Mathematics for Carpentry and the Construction Trades

Alfred P. Webster
Kathryn Bright Judy
Eastern Maine Vocational Technical Institute

Prentice Hall, Englewood Cliffs, New Jersey 07632

LIBRARY OF CONGRESS
Library of Congress Cataloging-in-Publication Data

Webster, Alfred P.
Mathematics for carpentry and the construction trades / Alfred P.
Webster, Kathryn Bright Judy.
p. cm.
Bibliography: p.
Includes index.
ISBN 0-13-562331-6
1. Carpentry--Mathematics. 2. Building--Mathematics. I. Judy,
Kathryn Bright. II. Title.
TH5612.W43 1989
513'.08694--dc19 88-15601
CIP

Editorial/production supervision
and interior design: Merrill Peterson
Cover design: Photo Plus Art
Manufacturing buyer: Bob Anderson
Page layout: Judy Winthrop

A Division of Simon & Schuster
Englewood Cliffs, New Jersey 07632

Printed in the United States of America
10 9 8 7 6 5 4 3 2 1

ISBN 0-13-562331-6

Prentice-Hall International (UK) Limited, *London*
Prentice-Hall of Australia Pty. Limited, *Sydney*
Prentice-Hall Canada Inc., *Toronto*
Prentice-Hall Hispanoamericana, S.A., *Mexico*
Prentice-Hall of India Private Limited, *New Delhi*
Prentice-Hall of Japan, Inc., *Tokyo*
Simon & Schuster Asia Pte. Ltd., *Singapore*
Editora Prentice-Hall do Brasil, Ltda., *Rio de Janeiro*

Contents

Preface

This text is intended to meet the needs of a two-semester course for students of carpentry and building construction.

Chapters 1 through 9 cover the fundamental mathematics necessary to a broad range of skills. Although problems in these chapters apply to a variety of areas, the emphasis has been given to applications in the building construction field. Material covering the metric system has been included in Appendix A and may be used when and if desired to meet the objectives of the course. Appendix B covers right-triangle trigonometry. This is an optional topic that can be useful to a builder and might be included after completion of Chapter 9.

It is suggested that calculators not be used in Chapters 1 through 3, allowing students to increase their mathematical skills in some basic areas. Thereafter, calculator usage is encouraged, with emphasis on efficiency and accuracy.

Chapters 10 through 24 cover matters of direct concern to the builder. The sequence of topics in these chapters follows the logical construction process insofar as is practical. Phases of the construction normally relegated to subcontractors (including masonry, plumbing, heating, electrical, among others) have not been covered. The occurrence of these phases of construction is of concern to the primary contractor, and their sequencing is alluded to in a summary chapter; however, the authors have made no attempt to include mathematics related to these areas.

Although it is not an objective of this text to be a complete "how-to" manual with respect to building techniques, a certain amount of instruction has been included. In many areas of building an understanding of the relevant mathematics is coupled with an understanding of how the construction is done. Furthermore, efficiency and accuracy (both highly desirable goals of the builder and estimator) are best achieved when an understanding of building methods has been achieved.

Users of this text will find the topic sequence logical with explanations clear and concise. Problems are realistic and practical. They are typical of the types of calculations that builders can expect to encounter in practice. Answers to the odd-numbered exercises have been included in the back of the book.

All of this is the result of the depth of experience the authors bring to users of this material. Their extensive backgrounds as teachers of applied mathematics and practitioners in the building construction field make this book valuable for its users.

The authors wish to express their appreciation to Sally Webster and Daisy Bright, both accomplished teachers of mathematics. Their identification of errors and suggestions for changes have been a valuable contribution to this effort.

A. P. Webster
K. B. Judy

chapter 1

Whole Numbers

1-1 PLACE VALUE AND ROUNDING

OBJECTIVES
Upon completing this section, the student will be able to:
1. Determine place value of numerals.
2. Round whole numbers to a given place value.

Place Value for Whole Numbers

To make sure that we understand the decimal system (the system we use), a brief review of place value is in order. A numeral such as 562 means 500 + 60 + 2. In other words, there are 5 hundreds, 6 tens, and 2 ones. Figure 1-1 indicates the place values for five hundred sixty-two (562), one thousand eight hundred fifty-three (1853), and five million, thirty thousand, four hundred six (5,030,406).

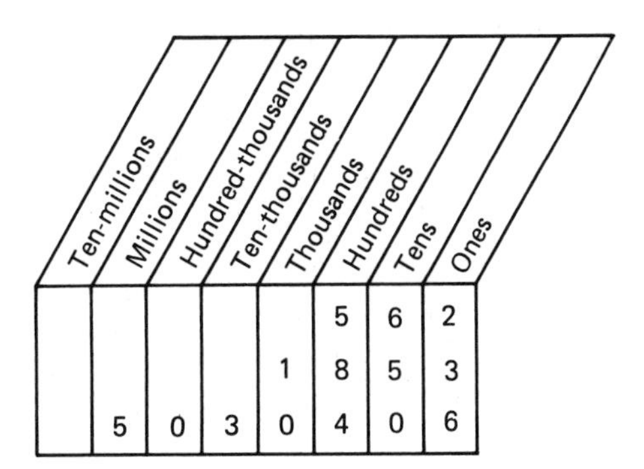

Figure 1-1

Exercise 1-1A

Determine the place value of the digits indicated, as in examples a and b.

	Number	Digit	Place value
a.	8**52**4	2	Tens place
b.	**63**,407	3	Thousands place
1.	42,5**27**	7	ONE PLACE
2.	165,**32**9	3	HUNDRED PLACE
3.	4,**06**1,755	0	HUNDRED THOUSAND PLACE
4.	8,**26**2,419	2	" " "
5.	**5**,317,204	5	million PLACE

Rounding Whole Numbers

Numbers can represent either exact quantities or approximate amounts. ''There are five people in the room.'' Five represents an exact number in this situation. We know that there cannot be $4\frac{1}{2}$ or $5\frac{1}{4}$ people in the room. On the other hand, a ''4-in. slab on grade'' represents an approximate number. All measured quantities are approximate since it is impossible to measure exactly without error. A 4-in. slab is not exactly 4 in. thick—it may be $3\frac{7}{8}$ in. thick in one place and $4\frac{1}{16}$ in. thick in another. Frequently, it is not possible or even desirable to state a quantity exactly. Estimating, a very important part of building construction, involves rounding numbers to approximate quantities. The following examples illustrate the proper method of rounding whole numbers.

Examples

A. Round 48,623 to the hundreds place.

48,**6**23

1. Locate the proper place. In this example the 6 is in the hundreds place.

48,**6**23 (arrow pointing to the 2)

2. Examine the digit to the right of the digit to be rounded. Examine the 2 in this case.

48,**6**23 ⇒ 48,600

3. If the digit to the right (the 2 in this example) is less than 5, that digit and all others to its right are replaced by zeros.

B. Round 48,623 to the thousands place.

4**8**,623

1. Locate the proper place. In this example the 8 is in the thousands place.

4**8**,623 (arrow pointing to the 6)

2. Examine the digit to the right of the digit to be rounded, a 6 in this example.

+1 (arrow pointing to the 8)
4**8**,623 ⇒ 49,000

3. If the digit to the right is equal to or greater than 5, add 1 to the digit to be rounded (the 8 in this example), and replace all digits to the right with zeros.

Exercise 1-1B

Using the rules for rounding whole numbers, round the following numbers to the place indicated.

1. 862,451 to the nearest ten 812,450; hundred 862,500.
2. 5899 to the nearest ten 5800; thousand 6000.
3. 541,722 to the nearest thousand 542,000; ten-thousand 541,000.
4. 5428 to the nearest ten 5430; thousand 5400.
5. 25,478,491 to the nearest hundred 25,478,500; million 25,000,000.

1-2 ORDER OF OPERATIONS

> **OBJECTIVES**
>
> Upon completing this section, the student will be able to:
>
> 1. Set up word problems involving more than one type of arithmetic operation.
> 2. Solve problems using the proper order of operations.

In solving problems with more than one operation, it is important to follow a certain order, as shown below. Arithmetic operations fall into several categories:

CATEGORY A: squares, square roots, and other exponents
CATEGORY B: multiplication and division
CATEGORY C: addition and subtraction

When solving a problem without parentheses, the following order should be followed:

Step 1. Perform operations in category A (squares, square roots, and other exponents).
Step 2. Perform operations in category B (multiplication and division).
Step 3. Perform operations in category C (addition and subtraction).

When parentheses are present, perform the operations in the parentheses first, then follow the order outlined above.

Examples

A. $8 - 6 \div 2 + 4 \times 2^2$

$8 - 6 \div 2 + 4 \times \underbrace{2^2}$ 1. Clear all exponents.

$8 - 6 \div 2 + 4 \times \mathbf{4}$

$8 - \underbrace{\mathbf{6} \div \mathbf{2}} + \underbrace{\mathbf{4} \times \mathbf{4}}$ 2. Perform all multiplication and division from left to right.

$8 - \mathbf{3} + \mathbf{16}$

$8 - 3 + 16 = 21$ 3. Perform all addition and subtraction from left to right.

B. $(8 - 6) \div 2 + (4 \times 2)^2$. Note that this problem is identical to Example A except for the parentheses.

$\underbrace{\mathbf{(8 - 6)}} \div 2 + \underbrace{\mathbf{(4 \times 2)}}^2$	1. Perform operations inside parentheses.
$\mathbf{2} \div 2 + \underbrace{\mathbf{8^2}}$	2. Evaluate exponents.
$2 \div 2 + \mathbf{64}$	
$\underbrace{\mathbf{2 \div 2}} + 64$	3. Perform all multiplication and division from left to right.
$\mathbf{1} + 64 = 65$	4. Add and subtract from left to right.

Note how different the answers are despite the fact that exactly the same numbers and operations occur in both problems. Order of operation is important!

Example

C. A contractor orders the following 2×4's: twenty-five 8-ft studs and thirty 12-ft studs. How many lineal feet of 2×4's has he ordered?

$$25 \times 8 + 30 \times 12 =$$

$$200 + 360 = 560 \text{ lineal feet}$$

These are frequently written with parentheses for clarification. Do the parentheses, as they are shown, change the answer?

$$(25 \times 8) + (30 \times 12) =$$

Exercise 1-2

Perform each operation in the correct order.

1. $5 + 8 \times 2$

2. $12 - 6 \times 2$

3. $8 + 6 \div 3 - 4$

4. $5 + 3 \times 2^2$

5. $8 \times (5 - 3) + 6$

6. $(3 + 2)^2 \div 5$

7. A trucker drives 30 mph for 1 hr and 55 mph for 2 hr to deliver prefabricated materials for a house. How far has he traveled?

8. Determine the total wattage in an electrical circuit with this load: five 175-watt (W) lamps, three 150-W lamps, two 75-W lamps, and four 60-W lamps.

REVIEW EXERCISES

1-1. Find the total cost if the following expenses are incurred by a contractor: framing lumber \$855, nails \$38, shingles \$152, siding \$278.

1-2. A table saw weighs 278 lb. If it is to be shipped in a wooden crate weighting 26 lb, what is the total weight of the saw when prepared for shipping?

1-3. An electrician uses the following lengths of Number 12 wire: 82 in., 185 in., and 1461 in. How much wire, in inches, did the electrician use?

1-4. A water meter installed at a construction site read 8351 ft^3 on July 1 and 15823 ft^3 on August 1. How many cubic feet of water were used in the month of July?

1-5. A total of 384 yd^3 of earth must be excavated for a basement. If 195 yd^3 have been removed, how many more cubic yards must be removed?

1-6. If holes spaced 17 in. center to center are to be drilled in a stud, how far apart are the two holes indicated on the diagram below?

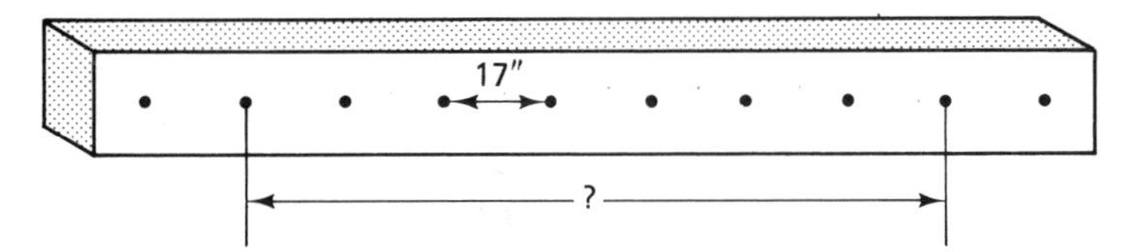

1-7. If lumber costs $439 per mbf (thousand board feet), what is the cost of 78 mbf?

1-8. A contractor's pickup truck averages 19 mpg. How far can he drive on a full tank of gas if the tank holds 27 gal?

1-9. A cabinetmaker can cut 325 wooden plugs from an 8-ft board. How many plugs can he cut from 71 boards of the same size?

1-10. Boards are cut 21 in. long to serve as fire stops between studs. If 386 fire stops are to be cut, how many inches of boards are needed? Ignore waste due to cutting.

1-11. A contractor purchases 42 mbf (thousand board feet) of lumber at a cost of $15,372. What is the price of 1 mbf?

1-12. How many joists spaced 16″ o.c. (on center) are required for a floor that is 32 ft long (384 in.). *Hint:* Note in the diagram that there is one more joist than the number of spaces. Since you are actually determining the number of spaces, you will need to add one joist.

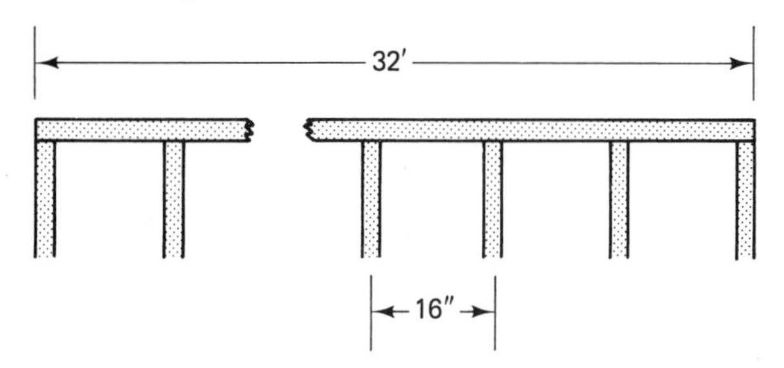

1-13. A set of basement stairs has an opening 81 in. long. How wide is each tread if there are nine steps? (Assume there is no overhang on the treads.)

1-14. A mason lays an average of 121 bricks per hour. How many hours will it take him to lay 4356 bricks?

1-15. A roll of copper tubing is 50 ft (600 in.) long. If gas range connectors must be 30 in. long, how many connectors can be cut from the roll of tubing? (Ignore cutting waste.)

1-16. The structural steel I-beam is to have 18 equally-spaced holes drilled in it as shown. How far apart are the holes center-to-center? (*Hint:* The number of spaces is one less than the number of holes.)

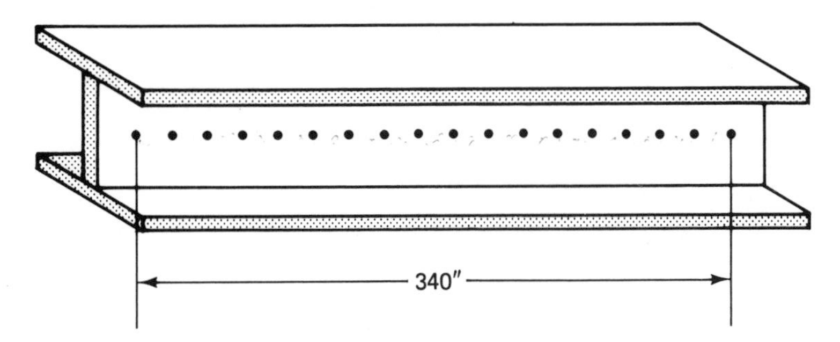

1-17. Bridging is to be cut 15 in. long. How many whole lengths of bridging (partial lengths can't be used) can be cut from a 1 × 3 piece of strapping that is 8 ft long?

1-18. How many lengths of shelving 26 in. long can be cut from an 8-ft 1 in. × 8 in. board?

1-19. A stairway has 13 risers. If the story height (distance from the top of the first to the top of the second floor) is 104 in., what is the height of each riser?

1-20. Bridging is to be 17 in. long. Ignoring cutting waste, how many pieces of bridging can be cut from a 12-ft length of 1 in. × 3 in. strapping? (Only whole pieces of bridging count!)

1-21. Shapers weighing 320 lb each are packed in shipping crates that each weigh 45 lb. What is the total shipping weight of 14 shapers?

1-22. Find the perimeter (distance around) of the floor plan shown.

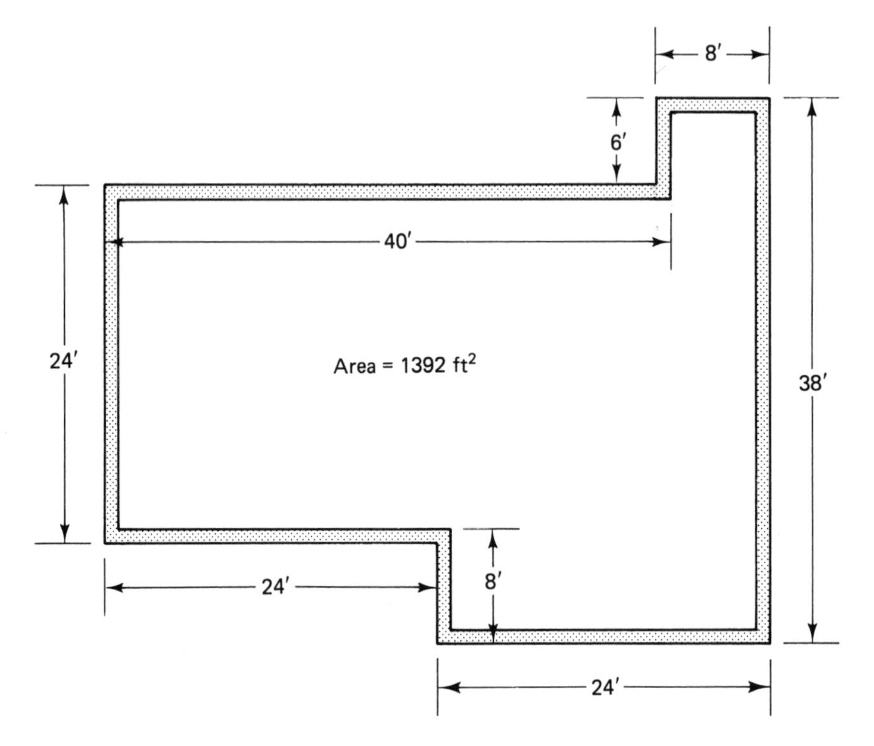

1-23. The floor plan below has the same area (1392 ft) as the one in problem 1-22. What is the perimeter of the floor plan? Which foundation and shell would be less expensive to erect? Why?

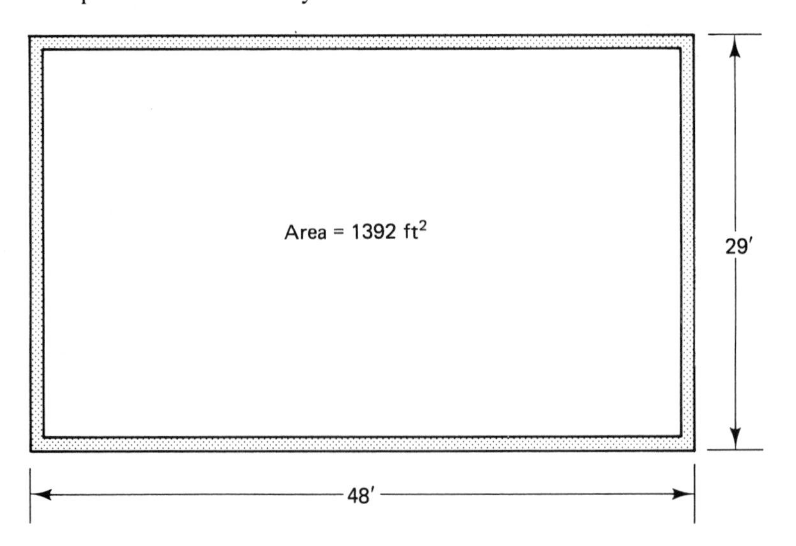

chapter 2

Fractions

2-1 EQUIVALENT FRACTIONS

OBJECTIVES

Upon completing this section, the student will be able to:

1. Reduce fractions to lowest terms.
2. Find equivalent fractions with the same denominators.
3. Change improper fractions to mixed numbers, and vice versa.

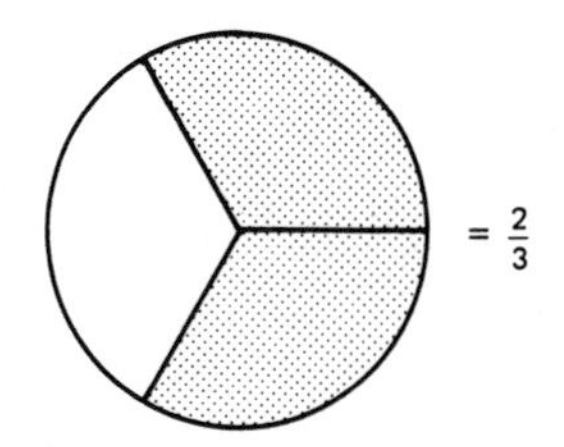

Figure 2-1

Fractions are used to express parts of a whole. The fraction $\frac{2}{3}$ indicates that 3 equal parts make up the whole and that 2 of these are being considered (Figure 2-1). The top number is the *numerator*, the *fraction bar* indicates division, and the bottom number is the *denominator*.

A **proper fraction** has a smaller numerator than denominator. Proper fractions represent less than a whole quantity. For example, in a quantity divided into 8 equal parts, if 3 are being used or considered, the fraction $\frac{3}{8}$ represents this concept. $\frac{8}{8}$ would represent a whole quantity in this case.

An **improper fraction** has a numerator larger than the denominator. Its value represents more than a whole quantity. $\frac{9}{5}$ is an example of an improper fraction (a misnomer, since there is really nothing improper about improper fractions). All improper fractions can be converted to mixed numbers.

If a quantity is divided into four equal parts and one of these parts is used, this is equivalent to dividing the quantity into 16 equal parts and using four of them (Figure 2-2). Notice that exactly the same portion of the square is darkened in both cases. $\frac{1}{4}$ and $\frac{4}{16}$ are equivalent fractions.

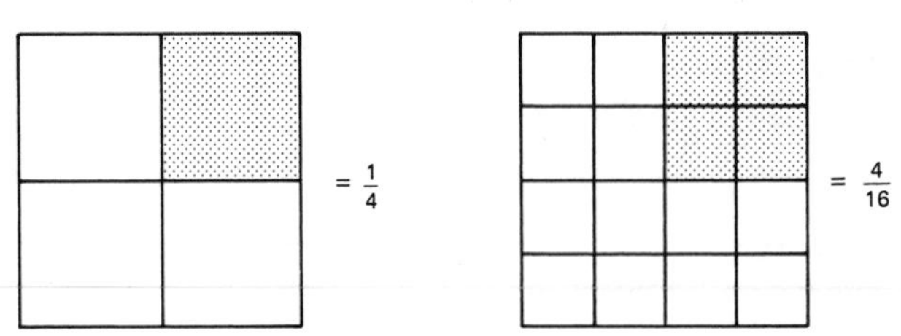

Figure 2-2

$$\frac{1}{4} = \frac{4}{16}$$

Another example of equivalent fractions is shown in Figure 2-3:

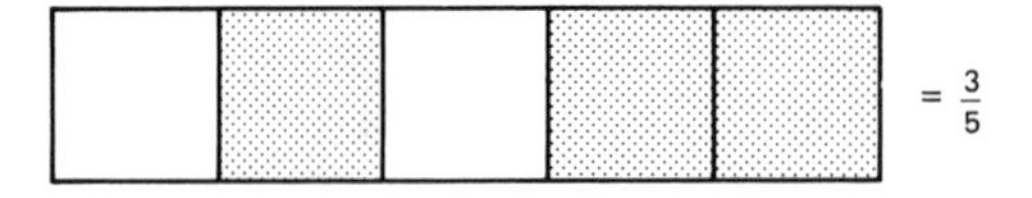

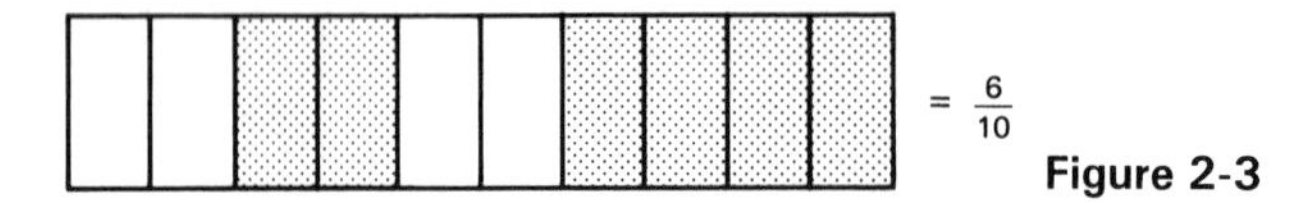

Figure 2-3

$$\frac{3}{5} = \frac{6}{10}$$

When an equivalent fraction is given with the smallest possible whole-number numerator and denominator, it is said to be **reduced to lowest terms.** For example:

$$\frac{25}{75} = \frac{1}{3}; \quad \frac{35}{49} = \frac{5}{7}$$

To reduce fractions to lowest terms, determine the largest number that can divide a whole number of times into both the numerator and the denominator, and reduce the fraction by that amount.

Examples

A. $\frac{25}{75} = \frac{25 \times 1}{25 \times 3} = \frac{\cancel{25} \times 1}{\cancel{25} \times 3} = \frac{1}{3}$

(divide numerator and denominator by 25)

B. $\frac{35}{49} = \frac{7 \times 5}{7 \times 7} = \frac{\cancel{7} \times 5}{\cancel{7} \times 7} = \frac{5}{7}$

(divide numerator and denominator by 7)

Exercise 2-1A

Reduce the following fractions to equivalent fractions in lowest terms.

1. $\frac{28}{36}$

2. $\frac{4}{30}$

3. $\frac{32}{64}$

4. $\frac{3}{39}$

5. $\frac{7}{21}$

6. $\frac{48}{64}$

7. $\frac{4}{17}$

8. $\frac{5}{45}$

9. $\frac{3}{24}$

10. $\frac{14}{26}$

Final answers should always be reduced to lowest terms. However, to perform intermediate steps, it is frequently necessary to change a reduced fraction into an equivalent fraction that is not in lowest terms.

Examples

C. Change $\frac{3}{4}$ into a fraction with a denominator of 36. Determine what number must be multiplied by the denominator 4, and then multiply both denominator and numerator by that number.

$$\frac{3}{4} = \frac{3 \times 9}{4 \times 9} = \frac{27}{36}$$

(numerator and denominator must be multiplied by 9)

D. What fraction with a denominator of 24 is equivalent to $\frac{5}{8}$?

$$\frac{5}{8} = \frac{5 \times 3}{8 \times 3} = \frac{15}{24}$$

(numerator and denominator must be multiplied by 3)

Exercise 2-1B

Find the equivalent fractions with the denominators indicated.

1. $\frac{5}{9} = \frac{?}{27}$

2. $\frac{5}{12} = \frac{?}{24}$

3. $\frac{3}{16} = \frac{?}{64}$

4. $\frac{5}{8} = \frac{?}{32}$

5. $\frac{1}{2} = \frac{?}{56}$

6. $\frac{5}{4} = \frac{?}{16}$

7. $\frac{9}{16} = \frac{?}{64}$

8. $\frac{35}{32} = \frac{?}{64}$

9. $\frac{5}{5} = \frac{?}{35}$

10. $\frac{2}{3} = \frac{?}{12}$

Changing Mixed Numbers to Improper Fractions

A mixed number is the sum of a whole number and a proper fraction. The mixed number $2\frac{1}{4}$ represents $2 + \frac{1}{4}$. Any improper fraction can be changed into a mixed number, and vice versa.

Example

E. Change the mixed number $2\frac{1}{4}$ into the equivalent improper fraction.

$$2\frac{1}{4} = 2 + \frac{1}{4} = \frac{2}{1} + \frac{1}{4}$$

1. Any whole number can be represented as a numerator over 1.

$$\frac{2 \cdot 4}{1 \cdot 4} + \frac{1}{4}$$

2. Change $\frac{2}{1}$ to its equivalent fraction with a denominator of 4.

$$\frac{8}{4} + \frac{1}{4} = \frac{9}{4}$$

3. Add the numerators to determine the improper fraction equivalent to the mixed number. *Note:* Do *not* add the denominators.

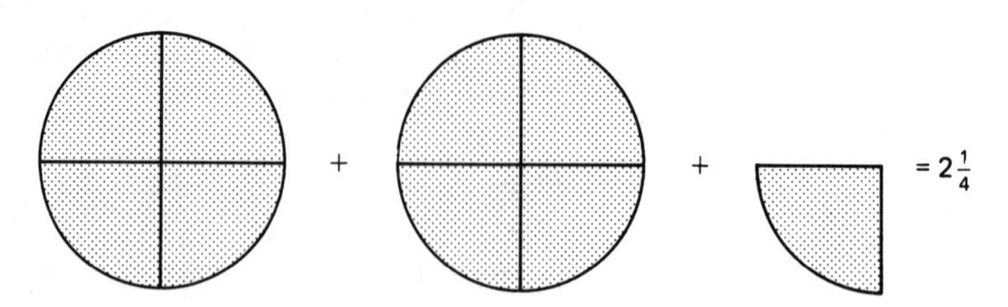

Here is a shortcut method for changing any mixed number to an improper fraction:

Examples

F. Change the mixed number $5\frac{2}{3}$ into an improper fraction.

$$5\frac{2}{3} \Rightarrow 3 \cdot 5 + \cdots$$

1. Multiply the denominator of the fraction times the whole number. In this example the denominator 3 is multiplied by the whole number 5.

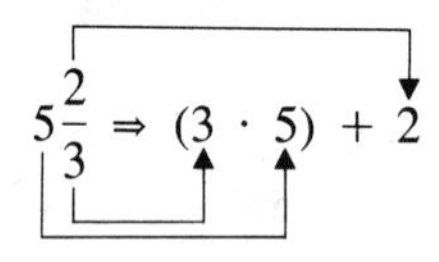

2. Add the original numerator to the product. In this example the numerator 2 is added to the product 15.

$$\frac{15 + 2}{3} = \frac{17}{3}$$

3. The resulting amount is the numerator of the improper fraction. The denominator is not changed.

G. Convert $3\frac{2}{3}$ to an improper fraction.

$$3\frac{2}{3} = \frac{(3 \cdot 3) + 2}{3} = \frac{9 + 2}{3} = \frac{11}{3}$$

Exercise 2-1C

Change the following mixed numbers to improper fractions.

1. $5\frac{1}{8}$ **2.** $3\frac{2}{5}$

3. $1\frac{7}{8}$ **4.** $7\frac{2}{3}$

5. $6\frac{3}{8}$ **6.** $4\frac{1}{5}$

7. $8\frac{3}{8}$ **8.** $7\frac{1}{9}$

9. $5\frac{2}{5}$ **10.** $3\frac{1}{4}$

Changing Improper Fractions to Mixed Numbers

An improper fraction can be changed to an equivalent mixed number by the following method.

Example

H. Change $\frac{14}{3}$ to a mixed number.

$$\frac{14}{3} = 3\overline{)14}$$

1. Remember: a fraction always indicates division. Divide the denominator into the numerator.

$$\begin{array}{r} 4\text{ R }2 \\ 3\overline{)14} \\ \underline{12} \\ 2 \end{array}$$

2. The whole number of times the denominator divides into the numerator is the whole-number part of the mixed number.

$$3\overline{)14}\ 4\text{ R }2 \Rightarrow 4\frac{2}{3}$$

3. The remainder becomes the numerator of the fraction part of the mixed number. Note that the denominator is the same in both the mixed number and the improper fraction.

Here is a shortcut method that works *only* in one special, but frequently occurring case:

Example

I. Convert $\frac{29}{21}$ to a mixed number.

$$\frac{29}{21} = 1\frac{?}{21}$$

1. By inspection, observe that the denominator 21 will divide into the numerator 29 only one time.

$$\begin{array}{r} 29 \\ -\ 21 \\ \hline 8 \end{array} \qquad \frac{29}{21} = 1\frac{8}{21}$$

2. Whenever the denominator divides *only one time*, the new numerator can be found by subtracting the denominator from the numerator of the improper fraction. *Note:* This method cannot be used if the whole-number part is any number other than 1.

Exercise 2-1D

Change the following improper fractions to mixed numbers and reduce to lowest terms where necessary. Use the shortcut method whenever possible.

1. $\frac{8}{5}$

2. $\frac{9}{4}$

3. $\frac{17}{16}$

4. $\frac{38}{32}$

5. $\frac{25}{16}$

6. $\frac{54}{8}$

7. $\frac{62}{15}$

8. $\frac{35}{32}$

9. $\frac{56}{8}$

10. $\frac{40}{16}$

Reading a Carpenter's Rule

The carpenter's rule shown is divided into inches, half-inches, $\frac{1}{4}$ inches, $\frac{1}{8}$ inches, and the smallest division on the ruler shown is $\frac{1}{16}$ of an inch. The following measurements are located on the ruler shown in Figure 2-4.

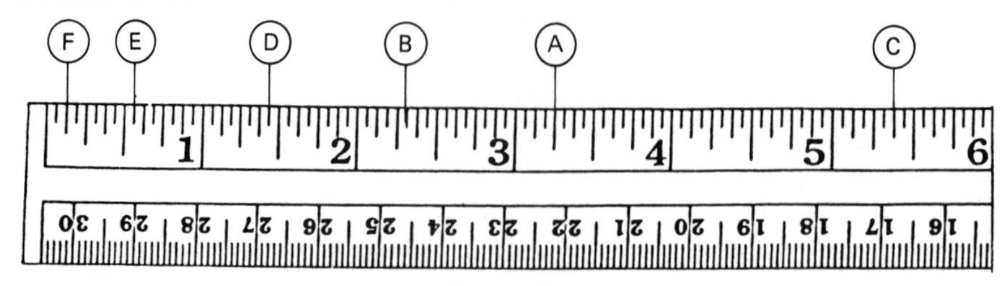

Figure 2-4

A $3\frac{1}{4}$ in.
B $2\frac{5}{16}$ in.
C $5\frac{3}{8}$ in.
D $1\frac{7}{16}$ in.
E $\frac{9}{16}$ in.
F $\frac{1}{8}$ in.

Exercise 2-1E

What measurement does each of the locations A to P on the ruler below indicate? Give each answer as a mixed number reduced to lowest terms.

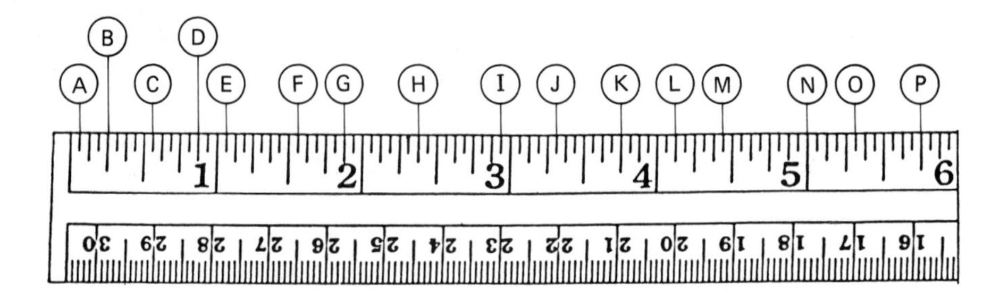

2-2 *FINDING THE LOWEST COMMON DENOMINATOR AND EQUIVALENT FRACTIONS*

OBJECTIVES

Upon completing this section, the student will be able to:

1. Find the lowest common denominator of fractions.
2. Determine the equivalent fractions with the lowest common denominators.
3. Order fractions according to size.

The **lowest common denominator,** usually referred to as the **LCD,** is the smallest number that is a multiple of each denominator.

Example

A. The LCD of 3/**8** and 5/**6** is 24. 24 is the smallest number that is a multiple of both 8 and 6.

$\mathbf{8} \times 3 = 24$

$\mathbf{6} \times 4 = 24$

Frequently, the LCD can be determined by inspection.

Examples

B. 8 is the LCD of 1/**4** and 3/**8**.

$\mathbf{4} \times 2 = 8$

$\mathbf{8} \times 1 = 8$

C. 6 is the LCD of 1/**2** and 2/**3**.

$\mathbf{2} \times 3 = 6$

$\mathbf{3} \times 2 = 6$

Here is one method that can be used when the LCD is not obvious by inspection. This method involves prime numbers. A **prime number** is any number that has factors of itself and 1 only. Here are the first 10 prime numbers:

2, 3, 5, 7, 11, 13, 17, 19, 23, 29

Examples

D. Find the LCD of the fractions $\frac{1}{5}$, $\frac{5}{9}$, and $\frac{4}{15}$.

$5 = 5$ $9 = 3 \cdot 3$ $15 = 3 \cdot 5$	1. Line up the denominators vertically and write them as prime factors.
→$5 = \mathbf{5}$ $9 = 3 \cdot 3$ $15 = 3 \cdot 5$ LCD $= \mathbf{5} \cdot$ ____	2. Starting at the top, include in the LCD every factor in the first denominator.
$5 = 5$ →$9 = \mathbf{3} \cdot \mathbf{3}$ $15 = 3 \cdot 5$ LCD $= \mathbf{5} \cdot \mathbf{3} \cdot \mathbf{3}$	3. Move to the next denominator (9 in this example). *All* factors *not already included* must be included in the LCD.
$5 = 5$ $9 = 3 \cdot 3$ →$15 = 3 \cdot 5$ LCD $= 5 \cdot 3 \cdot 3$	4. Neither the 5 nor the 3 in the denominator 15 should be included since both those factors are already in the LCD.
LCD $= 5 \cdot 3 \cdot 3 = 45$	5. Multiply the factors to determine the LCD.

E. Find the LCD and equivalent fractions of $\frac{5}{9}$, $\frac{11}{12}$, and $\frac{4}{15}$.

$9 = 3 \cdot 3$ $12 = 3 \cdot 2 \cdot 2$ $15 = 3 \cdot 5$	1. List the denominators and factor into prime factors.
→$9 = \mathbf{3} \cdot \mathbf{3}$ $12 = 3 \cdot 2 \cdot 2$ $15 = 3 \cdot 5$ LCD $= \mathbf{3} \cdot \mathbf{3} \cdot$ ____	2. Starting at the top, include in the LCD all factors in the first number (9).

$9 = 3 \cdot 3$
$\rightarrow 12 = 3 \cdot \mathbf{2} \cdot \mathbf{2}$
$15 = 3 \cdot 5$
$\text{LCD} = 3 \cdot 3 \cdot \mathbf{2} \cdot \mathbf{2} \cdot$ ____

3. Move to the next number (12). Include *only* those numbers that are not already included in the LCD. (Note that the 3 from the 12 is not included.)

$9 = 3 \cdot 3$
$12 = 3 \cdot 2 \cdot 2$
$\rightarrow 15 = 3 \cdot \mathbf{5}$
$\text{LCD} = 3 \cdot 3 \cdot 2 \cdot 2 \cdot \mathbf{5}$

4. Move to the last number (15). Include only the 5; the 3 is already included in the LCD.

$\text{LCD} = 3 \cdot 3 \cdot 2 \cdot 2 \cdot 5 = 180$

5. Multiply all the factors to determine the LCD.

$\frac{5}{9} = \frac{100}{180}$

$\frac{11}{12} = \frac{165}{180}$

$\frac{4}{15} = \frac{48}{180}$

6. Find the equivalent fractions using the LCD as the denominator.

Exercise 2-2A

Either by inspection or by the process of finding LCDs, determine the equivalent fractions with the lowest common denominator for the following pairs of fractions.

1. $\frac{3}{8}, \frac{5}{6}$

2. $\frac{3}{8}, \frac{5}{32}$

3. $\frac{2}{3}, \frac{3}{4}$

4. $\frac{3}{16}, \frac{5}{8}$

5. $\frac{5}{8}, \frac{2}{3}$

6. $\frac{1}{7}, \frac{2}{3}$

7. $\frac{9}{64}, \frac{3}{32}$

8. $\frac{5}{16}, \frac{1}{4}$

9. $\frac{5}{18}, \frac{2}{3}$

10. $\frac{7}{12}, \frac{11}{18}$

11. $\frac{3}{4}, \frac{1}{2}$

12. $\frac{5}{16}, \frac{5}{12}$

13. $\frac{1}{9}, \frac{2}{3}$

14. $\frac{3}{5}, \frac{2}{3}$

15. $\frac{11}{12}, \frac{5}{9}$

16. $\frac{3}{8}, \frac{5}{64}$

17. $\frac{5}{6}, \frac{1}{3}$

18. $\frac{2}{6}, \frac{1}{4}$

19. $\frac{3}{8}, \frac{5}{16}$

20. $\frac{5}{24}, \frac{2}{16}$

Most fractions that carpenters work with have denominators that are multiples of 2. These denominators can be found by inspection.

Examples

F. A $\frac{3}{8}$-in. drill bit is too small and a $\frac{1}{2}$-in. bit is too large for a particular job. Which bit should be tried next?

$\frac{3}{8} \quad \frac{1}{2}$

1. The next bit to be tried should be halfway between the $\frac{3}{8}$- and the $\frac{1}{2}$-in. bits.

$\frac{3}{8} = \frac{6}{16}$

$\frac{1}{2} = \frac{8}{16}$

2. Change both fractions to equivalent fractions with a denominator of 16. The new denominator should be twice the size of the larger denominator. In this example 8 is the larger denominator; therefore, the new denominator is $8 \times 2 = 16$.

$\frac{6}{16}, \frac{8}{16} \rightarrow \frac{7}{16}$

3. The numerator is halfway between the two numerators.

$\frac{7}{16}$ in.

4. The appropriate bit to try is $\frac{7}{16}$ in.

G. Given a 64-piece set of drill bits with every size from $\frac{1}{64}$ to $\frac{64}{64}$ in., what drill bit is one size larger than $\frac{5}{32}$ in.?

$\frac{5}{32} = \frac{10}{64}$

1. Change to the equivalent fraction with a denominator of 64.

$\frac{10}{64} + \frac{1}{64}$

2. Since the next larger size will be $\frac{1}{64}$ in. larger, add $\frac{1}{64}$ to the bit size. (Add the numerators only.)

$\frac{11}{64}$ in.

3. An $\frac{11}{64}$-in. bit is one size larger than a $\frac{5}{32}$-in bit.

H. Find the bit size that is halfway between a $\frac{3}{32}$-in. bit and a $\frac{1}{8}$-in. bit.

$\frac{3}{32} = \frac{6}{64}$

$\frac{1}{8} = \frac{8}{64}$

1. Find equivalent fractions with denominators twice the larger denominator $(32 \times 2 = 64)$.

$\frac{6}{64}, \frac{8}{64} \rightarrow \frac{7}{64}$

2. Find the number between the two numerators.

$\frac{7}{64}$ in.

3. This is the drill bit size halfway between a $\frac{3}{32}$-in. bit and a $\frac{1}{8}$-in. bit.

Exercise 2-2B

Assuming that all bits between $\frac{1}{64}$ and $\frac{64}{64}$ in. are available, determine which drill bit size is halfway between the following sizes.

1. $\frac{7}{32}$ and $\frac{1}{4}$ in.

2. $\frac{9}{32}$ and $\frac{5}{16}$ in.

3. $\frac{5}{8}$ and $\frac{3}{4}$ in.

4. $\frac{3}{8}$ and $\frac{1}{2}$ in.

5. $\frac{1}{2}$ and $\frac{3}{4}$ in.

6. $\frac{1}{16}$ and $\frac{1}{8}$ in.

7. $\frac{11}{16}$ and $\frac{3}{4}$ in.

8. $\frac{3}{32}$ and $\frac{1}{16}$ in.

9. $\frac{25}{32}$ and $\frac{13}{16}$ in.

10. $\frac{15}{32}$ and $\frac{1}{2}$ in.

Assuming that all bits between $\frac{1}{64}$ and $\frac{64}{64}$ in. are available, find the next size larger and the next size smaller than each given size.

	Size	Next size larger	Next size smaller
11.	$\frac{5}{32}$ (in.)		
12.	$\frac{3}{16}$ (in.)		
13.	$\frac{31}{32}$ (in.)		
14.	$\frac{9}{32}$ (in.)		
15.	$\frac{5}{8}$ (in.)		
16.	$\frac{1}{8}$ (in.)		
17.	$\frac{3}{4}$ (in.)		
18.	$\frac{1}{16}$ (in.)		
19.	$\frac{3}{8}$ (in.)		
20.	$\frac{9}{32}$ (in.)		

2-3 ADDITION OF FRACTIONS AND MIXED NUMBERS

OBJECTIVES

Upon completing this section, the student will be able to:

1. Add fractions and reduce to lowest terms.
2. Add mixed numbers and reduce to lowest terms.
3. Solve word problems involving addition of fractions and mixed numbers.

In order to be added, fractions must have the same denominators. The numerators are added, and the fraction is simplified if necessary.

Examples

A. $\frac{5}{8} + \frac{2}{8} = \frac{7}{8}$

1. Denominators are the same.
2. Add the numerators.

B. $\frac{1}{9} + \frac{2}{9} = \frac{3}{9} = \frac{1}{3}$

1. Denominators are the same.
2. Add the numerators and reduce the fraction to lowest terms.

C. $\frac{3}{4} + \frac{1}{8}$

1. Denominators are not the same.

$$\frac{3}{4} = \frac{6}{8}$$

$$+\frac{1}{8} = \frac{1}{8}$$

$$\frac{7}{8}$$

2. Determine the LCD and the equivalent fractions.
3. Add the numerators.

D. $\frac{2}{3} + \frac{5}{8}$

1. Denominators are not the same.

$$\frac{2}{3} = \frac{16}{24}$$

$$+\frac{5}{8} = \frac{15}{24}$$

$$\frac{31}{24} = 1\frac{7}{24}$$

2. Determine the LCD and find the equivalent fractions.
3. Add the numerators.
4. Convert to a mixed number.

E. $5\frac{5}{8} = 5 + \frac{5}{8}$

$$+\ 3\frac{1}{8} = 3 + \frac{1}{8}$$

$$8 + \frac{6}{8} = 8\frac{3}{4}$$

1. Add the whole numbers.
2. Add the fractions.
3. Reduce the fraction.

F. $3\frac{3}{4}$

$$+\ 1\frac{1}{4}$$

$$4\frac{4}{4} = 4 + 1 = 5$$

1. Add the whole numbers.
2. Add the fractions.
3. Simplify ($\frac{4}{4} = 1$).

G. $2\frac{5}{9} = 2\frac{5}{9}$

$$+\ 5\frac{2}{3} = 5\frac{6}{9}$$

$$7\frac{11}{9}$$

1. Determine the LCD and change to equivalent fractions.
2. Add the whole numbers.
3. Add the fractions.

$7\frac{11}{9} = 7 + 1\frac{2}{9}$ 4. Change the improper fraction to a mixed number.

$7 + 1\frac{2}{9} = 8\frac{2}{9}$ 5. Add the whole number and the mixed number.

Exercise 2-3

1. $\frac{2}{3} + \frac{1}{4}$

2. $\frac{5}{8} + \frac{1}{8}$

3. $\frac{9}{10} + \frac{4}{15}$

4. $\frac{3}{4} + \frac{5}{16}$

5. $2\frac{3}{16} + 5\frac{5}{16}$

6. $6\frac{1}{5} + 3$

7. $8\frac{3}{4} + 5\frac{24}{32}$

8. $3\frac{3}{4} + 7\frac{3}{16} + 5\frac{3}{8}$

9. $2\frac{5}{6} + 7\frac{1}{9}$

10. $3\frac{7}{8} + 4\frac{3}{4}$

11. A contractor acquires three adjoining lots with frontages of $131\frac{1}{3}$, $85\frac{3}{4}$, and $165\frac{5}{6}$ ft. What is the total frontage of the three lots?

12. A desk is made of $\frac{3}{4}$-in. particleboard and is covered on both sides with plastic laminate $\frac{1}{32}$ in. thick. What is the thickness of the desk top?

13. A 2 × 4 stud is actually $1\frac{1}{2} \times 3\frac{1}{2}$ in. How thick is a wall constructed of 2 × 4's that is covered on both sides by $\frac{3}{8}$-in. sheetrock? (The thickness is determined by the $3\frac{1}{2}$ in. dimension.)

14. Square ceramic floor tiles measure $3\frac{3}{8}$ in. on a side. If grout is to be $\frac{3}{16}$ in. wide, determine the total length needed to install four tiles as shown.

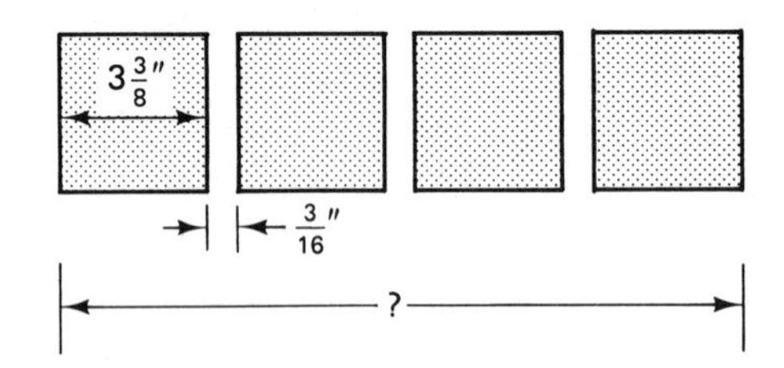

15. An outside wall is made up of the following: $\frac{3}{8}$-in. sheetrock, $\frac{15}{16}$-in. Styrofoam insulation board, $5\frac{1}{2}$-in. studs, and $\frac{3}{4}$-in. exterior sheathing. Determine the total thickness of the wall.

16. The top of a dresser is made of $\frac{3}{4}$-in. boards covered with a $\frac{1}{32}$-in. veneer on one side. What is the thickness of the top?

17. A partition between the living room and dining area is made of 2 × 4's covered with sheetrock on one side and sheetrock and paneling on the other. What is the total thickness of the wall if the 2 × 4's are actually $3\frac{1}{2}$ in., the sheetrock is $\frac{3}{8}$ in. thick, and the paneling is $\frac{1}{4}$ in. thick?

2-4 SUBTRACTION OF FRACTIONS AND MIXED NUMBERS

OBJECTIVES

Upon completing this section, the student will be able to:

1. Subtract fractions and express answers in simplest terms.
2. Subtract mixed numbers and express answers in simplest terms.
3. Solve word problems involving subtraction of fractions and mixed numbers.

Fractions and mixed numbers must have the same denominators in order to be subtracted. The numerators are subtracted, and fractions are simplified if necessary.

Examples

A.
$$\begin{array}{r} \frac{5}{9} \\ -\frac{1}{9} \\ \hline \frac{4}{9} \end{array}$$

Numerators are subtracted and the difference is written over the denominator.

B.
$$\begin{array}{rcl} \frac{3}{4} & = & \frac{6}{8} \\ -\frac{1}{8} & = & \frac{1}{8} \\ \hline & & \frac{5}{8} \end{array}$$

1. Determine the LCD and find the equivalent fractions.
2. Subtract the numerators.

C.
$$\begin{array}{rcl} 3\frac{7}{16} & = & 3\frac{7}{16} \\ -1\frac{1}{8} & = & 1\frac{2}{16} \\ \hline & & 2\frac{5}{16} \end{array}$$

1. Determine the LCD and the equivalent mixed numbers.
2. Subtract the fractions and subtract the whole numbers.

D.
$$\begin{array}{r} 3\frac{1}{4} \\ -1\frac{3}{4} \\ \hline \end{array}$$

In this example the fraction $\frac{3}{4}$ is too large to be subtracted from the fraction $\frac{1}{4}$.

$$\begin{array}{rcl} 3\frac{1}{4} & = 2 + 1 + \frac{1}{4} = & 2\frac{5}{4} \\ -1\frac{3}{4} & = 1 + \quad\;\; \frac{3}{4} = & 1\frac{3}{4} \\ \hline \end{array}$$

1. Rename $3\frac{1}{4}$ as $2 + 1 + \frac{1}{4}$.
2. Rename 1 as $\frac{4}{4}$ (any nonzero number over itself equals 1).
3. $2 + \frac{4}{4} + \frac{1}{4} = 2 + \frac{5}{4} = 2\frac{5}{4}$.

$$
\begin{array}{r}
2\frac{5}{4} \\
-\ 1\frac{3}{4} \\
\hline
1\frac{2}{4} = 1\frac{1}{2}
\end{array}
$$

4. Subtract.

E.

$$
\begin{array}{r}
5\frac{1}{8} \\
-\ 3\frac{5}{8} \\
\hline
\end{array}
$$

In this example the fraction $\frac{5}{8}$ is too large to be subtracted from the fraction $\frac{1}{8}$. Here is a **simplified method** for performing this subtraction:

In this example the fraction $\frac{5}{8}$ is too large to be subtracted from the fraction $\frac{1}{8}$. Here is a **simplified method** for performing this subtraction:

$$
\begin{array}{r}
\overset{-1}{\cancel{5}}\frac{\overset{1+8}{1}}{8} = 4\frac{9}{8} \\
-\ 3\frac{5}{8} = 3\frac{5}{8} \\
\hline
1\frac{4}{8} = 1\frac{1}{2}
\end{array}
$$

1. Reduce the whole number by 1.
2. *Add* the *denominator* and *numerator* of the fraction $\frac{1}{8}$ to obtain the *new numerator* 9.
3. Subtract and simplify the answer.

F.

$$
\begin{array}{rcrcr}
14\frac{3}{16} & = & \overset{-1}{\cancel{14}}\frac{\overset{3+16}{3}}{16} & = & 13\frac{19}{16} \\
-\ 9\frac{5}{8} & & -\ 9\frac{10}{16} & = & 9\frac{10}{16} \\
\hline
 & & & & 4\frac{9}{16}
\end{array}
$$

1. Find the LCD and determine the equivalent fractions.
2. Subtract 1 from the 14.
3. Add the denominator and numerator to determine the new numerator: 16 + 3 = 19.
4. Perform the subtraction.

G.

$$
\begin{array}{rcr}
8 & = & 7\frac{16}{16} \\
-\ 3\frac{5}{16} & = & -\ 3\frac{5}{16} \\
\hline
 & & 4\frac{11}{16}
\end{array}
$$

1. Change 8 into $7\frac{16}{16}$.
2. Subtract.

Exercise 2-4

Use the simplified method for subtracting mixed numbers where appropriate.

1. $\begin{array}{r} \frac{3}{4} \\ -\ \frac{5}{8} \\ \hline \end{array}$

2. $\begin{array}{r} \frac{5}{9} \\ -\ \frac{1}{6} \\ \hline \end{array}$

3. $\begin{array}{r} 3\frac{1}{8} \\ -\ 2\frac{15}{16} \\ \hline \end{array}$

4. $\begin{array}{r} 4\frac{1}{2} \\ -\ 2\frac{3}{8} \\ \hline \end{array}$

5. $6\frac{5}{9} - 2\frac{2}{3}$

6. $4\frac{5}{16} - 2$

7. $3\frac{1}{4} - \frac{5}{8}$

8. $9\frac{3}{16} - 3\frac{1}{8}$

9. $24\frac{5}{8} - 23\frac{25}{32}$

10. $14 - 5\frac{3}{8}$

11. A steel lintel above a fireplace is $46\frac{5}{16}$ in. long. From the diagram shown, determine the width of the fireplace opening.

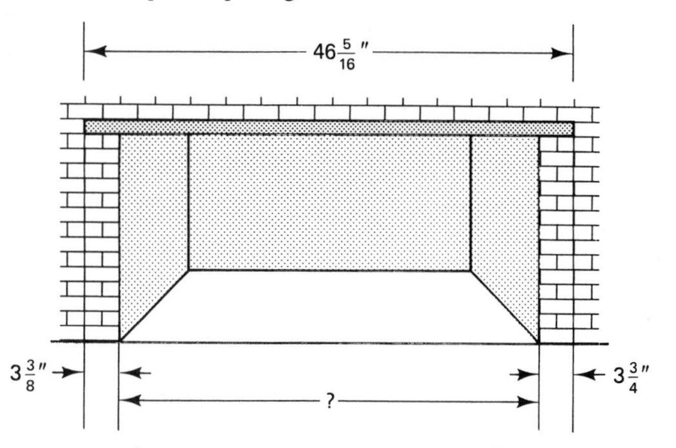

12. A planer takes a $\frac{3}{64}$-in. cut off a board that was $1\frac{7}{8}$ in. thick. What is the final thickness of the board?

13. A board 85 in. long is cut in two lengths. The saw kerf wastes $\frac{3}{16}$ in. If one length is $37\frac{7}{8}$ in. long, what is the other length?

14. A jointer is set to remove $\frac{5}{64}$ in. from the width of an oak board. If the board was $3\frac{7}{8}$ in. wide, what is its width after jointing once?

15. The outside width of a bookcase is $38\frac{5}{16}$ in. If the sides are made of $\frac{3}{4}$-in. stock, what is the inside width?

16. What is the final thickness of a $3\frac{1}{2}$-in. piece of stock if $\frac{3}{8}$ in. is planed off one side and $\frac{5}{16}$ in. is planed off the opposite side?

17. A board must be $46\frac{5}{16}$ in. long. How much should be cut from a 4-ft board?

18. The outside diameter (O.D.) of a pipe is $3\frac{1}{8}$ in. What is the inside diameter (I.D.) of the pipe if its walls are $\frac{3}{16}$ in. thick? (*Hint:* The wall thickness must be considered on both sides of the pipe.)

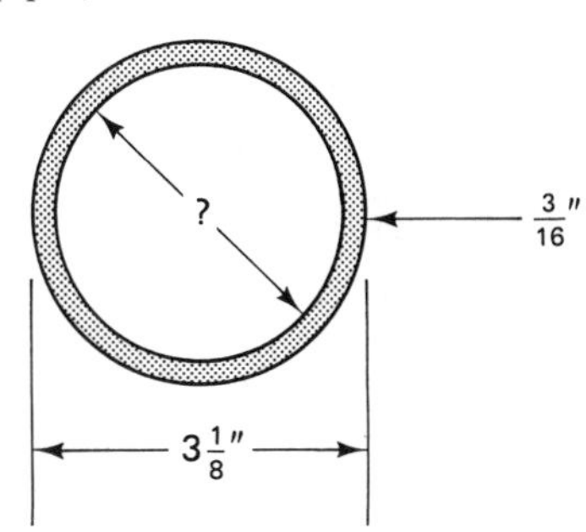

2-5 MULTIPLICATION OF FRACTIONS AND MIXED NUMBERS

OBJECTIVES

Upon completing this section, the student will be able to:

1. Multiply fractions.
2. Multiply mixed numbers.
3. Solve word problems involving multiplication of fractions and mixed numbers.

Multiplication of fractions is generally simpler than addition and subtraction of fractions. There is no need to find the LCD when multiplying. In multiplication, the numerators are multiplied together, and the denominators are multiplied together.

Examples

A. $\frac{3}{4} \times \frac{5}{8} = \frac{15}{32}$

1. Multiply the numerators together.
2. Multiply the denominators together.

B. $\frac{3}{8} \times \frac{16}{21} = \frac{3 \cdot 16}{8 \cdot 21} = \frac{48}{168}$

$\frac{48}{168} = \frac{2}{7}$

1. Multiply the numerators.
2. Multiply the denominators.
3. Reduce to lowest terms by dividing numerator and denominator by 24.

In Example B it would have been easier to reduce the fractions before multiplying the numerators and denominators.

$\frac{\overset{1}{\cancel{3}}}{8} \times \frac{16}{\underset{7}{\cancel{21}}}$

1. 3 divides into both the 3 in the numerator and the 21 in the denominator.

$\frac{1}{\underset{1}{\cancel{8}}} \times \frac{\overset{2}{\cancel{16}}}{7}$

2. 8 divides into both the 16 in the numerator and the 8 in the denominator.

$\frac{1}{1} \times \frac{2}{7} = \frac{2}{7}$

3. If all reducing is done before the multiplication, the answer will not need to be reduced further.

Examples

C. $5\frac{1}{4} \times \frac{3}{7}$

1. Mixed numbers cannot be multiplied directly.

$\frac{21}{4} \times \frac{3}{7}$

2. Change the mixed number to an improper fraction.

$\frac{\overset{3}{\cancel{21}}}{4} \times \frac{3}{\underset{1}{\cancel{7}}}$

3. Reduce by dividing the 21 in the numerator and the 7 in the denominator by 7.

$\frac{3}{4} \times \frac{3}{1} = \frac{9}{4}$

4. If no units of measurement are involved, it is best to leave the answer

as an improper fraction. When units are involved (inches, feet, pounds, etc.) the answer should be given as a mixed number.

$$\frac{9}{4} = 2\frac{1}{4}$$

5. Since most of the problems occurring in building construction involve units, in this book the answers will be converted to a mixed number.

D. $5\frac{3}{16} \times 4$

$$\frac{83}{16} \times \frac{4}{1}$$

1. Convert to an improper fraction.
2. A whole number should be put over a denominator of 1 when being multiplied by a fraction.

$$\frac{83}{\cancel{16}_{4}} \times \frac{\cancel{4}^{1}}{1}$$

3. Reduce.

$$\frac{83}{4} \times \frac{1}{1} = \frac{83}{4} = 20\frac{3}{4}$$

4. Multiply and convert to a mixed number.

Exercise 2-5

Give all answers in reduced form. Change improper fractions to mixed numbers.

1. $\frac{4}{5} \times \frac{5}{6}$

2. $\frac{17}{18} \times \frac{12}{17}$

3. $\frac{3}{4} \times \frac{5}{7}$

4. $5\frac{1}{2} \times 3\frac{1}{4}$

5. $14\frac{7}{8} \times 4$

6. $3\frac{2}{3} \times 4\frac{1}{2}$

7. $2\frac{3}{16} \times \frac{4}{5}$

8. $\frac{7}{8} \times 5\frac{5}{7}$

9. $8 \times \frac{3}{5}$

10. $7\frac{2}{9} \times \frac{3}{10}$

11. A beam is made of built-up 1 in. × 8 in. boards. If five boards are laminated together, what is the thickness of the beam? (1 × 8's are actually $\frac{3}{4}$ in. thick.)

12. Five boards each $7\frac{5}{8}$ in. long are needed. If they are to be cut from one board, how long must the board be? Allow 1 in. for waste.

13. A flight of stairs has 13 risers. If each riser is $8\frac{3}{16}$ in. high, what is the total height of the stairs? Give the answer in inches.

14. A stack of 4 ft × 8 ft sheets of $\frac{3}{8}$-in.-thick wallboard contains 55 sheets. What is the height of the stack?

15. A board is $86\frac{1}{4}$ in. long. What is $\frac{3}{4}$ of its length?

16. It takes a quarter of an hour ($\frac{1}{4}$ hour) to replace 1 ft of a special ceiling trim in a remodeling job. How many hours will it take to replace $37\frac{1}{2}$ ft of trim? Give the answer as a mixed number.

17. Cedar shingles are laid to expose $\frac{5}{12}$ ft per course. How high is a wall that requires 22 courses of shingles? Give the answer as a mixed number.

18. Seven lengths of board are to be removed from an 8-ft 2 × 4. Each length is to be

$5\frac{3}{8}$ in., and each cut wastes $\frac{1}{16}$ in. How many inches are removed from the board? (*Hint:* Since the entire board is not used, seven lengths will require seven cuts.)

2-6 DIVISION OF FRACTIONS AND MIXED NUMBERS

OBJECTIVES

Upon completing this section, the student will be able to:

1. Divide fractions.
2. Divide any combination of mixed numbers and fractions.
3. Solve word problems involving division of fractions and mixed numbers.

Division problems involving fractions are converted to multiplication problems. To convert the division problem to a multiplication problem, invert the divisor, and change the division sign to a multiplication sign.

Examples

A. $\frac{3}{4} \div \frac{5}{8}$

1. The divisor is always the number after the division sign. In this example the divisor is $\frac{5}{8}$.

$\frac{3}{4} \times \frac{8}{5}$

2. Invert (turn upside down) the divisor and change the division sign to a multiplication sign.

$\frac{3}{\not{4}_1} \times \frac{\not{8}^2}{5}$

3. Reduce the numerator and denominator. *Caution!* Do *not* attempt to reduce until the divisor has been inverted.

$\frac{3}{1} \times \frac{2}{5} = \frac{6}{5} = 1\frac{1}{5}$

4. Multiply and change to a mixed number.

B. Dividing a fraction by a whole number:

$\frac{3}{8} \div 5$

$\frac{3}{8} \div \frac{5}{1}$

1. Write the whole number as a fraction over 1.

$\frac{3}{8} \times \frac{1}{5}$

2. Invert the divisor.

$\frac{3}{8} \times \frac{1}{5} = \frac{3}{40}$

3. Multiply.

C. Dividing mixed numbers:

$4\frac{1}{8} \div 1\frac{5}{16}$

$\frac{33}{8} \div \frac{21}{16}$

1. Change the mixed numbers to improper fractions.

$\frac{33}{8} \times \frac{16}{21}$

2. Invert the divisor.

$\frac{\overset{11}{\cancel{33}}}{\underset{1}{\cancel{8}}} \times \frac{\overset{2}{\cancel{16}}}{\underset{7}{\cancel{21}}}$

3. Reduce the numerators and denominators.

$\frac{11}{1} \times \frac{2}{7} = \frac{22}{7} = 3\frac{1}{7}$

4. Multiply and convert to a mixed number.

D. $\frac{5}{9} \div \frac{3}{5}$

1. *Caution!* The numerators and denominators cannot be reduced as a division problem.

$\frac{5}{9} \times \frac{5}{3} = \frac{25}{27}$

2. Now no reducing can occur.

Exercise 2-6

1. $\frac{5}{8} \div 2\frac{3}{4}$

2. $\frac{4}{9} \div \frac{2}{3}$

3. $\frac{3}{5} \div \frac{5}{6}$

4. $1\frac{3}{4} \div 2\frac{5}{8}$

5. $6\frac{3}{4} \div 2\frac{5}{8}$

6. $9 \div \frac{3}{4}$

7. $\frac{5}{16} \div 15$

8. $11\frac{5}{16} \div 11\frac{5}{16}$

9. $1\frac{1}{2} \div 3\frac{3}{8}$

10. $4\frac{1}{16} \div 6\frac{1}{4}$

11. A wall is 8 ft high and $142\frac{3}{8}$ in. wide. It is to be covered by vertical boards 8 ft long and $3\frac{3}{4}$ in. wide. How many 8-ft boards are needed to cover the wall? (*Hint:* When determining the amount of material needed, always round UP.)

12. How many sheets of $\frac{1}{32}$-in. thick plastic laminate are in a pile $6\frac{3}{8}$ in. high?

13. Banister posts are cut from an 8-ft (96-in.) piece of cylindrical stock. How many posts $25\frac{3}{8}$ in. long can be cut from the stock? (*Hint:* When calculating the number of pieces that can be cut from a given length, always round *down*.)

14. The living room ceiling in a new house is to be taped and mudded. A $94\frac{1}{2}$-ft^2 area has already been done, and this represents $\frac{3}{8}$ of the job. How large is the ceiling?

15. An 8-ft board is to be sanded. If $2\frac{1}{2}$ ft has been done, what fraction of the board must still be sanded?

16. How many sheets of $\frac{3}{8}$-in. paneling are in a stack of sheets 3 ft high?

17. How many shelves for a bookcase can be cut from a board $15\frac{3}{4}$ ft long if each shelf is to be $3\frac{1}{4}$ ft?

18. How long does it take to cut through metal stock $4\frac{1}{2}$ in. thick if the cutting tool cuts at the rate of $\frac{3}{16}$ in./min?

REVIEW EXERCISES

2-1. A bookcase $56\frac{3}{8}$ in. high is made of $\frac{3}{4}$-in. stock with five equally spaced shelves. How much space is between each shelf?

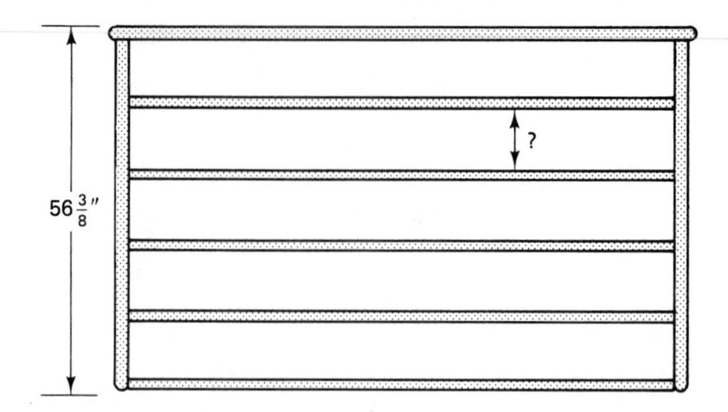

2-2. A $2\frac{1}{4}$-in.-diameter hole in a 2 in. × 6 in. stud must be enlarged to allow a pipe to fit through it snugly. If the pipe has a $3\frac{3}{16}$-in. diameter, by how much must the hole be enlarged?

2-3. A roof is covered with shingles weighing 255 lb per square. The nails used per square weigh $4\frac{1}{4}$ lb. What will be the weight of a roof requiring $3\frac{3}{4}$ squares?

2-4. The following colors are being mixed to produce a certain shade of paint: $4\frac{1}{4}$ gal of white, $1\frac{1}{8}$ gal of gray, $3\frac{5}{8}$ gal of blue, and $\frac{1}{4}$ gal of green. If each gallon of paint covers 320 ft^2, how many square feet will the mixture cover?

2-5. A cabinetmaker is cutting a large number of shelves $17\frac{1}{2}$ in. long. He has stock available in 2-, 4-, and 6-ft lengths. Which size should he choose in order to minimize waste?

2-6. What is the thickness of a pipe that is $3\frac{1}{16}$ in. O.D. and $2\frac{3}{8}$ in. I.D.?

2-7. A dimension stick is $21\frac{1}{4}$ in. long. If it is to be divided into five equal spaces, what will be the length of each space?

2-8. How many additional nails will be needed if $33\frac{1}{2}$ lb is sufficient to complete $\frac{3}{4}$ of a job?

2-9. An exterior wall has the following layers: $\frac{1}{4}$-in. paneling, $\frac{5}{8}$-in. firecode sheetrock, $5\frac{1}{2}$-in. studs, $\frac{1}{2}$-in. CDX plywood sheathing, $1\frac{1}{4}$-in. rigid insulation, and $\frac{5}{8}$-in. cedar shingles. What is the total thickness of the wall?

2-10. A construction corporation sells shares of stock for $12\frac{3}{8}$ dollars per share. If 11 investors jointly invest in 15 shares, what is the value of each investor's stock?

2-11. Order the following drill bits in size from smallest to largest: $\frac{7}{8}$, $\frac{11}{16}$, $\frac{43}{64}$, and $\frac{25}{32}$ in.

2-12. A hole is to be drilled $\frac{5}{64}$ in. oversized to accommodate a conduit with a $\frac{15}{32}$-in. diameter. What size of drill bit should be used?

2-13. A small pickup truck has $\frac{3}{8}$ of a tank of gas. If there are $4\frac{11}{16}$ gallons of gas left, how many gallons will the tank hold?

2-14. A 48-in. board is to be divided into four equal sections. If each cut wastes $\frac{1}{8}$ in., determine the length of each section. (*Hint:* Subtract the waste from the total length first. Remember, four sections require three cuts.) Give the answer to the nearest $\frac{1}{32}$ of an inch.

2-15. Four shims are used to raise a floor $1\frac{29}{32}$ in. on one corner. If two shims are each $\frac{13}{16}$ in. thick, and one shim is $\frac{5}{32}$ in., how thick must the fourth shim be?

chapter 3

Decimal Fractions

3-1 PLACE VALUE AND ROUNDING

OBJECTIVES

Upon completing this section, the student will be able to:

1. Identify decimal fractions and their fractional equivalents.
2. Identify the place value of a decimal fraction.
3. Round decimal fractions to a specified place.

Any common fraction with a denominator of 10, 100, 1000, and so on (known as powers of 10), can be written as a decimal fraction. Usually, a *decimal fraction* is called simply a *decimal*. Below are some common fractions and their decimal equivalents:

Common fraction	Decimal equivalent	Read as:
9/10	0.9	Nine-tenths
83/100	0.83	Eighty-three hundredths
521/1000	0.521	Five hundred twenty-one thousandths

Any common fraction with a denominator of 10 (or a power of 10) can be converted to a decimal in the following manner.

Examples

A. $\frac{9}{10} = \frac{9.}{10.} = \frac{.9.}{1.0.}$

1. Move the decimal point in the denominator to the left until the denominator equals 1.

$$\frac{.9}{1.0} = \frac{.9}{1}$$

2. Move the decimal point in the numerator to the left the same number of places. (When not shown, the decimal point is always understood to be to the right of the rightmost digit.)

$$\frac{.9}{1} = .9 \quad \text{or} \quad 0.9$$

3. Drop the denominator. (A denominator that equals 1 can always be omitted.) A leading zero is placed to the left of the decimal point if there is no whole number.

B. $\frac{521}{1000} = \frac{.521.}{1.000.}$

1. Move the decimal point in the denominator and numerator three places to the left.

$$\frac{.521}{1} = 0.521$$

2. Since the denominator is now equal to 1, it is dropped. *Note:* For each place the decimal point is moved to the left, the number is actually being divided by 10.

Any decimal can be written as a common fraction by reversing the process.

Examples

C. Convert 0.7 to a common fraction.

$$0.7 = \frac{.7}{1}$$

1. Any number can be written as a fraction with denominator of 1.

$$\frac{.7}{1.0} = \frac{.7.}{1.0.}$$

2. Move the decimal point in the numerator to the right as many places as necessary to make the numerator a whole number (one place to the right in this example).

$$\frac{7.}{10.} = \frac{7}{10}$$

3. Move the decimal point in the denominator the same number of places to the right, placing a zero for each move to the right.

D. Convert 0.41 to a common fraction.

$$0.41 = \frac{.41.}{1.00.} = \frac{41}{100}$$

Move the decimal point two places to the right in the numerator and the denominator. Place zeros in the denominator for each move to the right. (For each place the decimal point is moved to the right, the number is actually being multiplied by 10.)

Figure 3-1 indicates the place value for decimal fractions. A number such as 0.531 means 0.5 ($\frac{5}{10}$ or five tenths) + 0.03 ($\frac{3}{100}$ or three hundredths) + 0.001 ($\frac{1}{1000}$ or one thousandth). Shown in the figure are the numbers fifty-three thousandths (0.053 or $\frac{53}{1000}$), four hundred eighty-one thousandths (0.481 or $\frac{481}{1000}$), and two hundred fifteen and thirty-

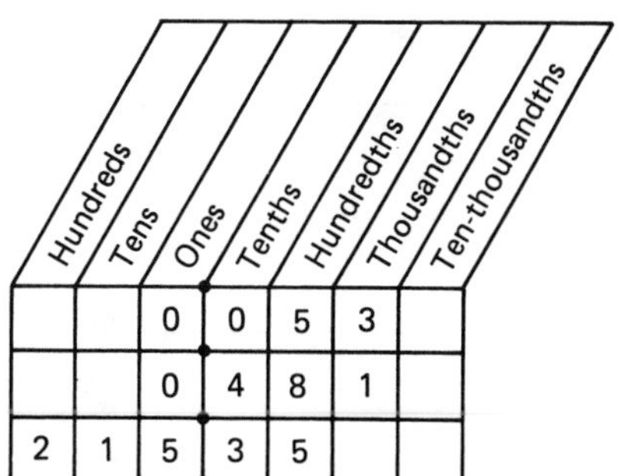

Figure 3-1

five hundredths (215.35 or $215\frac{35}{100}$). When reading a decimal number, the decimal point is always read as "and." The following examples will illustrate the method for rounding decimal fractions.

Examples

E. Round 0.2839 to two decimal places (to the nearest hundredths place).

0.2**8**39 — 1. Locate the digit to be rounded, in this example, 8.

0.2**8**39 (arrow at 3) — 2. Examine the digit just to the right of the digit to be rounded, 3 in this example.

0.2**8**~~39~~ — 3. If the digit to the right is less than 5, drop that digit and all other digits to the right of it.

0.28 — 4. Since the answer is rounded to two decimal places, only two digits are shown to the right of the decimal point.

F. Round 2.57381 to three decimal places (to the nearest thousandth).

2.57**3**81 — 1. Locate the digit in the thousandths place, in this example the 3.

2.57**3**81 (arrow at 8) — 2. Examine the digit just to the right of the digit to be rounded, in this case the 8.

2.57**3**81 (+1) — 3. If the digit to the right is 5 or greater, increase the digit to be rounded by 1, and drop all digits to its right.

2.57**4** — 4. *Note:* Do *not* replace the dropped digits with zeros. Replacement applies only to whole-number rounding.

Exercise 3-1

Round to the place value indicated.

1. 83.4067 to the nearest hundredths ____________, thousandths ____________

2. 0.0053 to the nearest hundredths ____________, thousandths ____________

3. 42.613 to the nearest ones ____________, tenths ____________

4. 2.4992 to the nearest hundredths ____________, ones ____________
5. 0.5038 to the nearest tenths ____________, hundredths ____________
6. 0.598 to the nearest ones ____________, hundredths ____________
7. 4.600 to the nearest tenths ____________, hundredths ____________
8. 0.8066 to the nearest ones ____________, hundredths ____________
9. 0.4251 to the nearest ones ____________, tenths ____________
10. 3.5012 to the nearest tenths ____________, thousandths ____________

3-2 ADDITION OF DECIMALS

> OBJECTIVES
>
> Upon completing this section, the student will be able to:
>
> 1. Add decimal fractions.
> 2. Solve word problems involving the addition of decimals.

Line up the decimal points. This is the key rule in addition (and subtraction) of decimals.

Examples

A. Add 5.82 + 30.6 + 0.005.

```
   5.82
  30.6
+  0.005
```

1. Position the decimal points in a vertical line. The decimal point in the answer will be in the same line.

```
   5.82
  30.6
+  0.005
  36.425
```

2. Add the numbers, positioning the digits properly on either side of the decimal point.

B. \$315 + \$0.82

```
  $315.00
+    0.82
  $315.82
```

Places without digits can be considered to be zeros.

Exercise 3-2

1. 8.25 + 23.4 + 0.15
2. 0.03 + 85 + 4.2
3. 4.45 + 0.02 + 12
4. 0.068 + 0.2 + 3.41
5. 5.612 + 62.62 + 3.84
6. 0.422 + 4.81 + 53
7. 6.72 + 67.2 + 0.672
8. 0.5 + 0.03 + 0.008
9. 2.03 + 0.003
10. 7.49 + 6.21 + 0.315
11. Find the thickness of a sheet of 7-ply plywood. The plies have thicknesses of 0.03125, 0.0625, 0.125, 0.25, 0.125, 0.0625, and 0.03125 in.

12. For a cabinet project, a carpenter purchases the following: plywood \$48.73, molding \$8.62, glue \$3.29, hinges \$9.37, drawer pulls \$19.49, finish nails \$2.16. What is the total cost of the purchases?
13. A shelf is covered with plastic laminate on both sides. The laminate is 0.03125 in. thick and the plywood shelf is 0.3125 in. thick. What is the total thickness of the shelf?
14. A survey of an odd-shaped lot indicates the five sides have the following dimensions: 225.62 ft, 78.25 ft, 145.62 ft, 138.45 ft, and 82.24 ft. Determine the distance around the lot.
15. A flyer for a lumberyard advertises the following: circular saws at \$38.99, utility cement mixers at \$189.95, cement mixing tubs at \$5.99, and exterior stain at \$11.97 per gallon. Find the total cost of purchasing a saw, a cement mixer, a tub, and 3 gal of exterior stain.
16. During construction of a house, the rainfall was 7.28 in. during April, 8.61 in. in May, and 7.42 in. in June. What was the total rainfall during the three months of construction?
17. A shim 0.0067 in. thick is placed under a plywood subfloor 0.75 in. thick. What is the thickness of the floor and shim?

3-3 SUBTRACTION OF DECIMALS

OBJECTIVES

Upon completing this section, the student will be able to:

1. Subtract decimal fractions.
2. Solve word problems involving subtraction of decimals.

Subtraction of decimals, like addition, requires that decimal points be lined up vertically.

Examples

A. 8.68 − 0.45

$$\begin{array}{r} 8.68 \\ -\ 0.45 \\ \hline 8.23 \end{array}$$

Line up decimal points and subtract.

B. 9.2 − 3.825

$$\begin{array}{r} 9.2\mathbf{00} \\ -\ 3.825 \\ \hline 5.375 \end{array}$$

When necessary, fill decimal places with zeros before subtracting. The same method is used for decimals and whole numbers when "borrowing" or "renaming."

C. \$19 − \$0.58

$$\begin{array}{r} \$19.\mathbf{00} \\ -\ \ \ 0.58 \\ \hline \$18.42 \end{array}$$

Line up decimal points, fill places with zeros, and subtract.

Exercise 3-3

1. 4.215 − 3.6
2. 93 − 14.21

3. 0.812 − 0.0053
4. 58.3 − 58.03
5. 0.297 − 0.0925
6. 4.62 − 3.1
7. 10.37 − 4.005
8. 8 − 7.99
9. 0.02 − 0.009
10. 13.51 − 4.11
11. A board 0.5625 in. thick has 0.0123 in. planed off it. What is the final thickness?
12. A lally column is positioned on a concrete footing as shown in the diagram. Determine the length of the lally column.

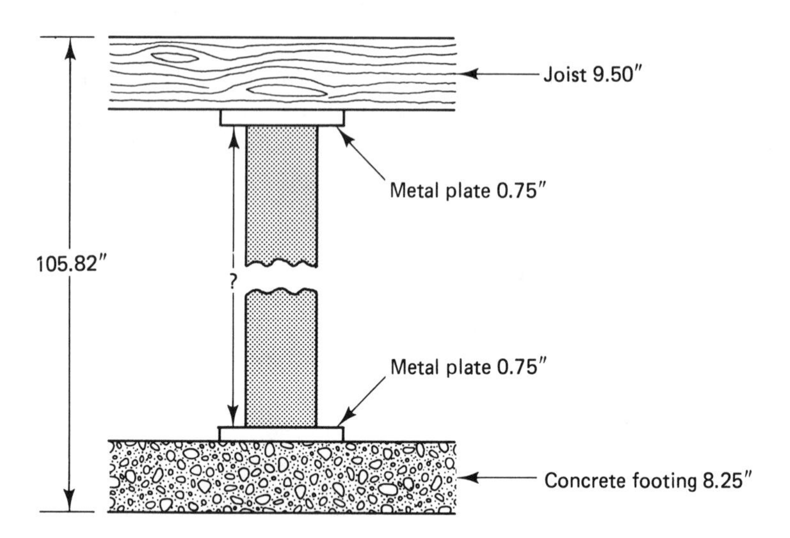

13. A pipe measures 1.0825 in. O.D. If the wall thickness is 0.052 in., what is the inside diameter of the pipe?
14. A template with a total length of 37.765 in. is divided into seven various lengths. Six of the lengths are: 4.285, 3.715, 4.155, 5.025, 5.135, and 7.325 in. What is the missing length?
15. A jointer takes off 0.128 in. from the side of an oak board. If the board was 5.500 in. wide, what is the width after jointing?
16. A carpenter charges a customer $125.80 for a job. How much of that price is for labor if the materials cost $57.92?
17. Two drill bits have diameters of 0.1875 and 0.625 in. What is the difference in the diameters of the two bits?
18. A carpenter charges $2855.50 to build a deck. If the cost of materials is $1125.35, and labor and overhead are $1225.50, what profit does the carpenter make on the job?

3-4 MULTIPLICATION OF DECIMALS

OBJECTIVES

Upon completing this section, the student will be able to:

1. Multiply decimal fractions.
2. Solve word problems involving multiplication of decimals.

When multiplying decimals, the decimal points are initially ignored and the numbers are multiplied as if they were whole numbers. The placement of the decimal point is then determined, and rounding is done where necessary.

Examples

A. 45.3 × 2.06

$$\begin{array}{r} 45.3 \\ \times\ \underline{2.06} \\ 2718 \\ +\ \underline{9060\ \ } \end{array}$$

1. Vertically line up the rightmost digits of each number.
2. Perform the indicated multiplication and addition, disregarding the decimal points.

$$\begin{array}{r} 45.3 \\ \times\ \underline{2.06} \\ 2718 \\ +\ \underline{9060\ \ } \\ 93.318 \end{array}$$

3. The placement of the decimal point is determined by the total number of decimal places in the two multipliers. 45.3 has one decimal place and 2.06 has two decimal places. Therefore, the decimal point in the answer must be positioned three places to the left.

B. 1.32 × 4.3

$$\begin{array}{rl} 1.32 & \leftarrow 2 \text{ decimal places} \\ \times\ \underline{4.3} & \leftarrow 1 \text{ decimal place} \\ 396 & \\ +\ \underline{528\ } & \\ 5.676 & \leftarrow 2 + 1 = 3 \text{ decimal places} \end{array}$$

Technically, it is not correct to leave the answer in Example B as 5.676, because that answer is more accurate than either of the multipliers. Since the answer cannot be more accurate than the least accurate number that produced it, the answer should be rounded to one decimal place. 5.676 rounds to 5.7 (the same accuracy as 4.3).

It is not the intent of this book to teach the sometimes confusing rules of precision and accuracy. Instead, ''reasonable rounding'' will be used. Especially with calculator usage, it will almost never be necessary for a carpenter to employ the accuracy that can be obtained on a calculator. Rounding to two, three, or sometimes four decimal places is generally sufficient.

Example

C. Variable-speed $\frac{1}{4}$-in. drills are on sale for \$29.59 each. What is the price of three drills? (Disregard sales tax.)

$$\begin{array}{rl} \$29.59 & \leftarrow 2 \text{ decimal places} \\ \times\ \underline{\qquad 3} & \\ \$88.77 & \leftarrow 2 \text{ decimal places} \end{array}$$

Here 3 is an exact number, so the accuracy is determined by the \$29.59.

Exercise 3-4

1. 4.28 × 5.3
2. 1.85 × 6
3. 0.43 × 2.1
4. 5.7 × 3.29
5. \$4.98 × 6
6. 3.81 × 4.2
7. 0.05 × 6.1
8. 3.3 × 4.8
9. 6.2 × 23
10. 31 × 6.2
11. A contractor purchases 14.853 gal of gas to be burned in a generator used to power construction equipment. Super-unleaded gas sells for \$1.089 per gallon. Determine

the cost of the gas for the generator. Round to two decimal places (the nearest penny).

12. A contractor hires five carpenters at $8.23 per hour. They are to be paid time and a half for all hours worked over 40. How much must the contractor pay in wages if the five carpenters each work 49.5 hours?

13. A sheet of plywood has five plies that are each 0.125 in. thick. What is the thickness of the plywood?

14. A furnace burns 0.83 gal of oil per hour at a certain setting. If No. 2 heating oil sells for $1.029 per gallon, what is the cost of running the furnace for 6.3 hr?

15. Find the total height of a seven-story building if each story measures 11.85 ft.

16. Scrap metal from a construction site sells for $0.035 per pound. What is a load of scrap metal worth that weighs 2380 lb?

17. A submersible pump in a well pumps 32.8 gal of water per minute. If water weighs 8.3 lb/gal, how many pounds of water does the pump deliver per minute?

18. A carpenter can lay a square of roofing shingles in 2.4 hr. How long would it take to lay 4.7 squares of shingles at that rate?

3-5 *DIVISION OF DECIMALS*

OBJECTIVES

Upon completing this section, the student will be able to:

1. Divide decimal fractions.
2. Solve word problems involving division of decimal fractions.

After manipulating the decimal points, division of decimals is similar to division of whole numbers.

Examples

A. Dividing a decimal by a whole number: $8\overline{)1.6}$.

$$\begin{array}{r} \text{quotient} \\ \text{divisor}\overline{)\text{dividend}} \end{array}$$

$$\begin{array}{r} . \\ 8\overline{)1.6} \end{array}$$

1. Position the decimal point in the quotient just above its location in the dividend.

$$\begin{array}{r} 0.2 \\ 8\overline{)1.6} \end{array}$$

2. Divide as whole numbers.

B. $14\overline{)0.395}$

$$\begin{array}{r} 0.0282 \\ 14\overline{)0.3950} \\ \underline{0\;0} \\ 39 \\ \underline{28} \\ 115 \\ \underline{112} \\ 30 \\ \underline{28} \end{array}$$

1. Position the decimal point in the quotient.
2. Place extra zeros after the decimal point in the dividend as necessary for placeholders.
3. Divide to one place beyond the desired accuracy.

```
     0.0282̸
14)0.3950  ⇒ 0.028
```

4. Round appropriately.

C. Dividing a decimal by a decimal: 4.23)30.4137.

```
         7.19
4.23.)30.41.37
      29 61
         80 3
         42 3
         38 07
         38 07
```

1. Move the decimal point in the divisor as many places as necessary to make it a whole number (two places in this example).
2. Move the decimal point in the dividend the same number of places.
3. Locate the decimal point in the quotient, and divide.

D. Dividing a whole number by a decimal to one decimal place accuracy: 8.3)57.

```
           .
8.3.)57.0.

        6.86
8.3.)57.0.00
     49 8
      7 2 0
      6 6 4
        5 60
        4 98
          62
```

1. Move the decimal point in the divisor one place to make the divisor a whole number.
2. Move the decimal point in the dividend the same number of places, using zeros to hold the place.
3. Place the decimal point in the quotient above the adjusted decimal point in the dividend.
4. For one-decimal-place accuracy, there should be two decimal places in the dividend.

6.86 rounds to 6.9

5. Round the answer to one decimal place.

Exercise 3-5

1. 42)4.158

2. 3.84 ÷ 0.6

3. 7.975 ÷ 2.75

4. 0.09)1.62

5. 5.3)48.336

6. 4.82 ÷ 0.01

7. 0.7 ÷ 35

8. 0.14)0.0154

9. 0.062)62

10. 48.618 ÷ 3.7

11. A pound of $1\frac{5}{8}$-in. galvanized sheetrock screws costs $4.32. If that amounts to $0.018 per sheetrock screw, how many are in a pound?

12. A carpenter paid $81.40 for lumber priced at $0.37 per board foot. How many board feet of lumber did he purchase?

13. Each tread in a stairway measures 9.3125 in. If the total run is 111.75 in., how many treads are there?

14. A carpenter uses 3.25 gal of paint to paint 1153.75 ft^2. How many square feet will 1 gal of paint cover?

15. A project requires 36 lineal feet of lumber. What is the price per lineal foot if the lumber costs $15.12?

16. A fill-up for a contractor's truck cost $27.54. If the tank took 28.13 gal of gas, what was the price per gallon? Give answer to nearest tenth of a penny.

17. Cut boards each measure 8.625 in. If the boards are set end to end, their total length is 163.875 in. How many cut boards are there?

18. How many pieces of oak flooring 2.25 in. wide are needed to cover a floor 38.25 in. wide?

19. A stack of 0.625-in.-thick plywood is piled 51.875 in. high. How many sheets of plywood are in the stack?

3-6 MULTIPLYING AND DIVIDING BY POWERS OF 10

OBJECTIVES

Upon completing this section, the student will be able to:

1. Mentally multiply numbers by powers of 10.
2. Mentally divide numbers by powers of 10.

Some common powers of 10 are as follows:

Greater than 1	*Less than 1*
10	0.1 or (1/10)
100	0.01 or (1/100)
1000	0.001 or (1/1000)
10,000	0.0001 or (1/10,000)
100,000	0.00001 or (1/100,000)
1,000,000	0.000001 or (1/1,000,000)

Note that each number differs from the number **1** only by the location of the decimal point. This fact can be used to easily multiply and divide numbers by a power of 10, by shifting the decimal point. The numerals in the original number will not be changed. Study the following examples to determine one method of mentally multiplying and dividing by powers of 10.

Examples

A. 5.83×100

1. First determine whether the answer will be larger or smaller than 5.83. Whenever the answer will be *larger*, the decimal point is moved to the *right*.

1.00.

2. Observe that the number 100 differs from the number 1 by *two decimal places*.

$5.83 \times 100 = 583$

3. Move the decimal point in the number 5.83 to the *right two places*.

B. 47.3×0.001

1. Determine whether the answer will be larger or smaller than 47.3. Whenever the answer will be *smaller*, the decimal point is moved to the *left*.

0.001.

2. Determine the number of decimal places 0.001 differs from the number 1, namely *three places*.

$47.3 \times 0.001 = 0.047$

3. Move the decimal point in the number 47.3 to the *left three places*.

C. 85 ÷ 1000	1. Determine whether the answer will be larger or smaller than 85. Remember for a smaller answer the decimal point is moved to the left.
1.000.	2. Note that the number 1000 is 3 decimal places from the number 1. (The decimal point is always to the right of the last digit if it is not shown.)
85. ÷ 1000 = 0.085	3. Move the decimal point in the 85 to the *left three decimal places*.
D. 47 ÷ 0.01	1. Determine whether the answer will be larger or smaller than 47. *Careful!* Dividing by a number between 0 and 1 makes the answer *larger*.
0.01.	2. Determine the number of decimal places 0.01 is from 1, namely 2.
47. ÷ 0.01 = 4700	3. Move the decimal point in the 47 to the *right two decimal places*.

Exercise 3-6

1. 83,000 ÷ 0.1
2. 4.87 × 1000
3. 0.05 ÷ 0.001
4. 0.05 × 0.01
5. 4800 ÷ 100
6. 853 × .01
7. 38 ÷ 1000
8. 0.005 × 100
9. 6200 ÷ 100
10. 4.72 × 10

3-7 DECIMAL AND FRACTION CONVERSIONS

OBJECTIVES

Upon completing this section, the student will be able to:

1. Convert common fractions to decimal fractions.
2. Convert mixed number fractions to mixed decimals.
3. Convert decimal fractions to common fractions.
4. Convert mixed decimals to mixed fractions.
5. Convert repeating decimals to fractions.
6. Convert decimal inches to the nearest $\frac{1}{2}$, $\frac{1}{4}$, $\frac{1}{8}$, $\frac{1}{16}$, $\frac{1}{32}$, and $\frac{1}{64}$ in.

Converting Common Fractions to Decimal Fractions

To convert a common fraction to its equivalent decimal, divide the numerator of the fraction by its denominator.

Examples

A. $\frac{3}{4}$ becomes $3 \div 4 = 0.75$.

$$\begin{array}{r} .75 \\ 4\overline{)3.00} \\ \underline{2\;8} \\ 20 \end{array}$$

B. $\frac{5}{8}$ becomes $5 \div 8 = 0.625$.

C. $\frac{1}{3}$ becomes $1 \div 3 = 0.3333. . . .$ This is called a *repeating decimal.* No matter how far this division is taken, the same pattern will repeat. Repeating decimals are usually shown as 0.333. . . or by putting a bar over the repeating digit(s), such as $0.\overline{3}$.

$$\begin{array}{r} .333... \\ 3\overline{)1.000...} \\ \underline{9} \\ 10 \\ \underline{9} \\ 10 \end{array}$$

$$0.333... = 0.\overline{3}$$

Converting Mixed Fractions to Decimals

When converting a mixed fraction to a decimal, keep in mind that the whole-number part of the mixed fraction will be exactly the same in the mixed decimal. Therefore, the whole number does not need to be converted.

Examples

D. $3\frac{7}{16}$ becomes $3 + (7 \div 16)$. Convert only the $\frac{7}{16}$ since the whole number 3 will be the same in both forms.

$$3 + \frac{7}{16} = 3 + (0.4375) = 3.4375$$

E. $4\frac{5}{12}$ becomes $4 + 5 \div 12$.

$$4 + (5 \div 12) = 4 + 0.416666...$$

$$4.41666... \quad \text{or} \quad 4.41\overline{6}$$

Note that the bar is over the 6 only, since that is the only digit that repeats.

Converting Decimals to Fractions

To convert a nonrepeating decimal to a fraction, put the decimal value over a denominator of 1, adjust the decimal point, and reduce, if necessary.

Examples

F. $0.375 = \frac{0.375}{1}$

1. Convert to a fraction with the denominator of 1.

$\frac{.375}{1.000} = \frac{.375.}{1.000.}$

2. Adjust the decimal point in the numerator to make it a whole number. Adjust the decimal point in the denominator the same number of places.

$\frac{375}{1000} = \frac{3}{8}$

3. Reduce where possible.

Converting Mixed Decimals to Mixed Fractions

Example

G. $121.35 = 121 + \frac{.35}{1}$

1. Separate out the whole number since only the fraction needs to be converted.

$\frac{.35}{1.00} = \frac{35}{100} = \frac{7}{20}$

2. Convert the decimal to a fraction and reduce.

$121 + \frac{7}{20} = 121\frac{7}{20}$

3. Don't forget to add on the whole number!

Converting Repeating Decimals to Fractions

Any repeating decimal can be converted to a fraction with a denominator of 9, 99, 999, and so on, depending on how many digits are repeating.

Examples

H. $0.\overline{5} = \frac{5}{9}$

1. Since 5 is the only repeating digit, the denominator is only one digit, a 9.

$\left(0.5 = \frac{5}{10}\right)$

2. *Careful!* 0.5 and $0.\overline{5}$ are not equal. The terminating decimal 0.5 equals $\frac{5}{10}$.

I. $0.83838383... = 0.\overline{83}$

1. This decimal has two repeating digits.

$0.\overline{83} = \frac{83}{99}$

2. The denominator is 99 because there are two repeating digits in the numerator.

J. $2.\overline{3} = 2 + .\overline{3}$

$.\overline{3} = \frac{3}{9}$

1. Separate the whole number and convert only the repeating decimal.

$\frac{3}{9} = \frac{1}{3}$

2. Reduce.

$2.\overline{3} = 2\frac{1}{3}$

3. Don't forget to include the whole number in your answer.

Note: The method described in Examples H to J works *only* if the repeating digits start just to the right of the decimal point (in the tenths place). For example, this method would *not* work for 0.8333. . . because the nonrepeating digit 8 is in the tenths place.

Converting Decimal Inches to Special Fractions

The most useful fractions for carpenters are those having denominators of 2, 4, 8, 16, 32, and 64, since these are the same divisions that are on a carpenter's rule. Especially with the common usage of calculators on the job, calculations are frequently done in decimal form and must be converted to the nearest useful equivalent fraction. Study the examples below carefully—this is a very important tool for the carpenter. The student is expected to use a calculator for these problems.

Examples

K. Convert 0.831 in. to the nearest $\frac{1}{16}$ in.

$$\frac{.831}{1} \times \frac{16}{16}$$

1. Multiply 0.831 times the fraction $\frac{16}{16}$.

$$\frac{.831}{1} \times \frac{16}{16} = \frac{13.296}{16}$$

2. Perform the multiplication in the numerator, leaving the denominator as 16.

$$\frac{13.296 \text{ in.}}{16} \doteq \frac{13}{16} \text{ in.}$$

3. Round the numerator to the nearest whole number.

L. Convert 6.894 in. to the nearest $\frac{1}{32}$ in.

$$\frac{.894}{1} \times \frac{32}{32}$$

1. Temporarily ignoring the whole number, multiply 0.894 times the fraction $\frac{32}{32}$.

$$\frac{.894}{1} \times \frac{32}{32} = \frac{28.608}{32}$$

2. Find the product of the numerators, leaving the denominator as 32.

$$\frac{28.608}{32} \doteq \frac{29}{32}$$

3. Round the numerator to the nearest whole number.

$$6.894 \text{ in.} \doteq 6\frac{29}{32} \text{ in.}$$

4. Add the whole number of inches to the fraction. (The symbol $\doteq$ means "approximately equal to.")

M. Convert 14.629 in. to the nearest $\frac{1}{64}$ in.

$$\frac{.629}{1} \times \frac{64}{64}$$

1. Multiply by $\frac{64}{64}$.

$$\frac{.629}{1} \times \frac{64}{64} = \frac{40.256}{64}$$

$$\frac{40.256}{64} \doteq \frac{40}{64}$$

2. Round numerator to the nearest whole number.

$$\frac{40}{64} = \frac{5}{8}$$

3. When possible, reduce the fraction to lowest terms.

$$14.629 \text{ in.} \doteq 14\frac{5}{8} \text{ in.}$$

4. Add the whole number to the reduced fraction.

Note: It is sometimes more useful to leave the fraction in the unreduced form. For instance, $\frac{40}{64}$ in. indicates to the carpenter that the measurement is accurate to the nearest $\frac{1}{64}$ in. Reducing to $\frac{5}{8}$ in. suggests that the measurement is accurate only to the nearest $\frac{1}{8}$ in. If knowing the degree of accuracy is important, the fraction should not be reduced. It should be reduced, however, before an actual measurement is made: It is much simpler to measure $\frac{5}{8}$ in. on a carpenter's rule than to measure $\frac{40}{64}$ in.

Exercise 3-7

Convert to equivalent decimals.

1. $\frac{5}{8}$

2. $\frac{3}{25}$

3. $\frac{2}{3}$

4. $\frac{8}{15}$

5. $1\frac{7}{8}$

6. $22\frac{25}{32}$

7. $8\frac{2}{3}$

8. $3\frac{5}{64}$

9. $19\frac{9}{16}$

10. $8\frac{3}{8}$

Convert to equivalent fractions, reduced to lowest terms.

11. 0.825

12. 0.135

13. 0.41

14. 4.825

15. 15.425

16. 7.82

17. 4.235

18. 9.17

19. 26.24

20. 11.006

Convert the repeating decimals to simplified fractions.

21. 0.666. . .

22. $0.\overline{936}$

23. $2.\overline{45}$

24. 6.18181818. . .

25. 25.4444. . .

26. $3.\overline{39}$

27. 3.66666. . .

28. $5.\overline{6}$

29. $35.\overline{2}$

30. 14.212121. . .

Convert these decimals to the nearest fraction equivalent, as specified, as shown in the example. Reduce to lowest terms whenever possible.

	Decimal	Half	Fourth	8th	16th	32nd	64th
a.	2.39 in.	$2\frac{1}{2}$ in.	$2\frac{1}{2}$ in.	$2\frac{3}{8}$ in.	$2\frac{3}{8}$ in.	$2\frac{3}{8}$ in.	$2\frac{25}{64}$ in.
31.	15.891 in.						
32.	35.319 in.						
33.	7.299 in.						
34.	121.814 in.						
35.	7.629 in.						
36.	38.79 in.						
37.	48.584 in.						
38.	134.327 in.						
39.	13.622 in.						
40.	58.05 in.						

41. Measurements of 28.365, 26.891, and 18.372 in. are to be totaled and converted to an equivalent fraction. What is the total of the three lengths to the nearest $\frac{1}{16}$ in.? (*Hint:* Add the decimal numbers and then convert the final answer.)

42. A blueprint specification calls for a steel rod to have a 0.429-in. diameter. If this rod is to fit snugly through a hole to be drilled in a stud, what size of drill bit, to the nearest $\frac{1}{16}$ in., should be used?

43. What is the smallest drill bit, to the nearest $\frac{1}{32}$ in., that can be used to drill a hole that must be at least 0.139 in. in diameter?

44. A length is measured on a decimal ruler (a ruler with inches divided into 10 equal units). If the decimal ruler reads 8.9 in., what is the closest equivalent value in $\frac{1}{8}$ in.?

45. The footing for a lally column is 8.235 in. thick. What does that equal to the nearest $\frac{1}{8}$ in.?

46. A desktop is made of $\frac{5}{8}$-in. plywood covered on one side with a laminate 0.03 in. thick. What is the thickness of the desktop to the nearest $\frac{1}{32}$ in.?

chapter 4

Weights, Measures, and Conversions

4-1 LINEAR MEASURE

OBJECTIVES

Upon completing this section, the student will be able to:

1. Solve problems involving units of length.
2. Convert from one unit of length to an equivalent unit.

The common units of length used in the United States are the inch, foot, yard, and mile. Rods, chains, fathoms, and hands are examples of other units of length which are used for specific purposes. Rods and chains are used in surveying, fathoms are used to measure nautical depths, and hands measure the height of horses. Following is a partial listing of linear equivalents.

1 ft = 12 in.

1 yd = 3 ft

1 mi = 5280 ft

1 rod = $16\frac{1}{2}$ ft

1 hand = 4 in.

1 chain = 66 ft

1 fathom = 6 ft

Every carpenter must be able to readily convert units from feet to inches, and vice versa. *To convert feet to inches: multiply the number of feet by 12.*

Examples

A. Convert 18 ft to inches.

$$18 \text{ ft} \times 12 \text{ in./ft} = 216 \text{ in.}$$

B. Convert 3.75 ft to inches.

$$3.75 \text{ ft} \times 12 = 45 \text{ in.}$$

To convert inches to feet: divide the number of inches by 12.

Multiply to convert to a smaller unit (*example:* converting feet to inches).
Divide to convert to a larger unit (*example:* converting inches to feet).

Examples

C. Convert 36 in. to feet.

$$36 \text{ in.} \div 12 \text{ in./ft} = 3 \text{ ft}$$

D. Change 18 in. to feet.

$$18 \text{ in.} \div 12 = 1.5 \text{ ft}$$

In Example D the answer is in decimal feet. In surveying, decimal feet are routinely used, but the carpenter generally works in feet and inches. Study the following examples for changing inches into feet and inches, and decimal feet into feet and inches.

Examples

E. Change 437 in. into feet and inches. It is recommended that these problems be worked on a calculator.

$437 \text{ in.} \div 12 = 36.41\overline{6} \text{ ft}$

1. Dividing inches by 12 yields feet.

$36.41\overline{6} \text{ ft} - 36 \text{ ft} = 0.41\overline{6} \text{ ft}$

2. Write down the 36 whole feet and then subtract longhand or with the calculator. The decimal part that is left is also in feet.

$0.41\overline{6} \text{ ft} \times 12 = 5 \text{ in.}$

3. Multiply the remaining decimal by 12. This converts the decimal feet back to inches.

$437 \text{ in.} = 36.41\overline{6} \text{ ft} = 36 \text{ ft } 5 \text{ in.}$

4. It is a good idea to write down the whole number of feet as they are found.

```
    feet    inches
    36  R  (5)
12)437
   36
    77
    72
    (5)
```

5. If this problem is worked without a calculator, the 36 in the quotient equals the number of feet, and the remainder 5 is inches.

36 R5 = 36 ft 5 in.

F. Change 19.58333. . . ft into feet and inches.

$19.58\overline{3} \text{ ft} - 19 \text{ ft} = 0.58\overline{3} \text{ ft}$

1. The 19 is the whole number of feet. Subtract the 19 and write it down as feet. *Note:* Key 19.583333 into the calculator. Rounding the number

when entering it in the calculator can cause a significant error in accuracy in some cases.

$0.58\overline{3}$ ft × 12 = 7 in. — 2. Multiply the decimal feet by 12 to convert to inches.

$19.58\overline{3}$ ft = 19 ft 7 in. — 3. Write down the inches.

Exercise 4-1A

Convert the following into feet and inches. Whenever a repeating decimal is to be converted, enter the repeating digits as many times as possible into the calculator. Rounding off can significantly alter the accuracy of the answer.

1. 2.75 ft
2. $14.58\overline{3}$ ft
3. 9.5 ft
4. $8.1\overline{6}$ ft
5. $7.08\overline{3}$ ft
6. 131 in.
7. 85 in.
8. 29 in.
9. 74 in.
10. 17 in.

Frequently, measurements do not come out to a whole number of inches. In such cases the carpenter will generally change decimal inches to the nearest $\frac{1}{8}$, $\frac{1}{16}$, etc., in., depending on the accuracy required. Study the following examples.

Examples

G. Find 15.671 in. to the nearest $\frac{1}{16}$ in.

15.671 in. − 15 in. = 0.671 in. — 1. Record and subtract 15, the whole number of inches.

$$\frac{0.671 \text{ in.}}{1} \times \frac{16}{16} = \frac{10.736 \text{ in.}}{16}$$

2. Multiply the decimal inches by $\frac{16}{16}$ since $\frac{1}{16}$ in. is the desired accuracy.

$$\frac{10.736 \text{ in.}}{16} \doteq \frac{11}{16} \text{ in.}$$

3. Round the numerator to the nearest whole number.

$$15.671 \text{ in.} \doteq 15\frac{11}{16} \text{ in.}$$

4. This is the fractional equivalent to the nearest $\frac{1}{16}$ in.

H. Surveyors' tapes are measured in feet and decimal feet to the nearest hundredth of a foot. If a surveyors' tape is used to measure a length of 7.43 ft, what is that length in feet and inches to the nearest $\frac{1}{8}$ in.?

7.43 ft − 7 ft = 0.43 ft — 1. Record and subtract the whole number of feet.

0.43 ft × 12 = 5.16 in. — 2. Change to inches.

5.16 in. − 5 in. = 0.16 in. — 3. Record and subtract the whole number of inches.

$$\frac{0.16 \text{ in.}}{1} \times \frac{8}{8} = \frac{1.28 \text{ in.}}{8}$$

4. Multiply by $\frac{8}{8}$ since that is the desired accuracy.

$$\frac{1.28 \text{ in.}}{8} \doteq \frac{1}{8} \text{ in.}$$

5. Round the numerator to the nearest whole number.

$$7.43 \text{ ft} \doteq 7 \text{ ft } 5\frac{1}{8} \text{ in.}$$

6. This is the fractional equivalent to the nearest $\frac{1}{8}$ in.

I. Determine 49.1782 in. in feet and inches to the nearest $\frac{1}{32}$ in. Reduce fraction if possible.

$$49.1782 \text{ in.} \div 12 = 4.0981833 \text{ ft}$$

1. Convert to feet.

$$4.0981833 \text{ ft} - 4 \text{ ft} = 0.0981833 \text{ ft}$$

2. Record and subtract the whole number of feet.

$$0.0981833 \text{ ft} \times 12 = 1.1782 \text{ in.}$$

3. Convert the decimal feet back to inches.

$$1.1782 \text{ in.} - 1 \text{ in.} = 0.1782 \text{ in.}$$

4. Record and subtract the whole inches.

$$0.1782 \text{ in.} \times \frac{32}{32} = \frac{5.7024}{32} \text{ in.}$$

5. Multiply by $\frac{32}{32}$ since that is the desired accuracy.

$$\frac{5.7024}{32} \doteq \frac{6}{32} = \frac{3}{16} \text{ in.}$$

6. Round the numerator to the nearest whole number and reduce to lowest terms.

$$49.1782 \text{ in.} \doteq 4 \text{ ft } 1\frac{3}{16} \text{ in.}$$

7. These are equivalent to the nearest $\frac{1}{32}$ in.

Exercise 4-1B

Change to feet and inches to the accuracy indicated.

1. 4.1785 ft to the nearest $\frac{1}{16}$ in.
2. 121.8316 in. to the nearest $\frac{1}{32}$ in.
3. 7.19932 ft to the nearest $\frac{1}{8}$ in.
4. 41.3167 in. to the nearest $\frac{1}{4}$ in.
5. 19.3751 in. to the nearest $\frac{1}{64}$ in.
6. 4.1923 ft to the nearest $\frac{1}{16}$ in.
7. 85.2914 in. to the nearest $\frac{1}{16}$ in.
8. 4.6281 ft to the nearest $\frac{1}{32}$ in.
9. 4.7522 ft to the nearest $\frac{1}{16}$ in.
10. 16.3343 ft to the nearest $\frac{1}{32}$ in.

4-2 OPERATIONS WITH MIXED UNITS

OBJECTIVES

Upon completing this section, the student will be able to:

1. Solve addition and subtraction problems involving mixed units.
2. Solve multiplication and division problems involving mixed units.

Addition and subtraction of measurements can be accomplished in either decimal feet or in feet and inches. Which method is simpler depends on the form of the measurements. Study the following examples.

Examples

A. Add 3 ft 8 in. + 5 ft 9 in. + 6 ft 3 in.

3 ft 8 in. 5 ft 9 in. + 6 ft 3 in. 14 ft 20 in.	1. Add the feet and inches separately.
14 ft + 1 ft 8 in.	2. If the total number of inches equals or exceeds 12, convert to feet and inches.
15 ft 8 in.	3. Combine the feet.

B. Subtract 5 ft 7 in. from 8 ft 3 in.

8 ft 3 in. − 5 ft 7 in.	1. The 7 in. cannot be subtracted directly from 3 in.
12 in. 7 ft 15 in. − 5 ft 7 in.	2. Borrow 1 ft from the 8 ft, convert it to 12 in., and add to the existing 3 in.
7 ft 15 in. − 5 ft 7 in.	3. Perform the subtraction.
2 ft 8 in.	4. 2 ft 8 in. is the difference.

C. Add 3.25 ft + $4.41\bar{6}$ ft and give the answer in feet and inches.

3.25 ft + $4.41\bar{6}$ ft $7.\bar{6}$ ft	1. Perform the addition. When using a calculator, do not round $4.41\bar{6}$ to 4.417. The answer will be more accurate if entered into the calculator as 4.41666666.
$7.\bar{6}$ ft − 7 ft = $0.\bar{6}$ ft	2. Record and subtract the whole number of feet.
$0.\bar{6}$ ft × 12 = 8 in.	3. Multiply the decimal feet by 12 to convert to inches.
$7.\bar{6}$ ft = 7 ft 8 in.	4. Give the answer in feet and inches.

Exercise 4-2A

Perform the indicated addition and subtraction. Give answers in feet and inches.

1. 8 ft 9 in. + 9 ft 2 in.
2. 3 ft 7 in. − 2 ft 10 in.
3. $4.\overline{6}$ ft − 2 ft 7 in.
4. 3.285 ft + 6.715 ft
5. 4 ft 11 in. + 5 ft 9 in. + 8 ft 11 in.
6. 16 ft 5 in. − 4 ft 10 in.
7. $3.41\overline{6}$ ft − 2 ft 10 in.
8. 7 ft 11 in. + 0 ft 9 in. + 3 ft 7 in.
9. $4.58\overline{3}$ ft − $2.91\overline{6}$ ft
10. 3 ft 5 in. + 1 ft 8 in.

Most addition, subtraction, and measuring is done with feet and inches. Multiplication and division, however, is best done with decimal feet or whole inches. The following examples demonstrate how to convert from feet and inches to decimal feet, and from feet and inches to whole inches.

Examples

D. Convert 5 ft 9 in. to decimal feet.

9 in. ÷ 12 = 0.75 ft	1. Divide the inches by 12 to convert to feet.
0.75 ft + 5 ft = 5.75 ft	2. Add the decimal feet to the whole number of feet.
5 ft 9 in. = 5.75 ft	

E. Change 18 ft 5 in. to decimal feet.

5 in. ÷ 12 = $0.41\overline{6}$ ft	1. Divide the inches by 12 to convert to feet. Do *not* round off.
$0.41\overline{6}$ ft + 18 ft = $18.41\overline{6}$ ft	2. Add the decimal feet to the whole number of feet.
18 ft 5 in. = $18.41\overline{6}$ ft	

F. Convert 9 ft 8 in. to inches.

9 ft × 12 = 108 in.	1. Multiply the feet by 12 to convert to inches.
108 in. + 8 in. = 116 in.	2. Add to the existing 8 in.
9 ft 8 in. = 116 in.	

G. A wall requires 19 studs each 7 ft 8 in. long. What would be the total length of the studs placed end to end? Give answer in feet and inches.

7 ft 8 in. × 19	1. This is a multiplication problem that will be solved by first converting the 7 ft 8 in. to decimal feet.
8 in. ÷ 12 = $0.\overline{6}$ ft	2. Convert feet and inches to decimal feet.
$0.\overline{6}$ ft + 7 ft = $7.\overline{6}$ ft	3. Add decimal feet to whole number of feet.

$7.\overline{6}$ ft × 19 = $145.\overline{6}$ ft

4. Multiply to obtain the total decimal feet. *Important:* When done on a calculator, do *not* remove these figures from the calculator or round. Accuracy will be lost.

$145.\overline{6}$ ft − 145 = $0.\overline{6}$ ft

5. Record and subtract the whole feet.

$0.\overline{6}$ ft × 12 = 8 in.

6. Convert the decimal feet to inches. *Note:* Some calculators will show the result as 7.999999. This should be interpreted as 8.

(7 ft 8 in.) × 19 = 145 ft 8 in.

7. This problem can also be solved by first changing feet to inches and adding to the existing inches.

H. Find $\frac{7}{8}$ of the length 5 ft 4 in.

5 ft 4 in. = $5.\overline{3}$ ft

1. Convert 5 ft 4 in. to decimal feet.

$5.\overline{3}$ ft × $\frac{7}{8}$ = $4.\overline{6}$ ft

2. Without removing 5.33333333 ft from the calculator, multiply by $\frac{7}{8}$ (multiply by 7 and then divide by 8).

$4.\overline{6}$ ft = 4 ft 8 in.

3. Convert 4.6 ft to feet and inches.

I. Divide 10 ft 6 in. by 9.

10 ft 6 in. = ? in.

1. Convert to inches before dividing.

10 ft × 12 = 120 in.

2. Convert 10 ft to inches.

120 in. + 6 in. = 126 in.

3. Add to the existing 6 in.

126 in. ÷ 9 = 14 in.

4. Divide.

14 in. = 1 ft 2 in.

5. Convert answer to feet and inches.

Exercise 4-2B

Perform the indicated operations, giving answers in feet and inches.

1. 8 ft 9 in. × 5

2. 4 ft 9 in. × $\frac{2}{3}$

3. 8 ft 2 in. ÷ 7

4. 4 ft 1 in. ÷ 7

5. 3 ft 8 in. ÷ 11

6. 2 ft 6 in. ÷ 5

7. 4 ft 6 in. ÷ 9

8. 5 ft 3 in. × 4

9. 16 ft 8 in. × $\frac{3}{5}$

10. 5 ft 1 in. × 6

4-3 UNIT CONVERSIONS

OBJECTIVES

Upon completing this section, the student will be able to:

1. Perform one-step conversions using the unit conversion method.
2. Perform two- and three-step conversions using the unit conversion method.

Because carpenters so frequently need to convert from feet to inches and vice versa, it is important to know the fast way to accomplish these conversions (covered in Section 4-1): feet to inches—multiply by 12; inches to feet—divide by 12. Many other conversions are also used in building construction, and sometimes it is not obvious whether the conversion factor should be multiplied or divided. The **unit conversion** method solves this problem. In the unit conversion method, the quantity to be converted is always multiplied by a fraction known as a *unity ratio*. In a unity ratio, the numerator and denominator are equivalent, but are different units of measurement. Here are some examples of unity ratios:

$$\frac{1\ \text{pound}}{16\ \text{ounces}} \qquad \frac{16\ \text{ounces}}{1\ \text{pound}} \qquad \frac{1\ \text{foot}}{12\ \text{inches}}$$

$$\frac{12\ \text{inches}}{1\ \text{foot}} \qquad \frac{1\ \text{yard}}{3\ \text{feet}} \qquad \frac{36\ \text{inches}}{1\ \text{yard}}$$

Multiplying by a unity ratio *does not change the original amount. It does change the type of units*. Study the following examples to determine how to convert from one unit to another using unity ratios.

Examples

A. Convert 9336 ft to miles.

$$1\ \text{mi} = 5280\ \text{ft}$$

$$\frac{1\ \text{mi}}{5280\ \text{ft}} \text{ or } \frac{5280\ \text{ft}}{1\ \text{mi}}$$

1. Using the conversion factor 1 mi = 5280 ft, set up the two unity ratios.

$$\frac{9336\ \text{ft}}{1} \times \frac{5280\ \text{ft}}{1\ \text{mi}}$$

2. This unity ratio will not work, because the final units would be (feet × feet)/mile or square feet/miles.

$$\frac{9336\ \cancel{\text{ft}}}{1} \times \frac{1\ \text{mi}}{5280\ \cancel{\text{ft}}}$$

3. This is the proper unity ratio to use. Units of feet cancel, leaving only the desired unit of miles.

$$\frac{9336}{1} \times \frac{1\ \text{mi}}{5280} \doteq 1.8\ \text{mi}$$

4. All units in the numerator are multiplied together. Divide by all units in the denominator. Multiplication and division by 1 can be ignored.

B. Convert 18.5 yards to rods. For this conversion, the following equivalents are used: 1 yd = 3 ft, and 16.5 ft = 1 rod.

$$18.5\ \text{yd} \times \frac{?}{?} = \frac{?\ \text{rods}}{1}$$

1. It is easiest to set up a problem like

this if the original units and final units are written down first.

$$\frac{18.5\,\cancel{yd}}{1} \times \frac{3\text{ ft}}{1\,\cancel{yd}} = \frac{?\text{ rods}}{1}$$

2. Since the direct conversion factor between yards and rods is not given, yards must first be converted to feet.

$$\frac{18.5}{1} \times \frac{3\,\cancel{ft}}{1} \times \frac{1\text{ rod}}{16.5\,\cancel{ft}} = \frac{?\text{ rods}}{1}$$

3. The resulting feet are converted to rods by multiplying by the unity ratio 1 rod/16.5 ft.

$$\frac{18.5}{1} \times \frac{3}{1} \times \frac{1\text{ rod}}{16.5} \doteq 3.36\text{ rods}$$

4. When the only unit left is the unit desired, perform the multiplication and division.

$$18.5\text{ yd} \doteq 3.36\text{ rods}$$

Exercise 4-3

Using the unit conversion method, convert to the units desired.

1. 8141 ft to miles
2. 121 chains to miles
3. 420 ft to fathoms
4. 8.3 ft to hands
5. 2.5 rods to yards
6. 82 ft to rods
7. 4.5 chains to feet

4-4 AREA AND VOLUME CONVERSIONS

OBJECTIVES

Upon completing this section, the student will be able to:

1. Convert from one unit of area to another.
2. Convert from one unit of volume to another.
3. Solve problems involving unit conversions.

Following are listed some common conversions for area and volume.

1 square foot = 144 square inches
1 square yard = 9 square feet
1 acre = 43,560 square feet
1 square mile = 640 acres
1 cubic foot = 1728 cubic inches
1 cubic yard = 27 cubic feet

1 gallon = 231 cubic inches

1 cubic foot = 7.5 gallons

Square units are units of area. The simplest types of areas to find are rectangular shapes, but the areas of triangles, circles, and irregular figures can also be measured and computed. For a rectangle, the area in square units is found by multiplying *length* $\times$ *width*. The length and width are in linear units, but the resulting area is in square units. This is because *inches* $\times$ *inches* = *square inches*, *feet* $\times$ *feet* = *square feet*, and so on.

Examples

A. How many square feet are there in a rectangular ranch house that measures 24 ft by 36 ft?

$$\text{area} = \text{length} \times \text{width}$$
$$= 36 \text{ ft} \times 24 \text{ ft} = 864 \text{ ft}^2$$

B. A certain square is 1 yd long by 1 yd wide. This means that it has an area of 1 yd^2. How many square feet of area does the square have? Study the figure to understand why 1 yd^2 = 9 ft^2.

$$1 \text{ yd}^2 = 1 \text{ yd} \times 1 \text{ yd}$$
$$= 3 \text{ ft} \times 3 \text{ ft}$$
$$= 9 \text{ ft}^2$$

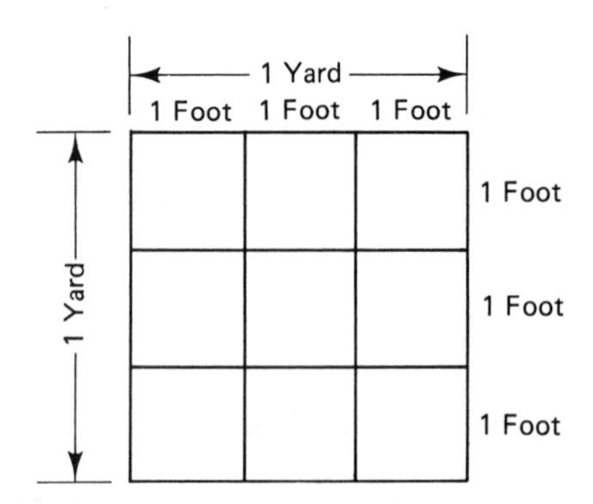

It is useful to know some of the more common square and cubic conversions. The conversion 1 yd^3 = 27 ft^3 is used frequently in building construction. Other commonly used conversions are 1 yd^2 = 9 ft^2, and 144 in.^2 = 1 ft^2. It is unnecessary to memorize all the square and cubic conversions. If the linear conversions are known, the square and cubic conversions can be found. Study the following examples.

Examples

C. Determine the number of square inches in 1 yd^2.

1 yd = 36 in.	1. Write down the linear conversion.
$(1 \text{ yd})^2 = (36 \text{ in.})^2$	2. Square each side of the conversion.
$1 \text{ yd}^2 = 1296 \text{ in.}^2$	3. 1 yd $\times$ 1 yd = 1 yd^2 and 36 in. $\times$ 36 in. = 1296 in.^2.

D. How many cubic feet are there in 1 yd^3?

1 yd = 3 ft	1. Write down the linear conversion.
$(1 \text{ yd})^3 = (3 \text{ ft})^3$	2. Cube both sides of the conversion since the cubic conversion is desired.

$1 \text{ yd}^3 = 27 \text{ ft}^3$	3. $1 \text{ yd} \times 1 \text{ yd} \times 1 \text{ yd} = 1 \text{ yd}^3$, and $3 \text{ ft} \times 3 \text{ ft} \times 3 \text{ ft} = 27 \text{ ft}^3$.

Caution! Be careful to recognize the difference between *4 square feet* and *4 feet square* (Figure 4-1). An area of 4 square feet can have any shape as long as it has a total area of 4 ft^2. An area 4 ft square has the specific shape of a square, and actually has an area of 4 ft × 4 ft or 16 ft^2.

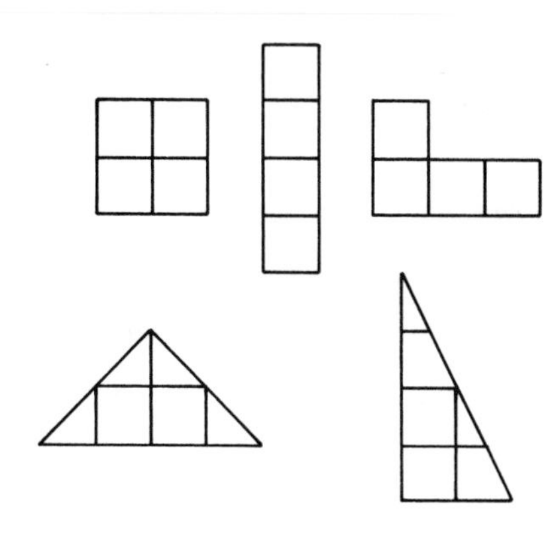

All of these figures have an area of *4 square feet*

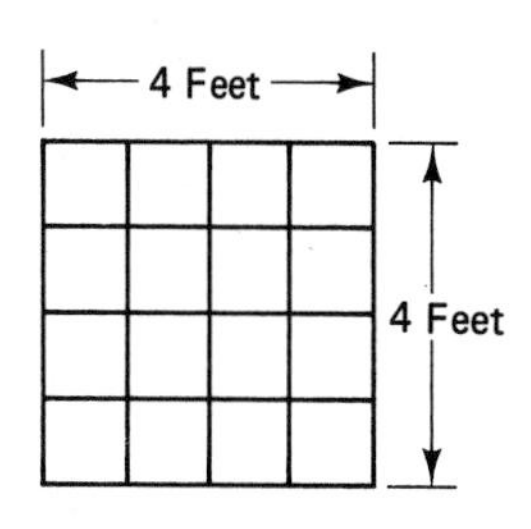

A figure *4 feet square* has an area of *16 square feet*

Figure 4-1

Examples

E. A rectangular cellar excavation is to be 40 ft long, 28 ft wide, and 8 ft deep. Determine the number of cubic yards of dirt to be excavated.

First the volume in cubic feet must be determined. For a rectangular solid, *volume* = *length* × *width* × *height* (or *depth*).

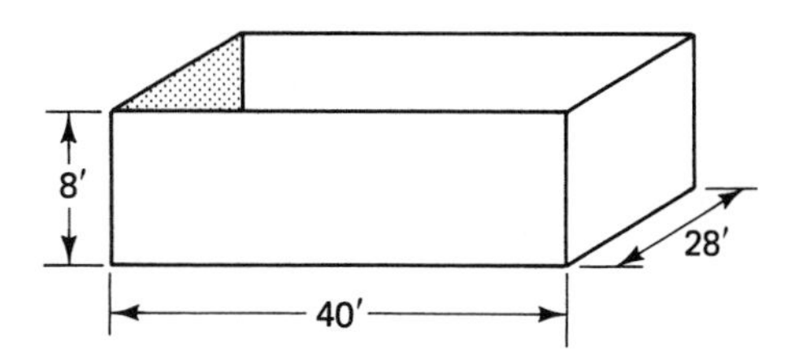

$40 \text{ ft} \times 28 \text{ ft} \times 8 \text{ ft} = 8960 \text{ ft}^3$	1. Determine the volume in cubic feet.
$\dfrac{8960 \cancel{\text{ft}^3}}{1} \times \dfrac{1 \text{ yd}^3}{27 \cancel{\text{ft}^3}}$	2. Using the unit conversion method, multiply by the conversion factor of $\frac{1}{27}$ to determine the equivalent number of cubic yards.
$\dfrac{8960}{1} \times \dfrac{1 \text{ yd}^3}{27} = 331.85 \text{ yd}^3$	3. Observe that the unwanted units cancel.

F. Convert 835.5 in.^2 into square feet.

$1 \text{ ft} = 12 \text{ in.}$	1. Assuming that the conversion between square feet and square inches is not known, write down the linear conversion.
$(1 \text{ ft})^2 = (12 \text{ in.})^2$	2. Square both sides.
$1 \text{ ft}^2 = 144 \text{ in.}^2$	3. This is the conversion in square units.

$$\frac{835.5 \cancel{\text{in.}^2}}{1} \times \frac{1 \text{ ft}^2}{144 \cancel{\text{in.}^2}}$$

4. Multiply by the unit conversion 1 ft^2/144 in.^2, canceling the units of square inches.

$$\frac{835.5}{1} \times \frac{1 \text{ ft}^2}{144}$$

5. The answer is in square feet, since that is the only unit that did not cancel.

$835.5 \text{ in.}^2 = 5.8 \text{ ft}^2$

G. One square mile contains how many square feet?

This problem can be worked in several ways. Two approaches are shown here.

Method 1: Squaring Linear Conversions

$$1 \text{ mi} = 5280 \text{ ft}$$
$$(1 \text{ mi})^2 = (5280 \text{ ft})^2$$
$$1 \text{ mi}^2 = 27{,}878{,}400 \text{ ft}^2$$

Method 2: Unit Conversion Method. For this method the following conversions must be known:

$$1 \text{ mi}^2 = 640 \text{ acres}$$
$$1 \text{ acre} = 43{,}560 \text{ ft}^2$$

$$\frac{1 \cancel{\text{mi}^2}}{1} \times \frac{640 \cancel{\text{acres}}}{1 \cancel{\text{mi}^2}} \times \frac{43{,}560 \text{ ft}^2}{1 \cancel{\text{acre}}} = 27{,}878{,}400 \text{ ft}^2$$

Observe that the unit fractions are set up in such a way as to cancel out all unwanted units. The problem started with square miles and ended up with square feet. Therefore, square miles and acres had to cancel out, leaving only the square feet.

Exercise 4-4

Convert to the units indicated.

1. 1 yd^3 to cubic inches
2. 8.56 yd^2 to square feet
3. 1426 acres to square miles
4. 6.8 ft^2 to square inches
5. 58,935 in.^3 to cubic yards
6. 141 ft^2 to square yards
7. 574 ft^2 to square rods (1 rod = 16.5 ft)
8. 3.4 mi^2 to acres
9. 84,000 yd^2 to acres
10. 165 ft^3 to cubic yards
11. A driveway requires 1500 ft^3 of gravel. How many cubic yards is that? Round to the nearest cubic yard.
12. A foundation wall is to be 7 ft high, 24 ft long, and 6 in. thick. How many cubic yards of concrete is needed for the wall? (*Hint:* First change the 6 in. to feet and find the number of cubic feet in the wall. Then convert to cubic yards.)
13. A building lot contains 35,200 ft^2. What part of an acre is this? Round to two decimal places.
14. A large meeting hall is 44 ft wide, 60 ft long, and has a 12-ft ceiling. How many

cubic yards of air are in the building? (The dimensions given are inside dimensions.)

15. A kitchen is 10 ft 5 in. long by 7 ft 9 in. wide. What is the area of the kitchen in square feet? (*Hint:* Change the units to decimal feet before multiplying.)

16. A square tile is 9 in. × 9 in. How many of these tiles are required to cover 1 yd^2?

17. A door opening measures 68 in. high by 32 in. wide. What is the area of the opening in square feet?

18. A pane of glass measures 1 ft 3 in. by 14 in. If the glass costs $0.65 per square foot, what is the price of a replacement pane?

4-5 RATE CONVERSION PROBLEMS

OBJECTIVES

Upon completing this section, the student will be able to:

1. Convert rates from one unit to another.
2. Solve word problems involving rate conversions.

Rates can frequently be thought of as "per" ratios: miles *per* hour, pounds *per* square inch, miles *per* gallon, dollars *per* pound, and so on. "Per" always can be written as a fraction bar with the unit before the "per" in the numerator and the unit after "per" in the denominator. Miles per hour can be written miles/hour. Pounds per square foot can be written pounds/square foot. Rates are unity ratios that are equal for a specific situation only. The unity ratio 12 in./1 ft is always true, because 12 in. always equals 1 ft. The rate 45 miles per hour is a unity ratio *only* for the specific instance when that is the speed. If it takes more or less time to travel 45 mi, then 45 mi does not equal 1 hr. Following are some unit conversions involving rates:

Examples

A. A car travels for 3.5 hr at 52 mph (miles per hour). How far has it gone?

$$\frac{3.5\text{ hr}}{1} \times \frac{?}{?} = \frac{?\text{ mi}}{1}$$

1. Write down the given units and those desired.

$$\frac{3.5\text{ \cancel{hr}}}{1} \times \frac{52\text{ mi}}{1\text{ \cancel{hr}}} = \frac{?\text{ mi}}{1}$$

2. The rate 52 mph means 52 miles/hours. "Per" always means a fraction bar.

$$\frac{3.5}{1} \times \frac{52\text{ mi}}{1} = 182\text{ mi}$$

3. When only the unit desired is left, perform the multiplication.

B. 3520 feet per minute = ? miles per hour.

$$\frac{3520\text{ ft}}{1\text{ min}} \times \frac{?}{?} = \frac{?\text{ mi}}{\text{hr}}$$

1. Write down the units given and those desired.

$$\frac{3520\text{ \cancel{ft}}}{1\text{ min}} \times \frac{1\text{ mi}}{5280\text{ \cancel{ft}}}\frac{?}{?} = \frac{?\text{ mi}}{\text{hr}}$$

2. Change feet to miles with the unit ratio 1 mi/5280 ft.

$$\frac{3520}{1\text{ \cancel{min}}} \times \frac{1\text{ mi}}{5280} \times \frac{60\text{ \cancel{min}}}{1\text{ hr}} = \frac{?\text{ mi}}{\text{hr}}$$

3. Convert minutes to hours using the unity ratio 60 min/1 hr.

$$\frac{3520}{1} \times \frac{1 \text{ mi}}{5280} \times \frac{60}{1 \text{ hr}} = \frac{40 \text{ mi}}{\text{hr}}$$

4. When only the desired units are left in the numerator and denominator, perform the multiplication and division.

3520 feet per minute = 40 mph

C. On a certain trip, a small plane used 8.9 gal of gas per hour while averaging 120 mph. What was the gas consumption in miles per gallon?

$$\frac{?}{?} \times \frac{?}{?} = \frac{\text{mi}}{\text{gal}}$$

1. Write down the ratio desired. Note that miles must end up in the numerator and gallons in the denominator.

$$\frac{120 \text{ mi}}{\cancel{\text{hr}}} \times \frac{1 \cancel{\text{hr}}}{8.9 \text{ gal}} = \frac{? \text{ mi}}{\text{gal}}$$

2. Notice that the 8.9 gal per hour ratio had to be inverted in order to achieve the desired units in the answer.

$$\frac{120 \text{ mi}}{1} \times \frac{1}{8.9 \text{ gal}} = \frac{13.5 \text{ mi}}{\text{gal}}$$

3. When only the units desired are left, perform the required division.

13.5 miles per gallon

This could have been set up and worked as a division problem without using unity ratios, but most students find this method easier to understand.

D. Dry mahogany weighs 53 pounds per cubic foot (53 lb/ft^3). What is its weight in ounces per cubic inch (oz/in.^{3})?

$$\frac{53 \text{ lb}}{1 \text{ ft}^3} \times \frac{?}{?} = \frac{? \text{ oz}}{\text{in.}^3}$$

1. Set up the units given and those desired.

$$\frac{53 \cancel{\text{lb}}}{1 \text{ ft}^3} \times \frac{16 \text{ oz}}{1 \cancel{\text{lb}}} = \frac{? \text{ oz}}{\text{in.}^3}$$

2. Multiply by the unity ratio 16 oz/lb. The units left at this point are in ounces/ft^3.

$$\frac{53}{1 \cancel{\text{ft}^3}} \times \frac{16 \text{ oz}}{1} \times \frac{1 \cancel{\text{ft}^3}}{1728 \text{ in.}^3} = \frac{? \text{ oz}}{\text{in.}^3}$$

3. Multiply by the unity ratio 1 ft^3/1728 in.^{3}.

$$\frac{53}{1} \times \frac{16 \text{ oz}}{1} \times \frac{1}{1728 \text{ in.}^3} = \frac{0.49 \text{ oz}}{\text{in.}^3}$$

4. When only the correct units are left, perform the multiplication and division.

53 lb/ft^3 = 0.49 oz/in.^{3}

Steps 2 and 3 can be reversed. The results would be the same.

Exercise 4-5

Convert to the units indicated.

1. 25 mph (miles per hour) to feet per hour
2. \$3.52 per pound to dollars per ounce
3. 25.8 gpm (gallons per minute) of water to pounds per minute (1 gallon of water weighs 8.35 lb)

4. 80 mph to yards per minute
5. 2.8 lb of nails per 400 board feet to ounces of nails per board foot
6. 55 mph to feet per second
7. The live-load rating in a building is 220 psf (pounds per square foot). What is this in psi (pounds per square inch)? (*Remember:* 1 ft^2 = 144 $in.^2$.)
8. A carpenter's truck averages 14 mpg. In most of the countries in the world this would be measured in kilometers per liter. Determine the truck's gas consumption in kilometers per liter. Use the conversions 1 gal = 3.785 liters and 1 mi = 1.61 km.
9. A water pump can pump 28 gpm. At that rate, how many pounds per minute can it pump? (1 gal of water = 8.35 lb.)
10. On a certain trip, a car averages 58 mph and gets 14.3 mpg. How much fuel does the car burn in gallons per hour?
11. A cast-iron pipe weighs 4.2 lb per foot of length. What does it weigh in ounces per inch of length?
12. A stack of dry maple boards weighs 225 lb. If dry maple weighs 49 lb/ft^3, how many cubic feet of maple are in the stack?
13. A certain insulation has an R factor of 4 per inch. What is the R factor per foot for this insulation?
14. A dry spruce 2 × 4 weighs approximately 1.5 oz per lineal inch. What is this equal to in pounds per lineal foot?

chapter 5

Ratio and Proportion

5-1 RATIO

OBJECTIVES

Upon completing this section, the student will be able to:

1. Identify and express ratios in several forms.
2. Express ratios in simplest terms.
3. Solve problems involving ratios.

A **ratio** is a comparison of (usually) two quantities. A gas–oil mix of 32 to 1 is a ratio of the amount of gas compared to oil to be mixed for a certain engine. The pitch of a roof is a ratio of the rise to the span. The Greek lowercase letter pi (π) represents the ratio of the circumference of a circle to its diameter. A blueprint may have a ratio of $\frac{1}{4}$ in. to 1 ft. Here are several common ways to express ratios.

Examples

A. 2 in./5 in.

1. Any ratio can be expressed as a fraction.

2 in./5 in. = 2/5

2. Whenever possible, reduce to a unitless ratio by canceling identical units in the numerator and denominator.

B. 2 : 5

1. Any ratio expressed as a fraction can also be expressed in this form, read "the ratio of two to five."

$2/5 \Rightarrow 2 : 5$	2. The fraction bar is replaced by a colon; the numerator is written before the colon; the denominator is written after the colon.

C. Express in simplest terms.

$7 : 14$	1. Any ratio written in this form can be reduced like a fraction.
$\frac{7}{7} : \frac{14}{7}$	2. 7 divides into both the first and second numbers (numerator and denominator in fraction form).
$1 : 2$	3. The simplified ratio is therefore 1 : 2.

D. Express 16 qt to 5 gal as a ratio in simplest terms.

16 qt : 5 gal	1. Write as a ratio.
4 gal : 5 gal	2. Change to the same units. (The 5 gallons could have been changed to quarts instead for the same final result.)
4 ~~gal~~ : 5 ~~gal~~	3. Cancel identical units.
4 : 5	4. This is a unitless ratio reduced to lowest terms. Removing the units makes the ratio more useful. Now it could be used for measuring 4 qt to 5 qt, or 4 cups to 5 cups, as well as 4 gal to 5 gal.

E. Simplify the ratio 2 ft 8 in. to 6 ft 8 in.

2 ft 8 in. : 6 ft 8 in.	1. Simplify to a reduced, unitless ratio.
32 in. : 80 in.	2. Change to inches.
32 : 80	3. Cancel the unit inches.
2 : 5	4. Reduce to lowest terms.
2 ft 8 in. : 6 ft 8 in. = 2 : 5	5. The ratio in simplest terms is 2 : 5.

F. A line 15 in. on a blueprint represents a length of 40 ft on a house. To what scale was the blueprint drawn?

15 in. = 40 ft	1. For scales such as maps and blueprints, the = sign is often used instead of a colon.
$\frac{15 \text{ in.}}{40} = \frac{40 \text{ ft}}{40}$	2. Blueprints are generally shown as a scale of X in. = 1 ft. Therefore, divide both sides by 40 to obtain 1 ft on the right-hand side.
$\frac{3}{8}$ in. = 1 ft	3. Simplify, leaving the blueprint in fractions of an inch compared to 1 ft of house length.

G. Simplify the ratio 8.25 : 3.16.

8.25 : 3.16	1. When a ratio cannot easily be

expressed as a ratio of two whole numbers, it is frequently expressed as a ratio of some quantity : 1.

$\frac{8.25}{3.16} : \frac{3.16}{3.16}$ — 2. Divide both sides by the second number to obtain the number 1 on the right side.

2.61 : 1 — 3. Divide and round where necessary.

H. An engine has a compression ratio of 4.6. What does this mean?

Compression ratio = 4.6 — 1. Whenever a ratio is written as one number, the comparison is considered to be to 1.

$4.6 = 4.6 : 1 = \frac{4.6}{1}$ — 2. In this instance, there is 4.6 times as much volume in the cylinder at its maximum as there is at its minimum.

I. A concrete mix uses a mixture of cement, sand, and crushed stone in the ratio of 1 : 2 : 5 by weight. Thus for every pound of cement used, 2 pounds of sand and 5 pounds of crushed stone are mixed with it. How much of each component is necessary if 4000 lb of concrete mix is needed for a job?

1 + 2 + 5 = 8 — 1. Add together to find the denominator. This means there are a total of 8 parts.

1/8 — 2. Since there is 1 part cement in the mixture, the fraction 1/8 represents the ratio of cement in the mixture.

2/8 = 1/4 — 3. There are 2 parts sand; therefore, 2/8 or 1/4 represents the ratio of sand in the mixture.

5/8 — 4. This represents the ratio of crushed stone in the concrete mixture.

1/8 × 4000 = 500 lb cement

1/4 × 4000 = 1000 lb sand

5/8 × 4000 = 2500 lb crushed stone

5. Multiply the fraction that represents each component by the total amount of the mixture.

The mixture needs:
500 lb cement
1000 lb sand
2500 lb crushed stone
4000 lb total mixture

6. The sum of the individual components must equal the weight of the mixture.

Exercise 5-1

Express as simplified, unitless ratios.

1. 3 ft : 6 in.
2. 25/80
3. 2 ft 5 in. : 29 in.
4. 3 ft 8 in. : 9 ft 2 in.
5. 22 in. : 5 ft 6 in.

6. 25 lb cement : 50 lb sand : 75 lb crushed rock
7. 3 rejects to 18 good plumbing joints
8. Problem 7 as a ratio of rejects to total attempts
9. 40 oz gas to 2.5 oz oil
10. 15 qt to 3 gal
11. The blueprint of a house is drawn to a scale of $\frac{1}{4}$ in. = 1 ft. If the outside dimensions of the house measure 6 in. × 10 in. on the blueprint, what are the dimensions of the house?
12. A board measures $3\frac{3}{8}$ in. wide and 2 ft 3 in. long. Express the ratio of length to width in simplest terms.
13. Two gears have 32 teeth and 20 teeth. What is the ratio of the larger gear to the smaller?
14. Paint is mixed using 1 gal of base white, 1 qt of light blue, and 1 cup of gray. Determine the simplified, unitless ratio of the paint mixture, in the order given. (1 gal = 4 qt; 1 qt = 4 cups)
15. Two partners in a construction company divide profits in the ratio 3 : 5. How much does each receive for a week when profits total $1152?
16. A concrete mix has a composition of cement to sand to crushed rock in the ratio 1 : 3 : 5. What fraction of the mixture is sand?

5-2 PROPORTION

OBJECTIVES

Upon completing this section, the student will be able to:

1. Solve for the unknown in a proportion.
2. Identify and set up direct and inverse proportions.
3. Solve word problems involving proportions.

Whenever two ratios can be set equal to each other, the equation formed is called a *proportion*. Many problems that can be solved by other means can be solved more easily using proportions.

Examples

A. Set up a proportion using the equal ratios (fractions) 2/4 and 3/6.

$$\frac{2}{4} = \frac{3}{6}$$

1. Since both 2/4 and 3/6 reduce to 1/2 this is a true statement.

$$\frac{2}{4} \times \frac{3}{6}$$

2. In a proportion, the products of the diagonal numbers are always equal. This is called *cross-multiplication*.

$$2 \times 6 = 4 \times 3$$

$$12 = 12$$

3. Cross-multiply.

B. $$\frac{X}{24} = \frac{7}{3}$$

1. Two ratios set equal to each other form a proportion. If three of the four quantities are known, the fourth can always be found.

$$\frac{X}{24} \times \frac{7}{3}$$

2. Cross-multiply by finding the products of the diagonal numbers.

$3X = (7)(24)$

3. These products are always equal to each other.

$3X = 168$

4. Perform the multiplication.

$X = 168 \div 3$

$= 56$

5. Divide by 3 to solve for X.

C. Solve for K in the following proportion.

$$\frac{14.87}{3.91} = \frac{5.12}{K}$$

1. The unknown can be in either the numerator or denominator on either side of the equal sign.

$$\frac{14.87}{3.91} \times \frac{5.12}{K}$$

2. Cross-multiply.

$14.87K = (3.91)(5.12)$

$14.87K = 20.0192$

3. Perform the multiplication.

$K = 20.0192 \div 14.87$

4. Divide by 14.87 to solve for K.

$= 1.35$

5. Divide and round.

D. Solve for J in the following proportion.

$$\frac{82.3}{5.2J} = \frac{19.5}{13.9}$$

1. Cross-multiply.

$(5.2)(19.5)(J) = (82.3)(13.9)$

2. Perform the multiplication.

$101.4J = 1143.97$

$J = 1143.97 \div 101.4$

3. Divide to solve for J.

$= 11.3$

4. Perform the division and round.

There are two types of proportions: direct proportions and inverse proportions. In a **direct proportion,** as one quantity increases, the corresponding quantity increases also. Or, if one quantity decreases, the other decreases also.

Examples

E. Three pounds of common bright 8d nails cost $1.19. What is the price of 7 lb of these nails?

As the pounds of nails increases, the cost of the nails will increase also. Therefore, this is a direct proportion.

F. A car averaging 55 mph can travel 225 miles in a certain length of time. How far can a car travel in the same length of time if it averages 62 mph?

This problem is also a direct proportion. As one quantity, miles per hour, increases, the other quantity, distance covered, will increase also.

If one quantity increases as the other quantity decreases, the proportion is an **inverse proportion.** Here are several examples of inverse proportions.

Examples

G. Two delivery trucks are traveling from a lumberyard to the same job site. Truck A reaches the site in 42 min, traveling an average speed of 38 mph.

Truck B can average only 29 mph. How long does it take truck B to reach the site?

Since both trucks are traveling the same distance, the one that travels faster is going to take less time. Hence, as the speed increases, the time decreases. This is an inverse proportion since increasing one quantity causes the other quantity to decrease.

H. A 12-in. pulley is belted to an 8-in. pulley. If the larger pulley is rotating at 500 rpm, how fast is the smaller pulley rotating?

As the size of the pulley decreases, the speed at which it rotates increases. Since a decrease in one quantity causes an increase in the other, this is an inverse proportion.

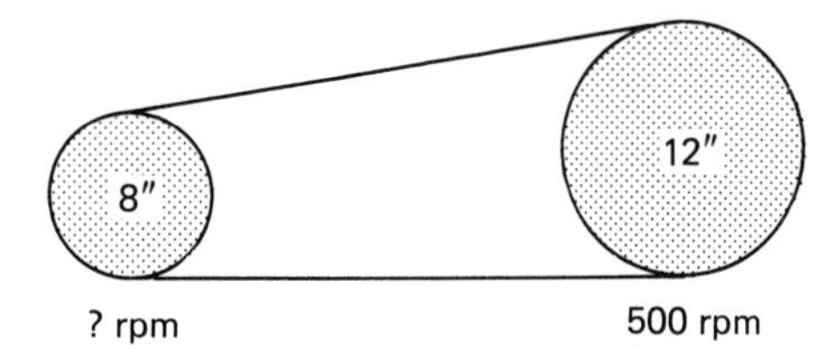

All proportions, whether direct or inverse, can be set up in a similar fashion as long as two rules are observed:

1. *Always set up the fraction so that the same units are over each other.* For example: 5 in./8 in.; 6 ft/11 ft; or 15 workmen/22 workmen. If the same units are always over each other in a fraction, the units will cancel.
2. *Always set the smaller unit over the larger unit:*

$$\frac{\text{small}}{\text{large}} = \frac{\text{small}}{\text{large}}$$

Examples

I. A blueprint has a scale of $\frac{1}{4}$ in. = 1 ft. If the length of a house on a blueprint is $7\frac{1}{2}$ in., what is the actual length of the house?

$\frac{1}{4}$ in. = 1 ft

$7\frac{1}{2}$ in. = ? ft

1. First determine what type of proportion is involved. This problem is a direct proportion since a longer line on the blueprint represents a longer wall on the house. Therefore, X, the unknown number of feet, is larger than 1 ft and must go in the denominator.

$$\frac{\frac{1}{4}\text{ in.}}{7\frac{1}{2}\text{ in.}} = \frac{1\text{ ft}}{X\text{ ft}}$$

2. Set up the ratios with inches/inches and feet/feet. The fractions are set up with the smaller number of inches and the smaller number of feet in the numerators.

$$\frac{\frac{1}{4}\text{ in.}}{7\frac{1}{2}\text{ in.}} \times \frac{1\text{ ft}}{X\text{ ft}}$$

3. Cross-multiply.

$$\frac{1}{4}X = (7\tfrac{1}{2})(1)$$

$$X = 7\tfrac{1}{2} \div \tfrac{1}{4}$$

4. Perform the multiplication and divide by $\frac{1}{4}$ to solve for X.

$$= 30 \text{ ft}$$

5. $7\frac{1}{2}$ in. on the blueprint represents 30 ft on the house.

J. ABC Construction Company can build four identical ranch houses in 9 weeks. How many weeks would it take to build 18 houses?

$$4 \text{ houses} = 9 \text{ weeks}$$
$$18 \text{ houses} = ? \text{ weeks}$$

1. This is a direct proportion since the number of weeks required to build the houses increases as the number of houses increases. Therefore X, the unknown number of weeks, is greater than 9 weeks and must go in the denominator.

$$\frac{4 \text{ houses}}{18 \text{ houses}} = \frac{9 \text{ weeks}}{X \text{ weeks}}$$

2. Houses are over houses and weeks are over weeks. The smaller number of houses and the smaller number of weeks are both in the numerator.

$$\frac{4 \text{ houses}}{18 \text{ houses}} \times \frac{9 \text{ weeks}}{X \text{ weeks}}$$

3. Cross-multiply.

$$4X = (9)(18)$$
$$X = (9)(18) \div (4)$$

4. Perform the multiplication and division to solve for X.

$$= 40.5 \text{ weeks}$$

5. It would take $40\frac{1}{2}$ weeks to build 18 ranch houses at the same rate of construction.

K. A construction crew of six carpenters (all working at the same rate) can build a house in 5 weeks. How long would it take a crew of 10 carpenters to build the same house?

$$6 \text{ carpenters} = 5 \text{ weeks}$$
$$10 \text{ carpenters} = ? \text{ weeks}$$

1. This is an inverse proportion, since a larger crew should be able to do the job in less time. Increasing carpenters causes a decrease in weeks. Therefore, X, the unknown number of weeks, is less than 5, and must go in the numerator.

$$\frac{6 \text{ carpenters}}{10 \text{ carpenters}} = \frac{X \text{ weeks}}{5 \text{ weeks}}$$

2. Like units are over each other: carpenters over carpenters and weeks over weeks. The 6 carpenters and the X weeks are in the numerators since they are the smaller units.

$$\frac{6 \text{ carpenters}}{10 \text{ carpenters}} \times \frac{X \text{ weeks}}{5 \text{ weeks}}$$

3. Cross-multiply.

$$10X = 30$$

4. Divide by 10.

$$X = 3 \text{ weeks}$$

5. It would take 3 weeks for 10 carpenters to do the same work that it would take 5 weeks for 6 carpenters to do.

L. Two pulleys are belted together. The larger pulley has a diameter of 24 in. and the smaller pulley has a 9-in. diameter. If the smaller pulley is rotating at 3600 rpm, how fast is the larger pulley rotating?

$$9 \text{ in.} = 3600 \text{ rpm}$$
$$24 \text{ in.} = ? \text{ rpm}$$

1. The larger pulley rotates slower than the smaller pulley; therefore, this is an inverse proportion. (As the

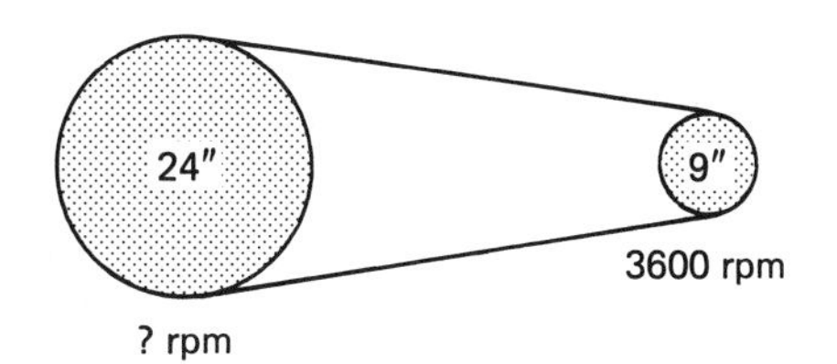

diameter of a pulley increases, its rotational speed decreases.) Thus the unknown rpm is less than 3600.

$$\frac{9 \text{ in.}}{24 \text{ in.}} = \frac{X \text{ rpm}}{3600 \text{ rpm}}$$

2. The units inches are over inches and rpm are over rpm. The smaller number of inches and rpm are in the numerator.

$$\frac{9 \text{ in.}}{24 \text{ in.}} \times \frac{X \text{ rpm}}{3600 \text{ rpm}}$$

3. Cross-multiply and solve for the unknown number of rpm.

$$X = (3600)(9) \div (24)$$

4. Divide by 24 to solve for X.

$$= 1350 \text{ rpm}$$

5. The larger pulley rotates at 1350 rpm.

Exercise 5-2

1. $\frac{3}{7} = \frac{X}{28}$

2. $\frac{2Y}{5} = \frac{8}{15}$

3. $\frac{3.08}{J} = \frac{1.45}{7.29}$

4. $\frac{3.6}{4.7} = \frac{6.3}{5.1X}$

5. $\frac{6.1}{2X} = \frac{8.3}{5.7}$

6. $\frac{4\frac{1}{4}}{X} = \frac{8\frac{1}{2}}{5\frac{1}{2}}$

7. $\frac{3\frac{3}{5}}{8\frac{2}{5}} = \frac{2X}{4\frac{1}{5}}$

8. $\frac{5.27}{3.1X} = \frac{8.64}{17.9}$

9. $\frac{2X}{5} = \frac{8}{27}$

10. $\frac{2\frac{1}{2}}{15} = \frac{X}{4}$

11. A lumber truck can travel 183 mi on 15.5 gal of gas. How far can it go on 12.2 gal?

12. $3\frac{1}{2}$ board feet of dry white maple weighs approximately 4.1 lb. What is the weight of 19 board feet of the same wood? Round to one decimal place.

13. The scale on a map is 1 in. = 25 mi. What is the distance from town A to town B if they are $2\frac{1}{8}$ in. apart on the map?

14. A room is 12 ft 6 in. wide by 18 ft 3 in. long. What would be the length and width of the blueprint of this room if a scale of $\frac{3}{8}$ in. = 1 ft is used? Give answer to nearest $\frac{1}{32}$ in.

15. A carpenter works 37 hr one week and makes $306.73. At that same rate, how much would he earn in a workweek of 29 hr?

16. A 42-gal hot-water tank holds 351 lb of water. How many pounds of water can a 55-gal tank hold? (Round to the nearest whole pound.)

17. There are 27 lb of sand in 108 lb of concrete mix. How many pounds of sand is needed for 800 lb of the concrete mix?

18. A 1 in. × 4 in. dry white pine board weighs 0.7 lb per lineal foot. What is the weight of an 8-ft board?

19. A rectangular pattern $3\frac{1}{2}$ in. × $5\frac{1}{2}$ in. is to be enlarged such that the width is $8\frac{3}{4}$ ft. What will the corresponding length be?

20. Seven masons can complete a job in $8\frac{1}{2}$ days. How many masons, working at the same rate, would be needed to do the job in $3\frac{1}{2}$ days?

21. A concrete wall 28 ft long weighs 23,000 lb. What is the weight of a concrete wall 41 ft long? (Assume that other dimensions are equal.)

22. A painter uses 3 qt of paint to cover 285 ft^2 of drywall. How many square feet of drywall can $5\frac{1}{2}$ gal cover?

23. A large gear with 520 teeth meshes with a smaller gear with 280 teeth. What is the speed of the smaller gear if the larger gear is rotating at 258 rpm?

24. Lumber costs $13.02 for 14 board feet. At that rate, what is the price of 6 board feet?

25. Four carpenters can build storage units for a school in 78 hr. How long would it take three carpenters, working at the same rate, to build the units?

26. A lumber truck averaging 38 mph delivers building materials to a job site in 1 hr 15 min. What speed must the truck average to get the materials to the site in 55 min? (Round to the nearest whole number.)

27. Interior trim boards cost $28.80 for 80 ft. What would 228 ft of trim boards cost?

28. A picture frame has its width and length in the ratio 89 : 144. (The ratio 89 : 144 is considered to be the most pleasing rectangular proportion to the human eye.) What is the width of the frame if the length is 4 ft 6 in.? Give answer in feet and inches to the nearest $\frac{1}{8}$ in.

chapter 6

Percents

6-1 CONVERSION OF FRACTIONS AND DECIMALS

OBJECTIVES

Upon completing this section, the student will be able to:

1. Demonstrate an understanding of the meaning of percent.
2. Convert decimals to percents.
3. Convert fractions to percents.
4. Convert percents to decimals.
5. Convert percents to fractions.

Percent or **per cent** can literally be translated as "*out of 100.*" When quantities are written as percents, they are being compared to a base of 100. The symbol for percent is **%**. Note the similarity between the symbols **%** and **/100**. Indeed, % is a shortcut way of writing /100. Thus 15% means 15/100 or 15 out of 100.

Percents are used daily. A power tool is on sale for 15% off. A sales tax of 5% is added to purchases. A waste allowance of 20% is figured for flooring. It will cost 35% more to build a particular house in New York than in Maine. The interest rate for home mortgages is $13\frac{1}{2}$%. These are a few examples of the types of percent problems carpenters and contractors deal with regularly.

Any decimal or fraction can be changed to a percent, and vice versa. Study the following examples carefully to determine the methods used for the various conversions.

Changing a Decimal to a Percent

Examples

A. Change 0.82 to a percent.

$0.82 = \frac{82}{100}$

1. Because % means "over 100," the % sign can be substituted for two decimal places or for the denominator /100.

$0.82. \Rightarrow 82\%$

2. Move the decimal point two places to the *right* and place the % sign after the number.

$\frac{82}{100} \Rightarrow 82\%$

3. Alternately, replace the fraction bar and the denominator 100 with the % sign after the number.

B. Change 1.4 to a percent.

$1.4 = 1.40$

1. Position zeros after the decimal point as needed.

$1.40. \Rightarrow 140\%$

2. Move the decimal point two places to the *right* and include the percent sign.

C. Change 0.025 to a percent.

$0.02.5 \Rightarrow 2.5\%$ or $2\frac{1}{2}\%$

Move the decimal point two places to the *right* and include the percent sign. The percent can be written as a decimal or a fraction.

D. $1 = 100\%$

1. The number 1 is equal to 100%.

$0.8 = 80\%$

2. Any number smaller than 1 equals a percent less than 100%.

$2.3 = 230\%$

3. Any number greater than 1 equals a percent greater than 100%.

Changing a Fraction to a Percent

Examples

E. Change $\frac{5}{8}$ to a percent.

$\frac{5}{8} = 0.625$

1. A fraction should first be changed to a decimal.

$0.625 = 62.5\%$

2. Convert the decimal to a percent.

F. Change $1\frac{3}{4}$ to a percent.

$1\frac{3}{4} = 1.75$

1. Convert to a decimal.

$1.75 = 175\%$

2. Convert the decimal to a percent.

G. Change $\frac{1}{3}$ to a percent.

$\frac{1}{3} = .3333...$

1. Convert to a repeating decimal.

$0.333... = 33.333...\%$

2. Convert to a percent.

$33.333\ldots\% = 33\frac{1}{3}\%$

3. A common repeating decimal such as $0.333\ldots$ ($\frac{1}{3}$) or $0.666\ldots$ ($\frac{2}{3}$) is usually converted to the equivalent fraction.

Changing Percents to Decimals

Examples

H. Change 85% to a decimal.

$85\% = 85.\%$

$0.85. = 0.85$

To convert a percent to a decimal, move the decimal point two places to the *left* and drop the % sign.

I. Change $22\frac{1}{4}\%$ to a decimal.

$22\frac{1}{4}\% = 22.25\%$

1. Change the fraction portion to a decimal.

$.22.25 = 0.2225$

2. Move the decimal point two places to the *left* and drop the % sign.

J. Change 0.05% to a decimal.

$0.05\% = 00.05\%$

1. Position zeros to the left of the decimal point, as needed. *Careful!* There is a tendency to move the decimal point incorrectly to the right when the percent itself is a decimal.

$.00.05 = 0.0005$

2. Always move the decimal point to the *left* to convert a percent to a decimal.

Converting a Percent to a Fraction

Examples

K. Convert 42% to the equivalent common fraction.

$42\% = \frac{42}{100}$

1. Drop the percent sign and show as a fraction over 100. (*Remember:* % means /100.)

$\frac{42}{100} = \frac{21}{50}$

2. Reduce where possible.

L. Change 150% to a fraction.

$150\% = \frac{150}{100}$

1. Show as a fraction over 100.

$\frac{150}{100} = \frac{3}{2} = 1\frac{1}{2}$

2. Reduce and convert to a mixed number.

M. Change $166\frac{2}{3}\%$ to a fraction.

$166\frac{2}{3}\% = \frac{500}{3}\%$

1. Convert the mixed number to an improper fraction. Note that the improper fraction is still a percent.

$\frac{500\%}{3} \times \frac{1}{100} =$

2. Multiply by 1/100 and drop the % sign. (This is equivalent to setting $\frac{500}{3}$ over a denominator of 100.)

$\frac{\not{5}\not{0}\not{0}}{3} \times \frac{1}{\not{1}\not{0}\not{0}} = \frac{5}{3}$	3. Reduce and multiply.
$166\frac{2}{3}\% = \frac{5}{3}$ or $1\frac{2}{3}$	4. Leave as an improper fraction or change to a mixed number.

N. Change $\frac{1}{4}\%$ to a fraction.

$\frac{1}{4}\% = \frac{1}{4} \times \frac{1}{100}$	1. Replace the % sign by multiplying by $\frac{1}{100}$.
$\frac{1}{4} \times \frac{1}{100} = \frac{1}{400}$	2. Multiply.
$\frac{1}{4}\% = \frac{1}{400}$	3. $\frac{1}{4}\%$ is equivalent to $\frac{1}{400}$.

Exercise 6-1

Complete the following table, converting all improper fractions to mixed numbers.

	FRACTION	DECIMAL	PERCENT
1.	$1\frac{4}{5}$		
2.		0.3	
3.			182%
4.	$\frac{3}{8}$		
5.		0.25	
6.			0.04%
7.	$\frac{1}{1000}$		
8.		1.25	
9.			$\frac{1}{2}\%$
10.	$2\frac{2}{3}$		
11.		0.025	
12.			24%
13.	$\frac{2}{3}$		
14.		1.00	
15.			2.5%
16.	$\frac{3}{5}$		
17.		0.62	
18.			143%
19.	$\frac{3}{5000}$		
20.			$266\frac{2}{3}\%$

6-2 PERCENT PROBLEMS

OBJECTIVES

Upon completing this section, the student will be able to:

1. Identify the base, rate of percent, and amount in percent problems.
2. Find the amount when given the base and rate.
3. Find the rate when given the base and amount.
4. Find the base when given the rate and amount.
5. Solve ''more than'' and ''less than'' problems.

Every percent problem has three parts: the **base,** the **rate,** and the **amount.** In order to set up and solve percent problems, we must first be able to identify the components. Study the following examples.

Examples

A. What is 18% of 52?

What is *18%* of 52?	1. The *rate* is the percent; therefore, it can be identified by the % sign.
What is 18% *of 52*?	2. The *base* is the quantity after the word *of*.
What is 18% of 52?	3. Since both the rate and base have been identified, *what* refers to the *amount*. Hence we are looking for the amount in this example.

In this problem, the **amount** is the unknown.

B. 25 is what percent of 38?

25 is *what percent* of 38?	1. *What percent* indicates that we are looking for the *rate* in this example.
25 is what percent *of 38*?	2. The *base* is 38, since it is the quantity that comes after the word *of*.
25 is what percent of 38?	3. Since the rate and base have both been identified, 25 represents the *amount*.

In this problem, the **rate** is the unknown.

C. 22 is 62% of what number?

22 is *62%* of what number?	1. 62% is the *rate* since % identifies the rate.
22 is 62% *of what number*?	2. *Of* identifies *what number* as the *base*. Therefore, we are looking for the base.
22 is 62% of what number?	3. Since the rate and base have been identified, 22 is the *amount*.

In this problem, the **base** is the unknown.

D.

15 is *what percent* of 32?	The *rate* is the unknown.
35 is 82% of *what number*?	The *base* is the unknown.
What is 21% of 85?	The *amount* is the unknown.

Exercise 6-2A

Identify which of the three parts (base, rate, or amount) is the unknown.

1. 22% of 58 equals what number?
2. 48 is what percent of 19?
3. 45 is 14% of what number?
4. 38 out of 200 represents what percent?
5. 125 is 20% of what number?
6. 25 is what percent of 125?
7. 42 is 18% of what number?
8. What is 155% of 24?
9. 150 is 25% of what number?
10. What is $\frac{1}{2}$% of 1000?

There are several ways to approach percent problems. The **percent proportion** method is discussed here because it works equally well for all types of percent problems.

Percent Proportion Method

The problem is set up as a proportion using the following form:

$$\frac{\text{rate (\%)}}{100\%} = \frac{\text{amount}}{\text{base}}$$

This method will work regardless of which part of the problem is unknown. The following examples refer back to Examples A, B, and C.

Examples

E. Set up and solve Example A: What is 18% of 52?

rate → 18%; This is always 100% → 100%; unknown amount → amount; base → 52

$$\frac{18\%}{100\%} = \frac{\text{amount}}{52}$$

1. 18% is the rate, 52 is the base, and the amount is unknown.

$$\frac{18}{100} \times \frac{X}{52}$$

2. Cross-multiply.

$$100X = (18)(52)$$

3. Solve for the unknown amount X.

$$X = \frac{936}{100} = 9.36$$

9.36 is 18% of 52

4. 9.36 is the amount.

F. Set up and solve Example B: 25 is what percent of 38?

unknown rate → X%; amount → 25; base → 38

$$\frac{X\%}{100\%} = \frac{25}{38}$$

1. Here the rate is the unknown.

$$\frac{X}{100} \times \frac{25}{38}$$

2. Cross-multiply.

$$38X = (25)(100)$$

3. Solve for X and round.

$$X = \frac{2500}{38} \doteq 65.8$$

4. 65.8% is the rate.

25 is 65.8% of 38

G. Set up and solve Example C: 22 is 62% of what number?

$$\frac{62\%}{100\%} = \frac{22}{X}$$ (rate: 62%; amount: 22; unknown base: X)

1. Here the base is the unknown.

$$\frac{62}{100} \times \frac{22}{X}$$

2. Cross-multiply.

$$62X = (22)(100)$$

$$X = \frac{2200}{62} \doteq 35.5$$

3. Solve for X and round.

22 is 62% of 35.5

4. 35.5 is the base.

Exercise 6-2B

These are the same problems as Exercise 6-2A. Set up and solve for the unknown.

1. 22% of 58 equals what number?
2. 48 is what percent of 19?
3. 45 is 14% of what number?
4. 38 out of 200 represents what percent?
5. 125 is 20% of what number?
6. 25 is what percent of 125?
7. 42 is 18% of what number?
8. What is 155% of 24?
9. 150 is 25% of what number?
10. What is $\frac{1}{2}$% of 1000?

''More Than'' and ''Less Than'' Problems

Frequently, percent problems involve ''more than'' and ''less than.'' Here are two examples of more than/less than problems:

1. 25 is 14% *more than* what number?
2. What is 18% *less than* 52?

More than/less than problems are usually changed to ''of'' problems and then worked as percent proportions.

Examples

H. What is 19% more than 83?

What is 19% *more than* 83?

1. Convert this ''more than'' into an ''of'' problem.

What is + $\begin{array}{r} 100\% \\ \underline{19\%} \\ 119\% \end{array}$ of 83?

2. For a ''more than'' problem, *add* the percent given to 100%, and change *more than* to *of*. ''More than'' implies that the entire 100% is to be used plus 19% more. In this example, the entire 83 *plus* 19% of 83 is desired.

What is 119% of 83?

3. This is now similar to other percent problems in which the amount is unknown.

$$\frac{119\%}{100\%} = \frac{X}{83}$$

4. Solve by percent proportion.

$$100X = (119)(83)$$

5. In a "more than" problem, the amount is greater than the base.

$$X = \frac{9877}{100}$$

98.77 is 19% more than 83

I. 225 is 28% less than what number?

225 is 28% *less than* what number?

1. Convert this "less than" problem to an "of" problem.

225 is $\begin{array}{r} 100\% \\ -\ \ 28\% \\ \hline 72\% \end{array}$ of what number?

2. For a "less than" problem, *subtract* the percent given from 100% and change the *less than* to *of*.

225 is 72% of what number?

3. Set up and solve by percent proportion.

$$\frac{72\%}{100\%} = \frac{225}{X}$$

4. Cross-multiply.

$$72X = (225)(100)$$

$$X = \frac{22{,}500}{72} = 312.5$$

5. In a "less than" problem, the amount is less than the base.

225 is 28% less than 312.5

J. Here are several examples of "more than" and "less than" problems rewritten as "of" problems.

1. *Original:* What is 12% more than 55?

   Rewritten: What is 112% of 55?

   (Add to 100% for a "more than" problem.)

   Setup: $\frac{112\%}{100\%} = \frac{X}{55}$

2. *Original:* What is 22% less than 39?

   Rewritten: What is 78% of 39?

   (Subtract from 100% for a "less than" problem.)

   Setup: $\frac{78\%}{100\%} = \frac{X}{39}$

3. *Original:* 15 is 32% more than what number?

   Rewritten: 15 is 132% of what number?

   ("More than" : add to 100%)

   Setup: $\frac{132\%}{100\%} = \frac{15}{X}$

4. *Original:* 88 is 10% less than what number?

Rewritten: 88 is 90% of what number?

(''Less than'' : subtract from 100%)

Setup: $\frac{90\%}{100\%} = \frac{88}{X}$

K. 139 is what percent more than 121?

$\frac{X\%}{100\%} = \frac{139}{121}$

1. If the rate is the unknown, the adjustment is made *after* the rate is determined.

$121X = 13900$

$X\% = 114.9\%$

139 is 14.9% more than 121

2. 139 is 114.9% of 121. Since 121 = 100%, 139 is 14.9% *more than* 121.

Exercise 6-2C

Solve the following more than/less than problems.

1. What is 32% more than 130?
2. 128 is what percent less than 155?
3. 392 is 18% less than what number?
4. What is 4% less than 226?
5. 553 is what percent more than 421?
6. 15 is 25% more than what number?
7. What is 47% more than 56?
8. What is 20% less than $89.95?
9. $455.92 is 14% less than what number?
10. What is $21\frac{1}{2}$% more than 162?

6-3 WORD PROBLEMS INVOLVING PERCENT

OBJECTIVES

Upon completing this section, the student will be able to:

1. Identify the three parts of a word problem involving percent.
2. Set up and solve for the unknown.

Word problems involving percent frequently do not have the words ''of,'' ''more than,'' or ''less than'' to identify the base. Therefore, other ways of determining the base must be considered. Study the following examples.

Examples

A. A large construction firm has 150 workers. If 20% of the workers have at least 15 years of carpentry experience, how many workers have at least 15 years of experience?

20% is the rate	1. Rate is determined by the % sign.
150 is the base	2. If there is a number that represents the *total* (in this case total workers), that is the base.
The unknown is the amount	3. The amount is the number (unknown in this example) that represents *part* of the total. Part of the workers have 15 or more years of experience.
$\frac{20\%}{100\%} = \frac{X}{150}$	4. Solve as a percent proportion.
$X = 30$ workers	5. 30 out of 150 workers is equivalent to 20% of 150 workers.

B. A carpenter is making $10.78 per hour after an increase of 10%. What was he making per hour before the raise?

10% + 100% = 110% The rate is 110%	1. A 10% raise represents a "more than" problem. (Any increase can be thought of as "more than.") Therefore, the rate is 110%.
The base is the unknown	2. Whenever there is a time difference, the *original quantity*, before any increase or decrease, is the base. The carpenter's wage before the raise is the original wage; therefore, it is the base.
$10.78 is the amount	3. The present quantity is the amount.
$\frac{110\%}{100\%} = \frac{\$10.78}{X}$	4. Solve as a percent proportion.
$X = \$9.80$	5. The carpenter was making $9.80 before the raise.

C. A builder charges $3850 to build a deck. If the lumber for the deck cost $1225, what percent of the price was for lumber?

$3850 is the base	1. The base represents the *total* price.
$1225 is the amount	2. The amount represents a specific *part* of the total.
The unknown is the rate	3. We are asked to find the percent.
$\frac{X\%}{100\%} = \frac{\$1225}{\$3850}$	4. Set up and solve as a percent proportion.
$X = 31.81\% \doteq$ 31.8% or 32%	5. Approximately 32% of the total price charged for the deck is to cover the lumber.

Exercise 6-3

Set up and solve the following word problems.

1. On a certain job, 53% of the expenses incurred by a contractor are for lumber and materials. If the expenses total $8250 for the job, what is the cost of the lumber and materials?

2. $2\frac{1}{4}$-in. red oak flooring boards are to be used in a living room. 1350 lineal feet is required to cover the floor. How many lineal feet should be ordered if a 15% allowance is made for waste? (*Hint:* The order should include 15% more than the amount needed to cover the floor.)

3. Circular saws are on sale for $47.96. If this is 20% less than the regular price, what is the regular price of the saw?

4. The price of $\frac{3}{4}$-in. CDX plywood increased from $23.95 to $25.98 per sheet at a certain lumberyard. What percent increase is this? Round the percent to one decimal place.

5. An architect charges 8% of the estimated cost of the house for her services. If the architect's fees are $10,000, what is the estimated cost of the house?

6. A contractor charges $12,520 for a job. If he estimates that his total cost will be $9850, what percent profit does he expect to earn on the job?

7. Labor costs on a house amount to $42,520. If this represents 43% of the total cost to the contractor, what is his total cost?

8. A contractor gets a $5\frac{1}{2}$% discount from retail price at a building supplies company. If the retail price of his purchase totals $8520.50, what does he pay for the materials? Round to the nearest cent.

9. A keg of nails costs $55.00. If there is a $4\frac{1}{2}$% sales tax, what must the purchaser pay for the nails? Round to the nearest cent.

10. A builder made a profit of $12,500 on the sale of a new house. If the house sold for $131,450, what percent profit did the builder make?

11. A 7.25-ft^2 door is cut from a 32-ft^2 sheet of cabinet-grade plywood. Ignoring waste, what percent of the plywood is used in the door?

12. 65% of all belt sanders manufactured by a certain company are still working 8 years later. What fraction of these sanders are still working after 8 years?

13. The width of a cupboard door is 40% of its height. If the width of the door is 16 in., what is its height? Give answer in feet and inches.

14. A carpenter's income for July is $2534. If he deposits $6\frac{1}{2}$% of that in a savings account, how much does he save during July?

15. A builder receives a $7\frac{1}{2}$% discount on plumbing and electrical supplies. If he purchases supplies that would retail for $5846.00, how much money does the discount save him?

16. Allowing 15% for waste, how many board feet of lumber must be ordered for a project requiring 450 board feet?

17. A building lot is exactly 1 acre (43,560 ft^2). A one-story house has 1750 ft^2, and a detached garage has 672 ft^2. What percent of the lot is taken up by buildings? Round the percent to one decimal place.

18. A bank requires a 12% down payment on a new house costing $143,550. How much is the down payment?

19. A new house cost $82,530 to construct. If the foundation cost $12,225, what percent of the total cost was the foundation?

20. A certain lot of lumber is 65% *clear* (that is, free of defects and knots). If there is 85,000 lineal feet of lumber in the lot, how much clear lumber is in the lot?

chapter 7

Angles and Triangles

7-1 ANGLE MEASURE

OBJECTIVES
Upon completing this section, the student will be able to: 1. Convert decimal degrees to DMS (degrees, minutes, and seconds). 2. Convert DMS to DD (decimal degrees). 3. Add and subtract DMS. 4. Multiply and divide DD and convert to DMS.

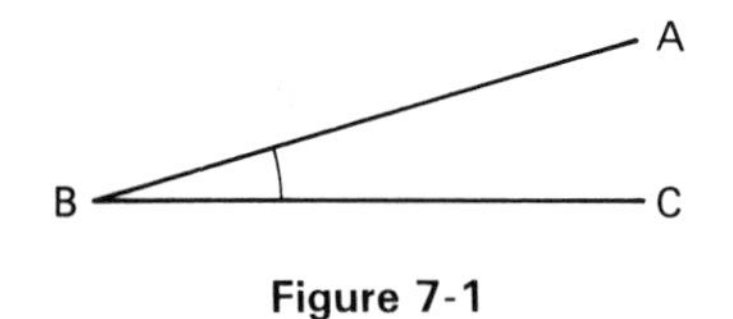

Figure 7-1

Angles are formed when two lines meet or intersect. The point at which they meet is called the **vertex** of an angle. The angle shown in Figure 7-1 can be written either as $\angle ABC$, $\angle CBA$, or simply as $\angle B$. If there is no possibility of confusing it with other angles, an angle is usually indicated by just the letter at its vertex. (B is at the vertex in this example.)

The designation $\angle J$ is not sufficient in Figure 7-2. The three-letter $\angle MJN$ or $\angle NJM$ must be given to distinguish this angle from $\angle NJK$, $\angle KJL$, and so on.

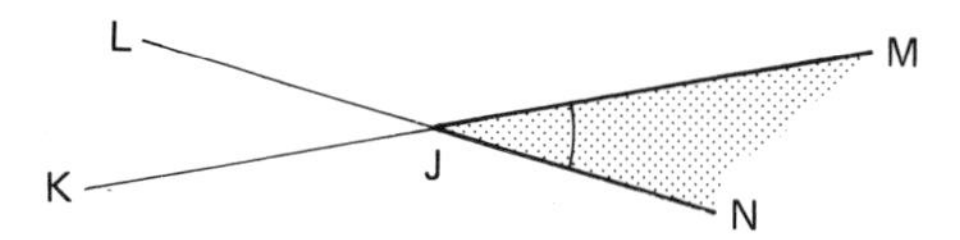

Figure 7-2

The **size** of an angle is measured in **degrees.** A degree is a unit of rotation. There are 360° in one full rotation. A small raised circle, °, is the symbol for degree. A full circular rotation is 360° (see Figure 7-3a).

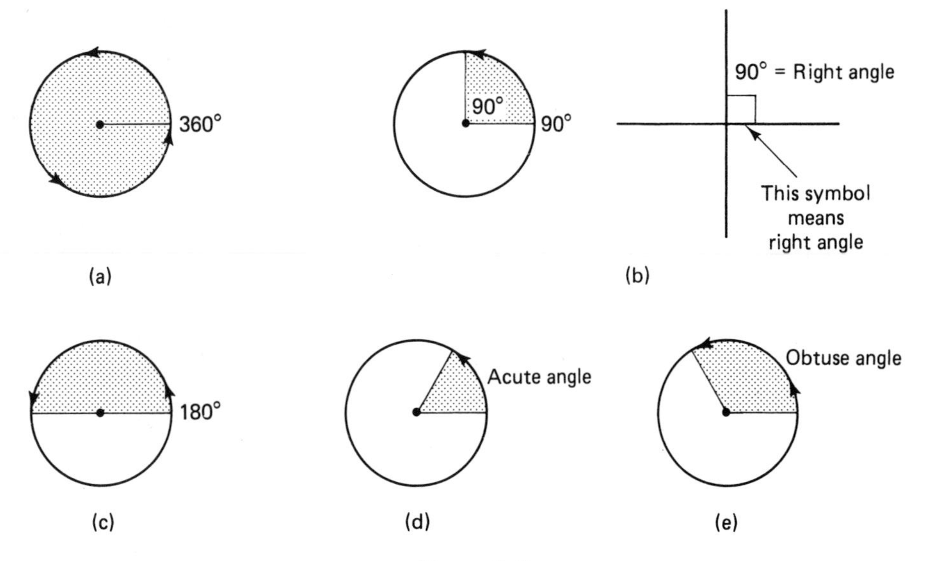

Figure 7-3

A quarter circular rotation is 90° (Figure 7-3b). The angle formed is called a **right angle.** Whenever two lines meet in such a way that they form a right angle, the lines are said to be *perpendicular*.

A half circular rotation is 180° ($\frac{1}{2}$ of 360°). A 180° angle, called a **straight angle,** does not look like an angle at all, but like a straight line. This is illustrated in Figure 7-3c.

Any angle less than 90° is called an **acute angle** (Figure 7-3d). Angles of 10°, 45°, 50°, and 85° are examples of acute angles.

Any angle greater than 90° but less than 180° is called an **obtuse angle** (Figure 7-3e). Angles of 120°, 150°, 93°, and 175° are examples of obtuse angles.

The size of an angle is not determined by the length of its sides. The two angles shown in Figure 7-4a are both 30°. Because the degree of rotation is the same for both of these angles, $\angle B = \angle E$. The sides of $\angle B$ could be extended without changing its size (Figure 7-4b).

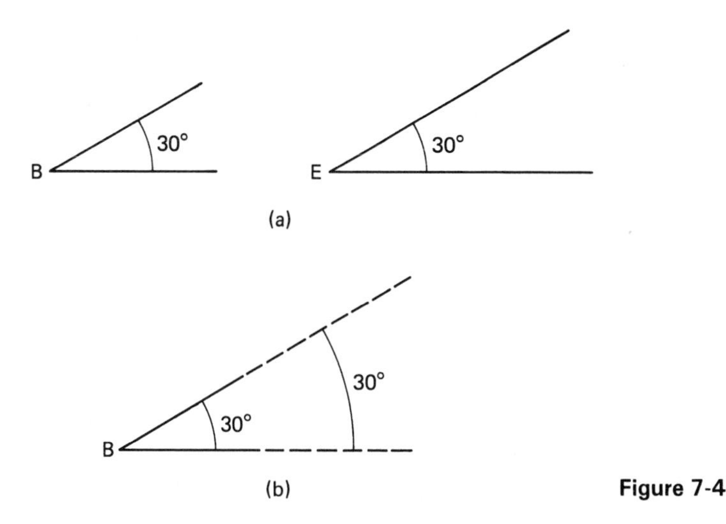

Figure 7-4

In measuring lengths it would rarely be acceptable for a carpenter to measure to the nearest foot. In most cases he or she would need to measure a length to the nearest inch (for the width of a driveway), to the nearest $\frac{1}{8}$ in. (for a rafter), or to the nearest $\frac{1}{32}$ in. (for a cabinet door). Similarly, measuring an angle to the nearest degree is not always accurate enough. Just as feet can be subdivided into inches and fractions of an inch, degrees can be subdivided into units called minutes and seconds. Each degree is divided into 60 minutes, and each minute is divided into 60 seconds.

$$1 \text{ degree} = 60 \text{ minutes} \qquad 1^\circ = 60'$$

$$1 \text{ minute} = 60 \text{ seconds} \qquad 1' = 60''$$

An angle that measures 62 degrees, 14 minutes, and 38 seconds would be written as $62^\circ\ 14'\ 38''$. The symbols for minutes and seconds are the same as those for feet and inches.

Addition of DMS (Degrees, Minutes, and Seconds)

To determine the method used for adding degrees, minutes, and seconds, study the following examples.

Examples

A. Add $42^\circ\ 18'\ 32''$ and $35^\circ\ 22'\ 25''$.

$$\begin{array}{r} 42^\circ\ 18'\ 32'' \\ +\ \underline{35^\circ\ 22'\ 25''} \end{array}$$

1. Line up degrees under degrees, minutes under minutes, and seconds under seconds.

$$\begin{array}{r} 42^\circ\ 18'\ 32'' \\ +\ \underline{35^\circ\ 22'\ 25''} \\ 77^\circ\ 40'\ 57'' \end{array}$$

2. Add the degrees to degrees, minutes to minutes, and seconds to seconds.

B. Find the sum of $41^\circ\ 28'\ 35''$ and $52^\circ\ 51'\ 43''$.

$$\begin{array}{r} 41^\circ\ 28'\ 35'' \\ +\ \underline{52^\circ\ 51'\ 43''} \\ 93^\circ\ 79'\ 78'' \end{array}$$

1. Line up and add the units individually.

$$\begin{array}{r} 78'' = 60'' + 18'' \\ 78'' = 1' + 18'' \\ 93^\circ\ 79'\ 78'' = 93^\circ\ 80'\ 18'' \end{array}$$

2. Convert $78''$ into minutes and seconds and add the 1 minute to the existing $79'$.

$$\begin{array}{r} 80' = 60' + 20' \\ 80' = 1^\circ + 20' \\ 93^\circ\ 80'\ 18'' = 94^\circ\ 20'\ 18'' \end{array}$$

3. Convert $80'$ to degrees and minutes and add to the existing 93°.

4. All minutes and seconds 60 or greater should be converted to the next higher unit.

If the sum of two angles is 90°, the angles are said to be **complementary.** $\angle B$ and $\angle C$ shown in Figure 7-5a are complementary; $\angle XYZ$ and $\angle ZYW$ (Figure 7-5b) are complementary.

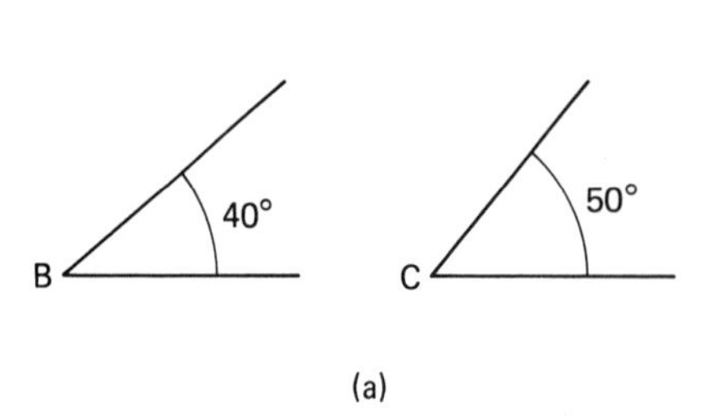

Figure 7-5

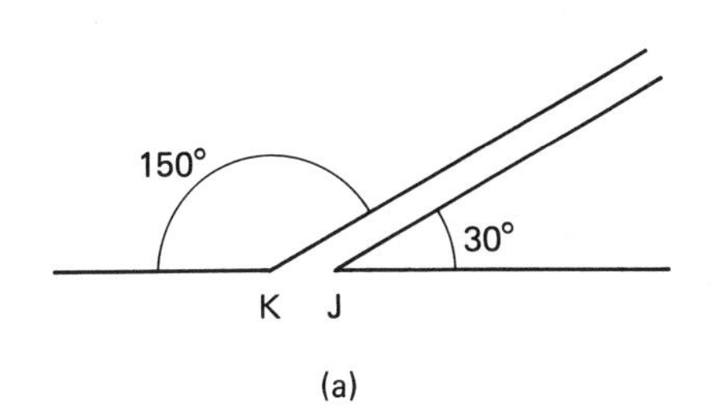

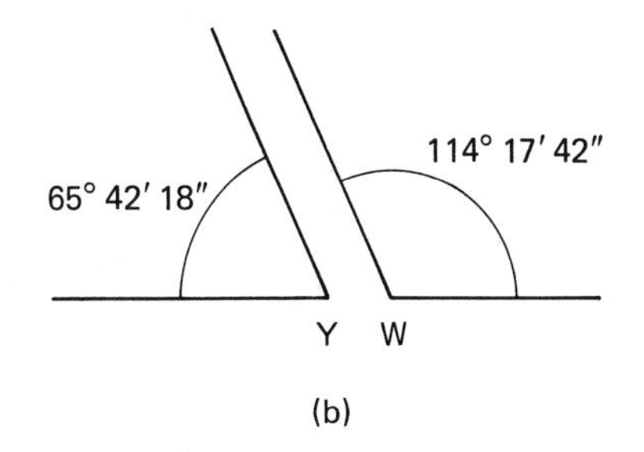

Figure 7-6

If the sum of two angles is 180°, the angles are said to be **supplementary** (see Figure 7-6). ∠*K* is supplementary to ∠*J*. Similarly, ∠*Y* and ∠*W* are supplementary.

Subtraction of DMS

Study the following examples that illustrate subtraction of angles.

Examples

C. Subtract 18° 43′ 22″ from 62° 51′ 45″.

$$\begin{array}{r} 62^\circ\ 51'\ 45'' \\ -\ \underline{18^\circ\ 43'\ 22''} \\ 44^\circ\ 08'\ 23'' \end{array}$$

Line up and subtract similar units.

D. ∠*X* is complementary to ∠*Y*. If ∠*X* measures 32° 15′ 41″, what is ∠*Y*?

$$\begin{array}{r} 90^\circ \\ -\ \underline{32^\circ\ 15'\ 41''} \end{array}$$

1. Since the angles are complementary, subtract ∠*X* from 90° to determine ∠*Y*.

$$\begin{array}{r} 89^\circ\ 60' \\ -\ \underline{32^\circ\ 15'\ 41''} \end{array}$$

2. Convert 90° to 89° 60′.

$$\begin{array}{r} 89^\circ\ 59'\ 60'' \\ -\ \underline{32^\circ\ 15'\ 41''} \end{array}$$

3. Convert 60′ to 59′ 60″.

$$\begin{array}{r} 89^\circ\ 59'\ 60'' \\ -\ \underline{32^\circ\ 15'\ 41''} \\ 57^\circ\ 44'\ 19'' \end{array}$$

4. Subtract.

$Y = 57^\circ\ 44'\ 19''$

E. Subtract 38° 15′ 46″ from 62° 12′ 22″.

$$\begin{array}{r} 62^\circ\ 12'\ 22'' \\ -\ \underline{38^\circ\ 15'\ 46''} \end{array}$$

1. Line up units for subtraction. Note that 46″ is too large to subtract from 22″, and 15′ is too large to subtract from 12′.

$$\begin{array}{r} 62^\circ\ 11'\ 82'' \\ -\ \underline{38^\circ\ 15'\ 46''} \end{array}$$

2. Convert 12′ to 11′ 60″, and add the 60″ to the existing 22″ for a total of 82″.

$$\begin{array}{r} 61^\circ\ 71'\ 82'' \\ -\ \underline{38^\circ\ 15'\ 46''} \end{array}$$

3. Convert 62° to 61° 60′, and add the 60′ to the existing 11′ for a total of 71′.

$$\begin{array}{r} 61^\circ\ 71'\ 82'' \\ -\ \underline{38^\circ\ 15'\ 46''} \\ 23^\circ\ 56'\ 36'' \end{array}$$

4. Subtract.

Exercise 7-1A

Perform the addition and subtraction indicated.

1. 48° 15′ 37″ + 51° 9′ 22″
2. 51° 13′ 45″ + 18° 22′ 55″
3. 61° 19′ 38″ − 41° 16′ 15″
4. 82° 42′ 21″ − 51° 52′ 25″
5. 62° 14′ 38″ − 14° 14′ 51″
6. What angle is complementary to 42° 14′ 35″?
7. What angle is supplementary to 121° 19′ 25″?
8. 52° 19′ is complementary to what angle?
9. 27° + 52° 19′
10. What is the supplement of 82°?

Decimal Degrees

In some situations decimal degrees (hereafter referred to as DD) must be used instead of DMS. Study the following examples to determine how to convert from DD to DMS and from DMS to DD. It is recommended that a calculator be used for these conversions.

Examples

F. Convert 82.5361° from decimal degrees to DMS.

82.5361° — 1. DD are usually given to four decimal places.

82.5361° − 82° = 0.5361° — 2. Record the 82° and subtract from the calculator. The decimal value 0.5361 left in the calculator is still in degrees.

0.5361° × 60 min/deg = 32.166′ — 3. To convert to minutes, multiply the DD 0.5361 by 60 (since 60′ = 1°).

32.166′ − 32′ = 0.166′ — 4. Record the 32′ and subtract from the calculator.

0.166′ × 60 sec/min = 9.96″ — 5. Multiply the remaining 0.166′ by 60 to convert to seconds (since 60″ = 1′).

9.96″ ≐ 10″ — 6. Round to the nearest whole second.

82.5361° = 82° 32′ 10″

7. These are equivalent, to the nearest second.

G. Convert 41.0066° to DMS.

41.0066° − 41° = 0.0066° — 1. Record the 41° and subtract from the calculator.

$0.0066° \times 60 = 0.396'$	2. Convert to minutes by multiplying by 60. Notice that this does not yield a whole number of minutes.
$0.396' \times 60 = 23.76''$	3. Since there are no full minutes, write down 0′ and convert the decimal minutes to seconds by multiplying 0.396′ by 60.
$23.76'' \doteq 24''$	4. Round to the nearest second.
$41.0066° = 41°\ 0'\ 24''$	5. These are equivalent to the nearest second.

H. Convert 25° 18′ 32″ to decimal degrees.

$25°\ 18'\ 32''$ $32'' \div 60 \text{ sec/min} = 0.5\overline{3}'$	1. When converting DMS to DD, convert from *right* to *left*. To convert 32″ to minutes, *divide* by 60.
$0.5\overline{3}' + 18' = 18.5\overline{3}'$	2. Add the decimal minutes to the existing 18′ *without* removing the numbers from the calculator.
$18.5\overline{3}' \div 60 \text{ min/deg} = 0.30\overline{8}°$	3. Convert the total minutes to degrees by dividing by 60.
$0.30\overline{8}° + 25 = 25.30\overline{8}°$	4. Add the decimal degrees to the existing 25° for the total DD.
$25.30\overline{8}° \doteq 25.3089°$	5. Round to four decimal places.
$25°\ 18'\ 32'' = 25.3089°$	6. These are equivalent to the nearest second. No intermediate values should be removed from the calculator until the final answer is obtained. Otherwise, accuracy will be lost.

I. Convert 62° 0′ 45″ to DD.

$62°\ 0'\ 45''$	1. Note there are no minutes in this example.
$45'' \div 60 \text{ sec/min} = 0.75'$	2. Divide 45″ by 60 to convert to minutes.
$0.75' \div 60 \text{ min/deg} = 0.0125°$	3. Since there are no whole minutes to add to the decimal minutes, divide by 60 again to convert to decimal degrees.
$0.0125° + 62° = 62.0125°$	4. Add the decimal degrees to the 62° to get the resultant DD.
$62°\ 0'\ 45'' = 62.0125°$	5. These are equivalent to the nearest second.

Here is an easy way to remember when to multiply by 60 and when to divide by 60.

To convert DD to D**MS**, **M**ultiply by 60.
To convert DMS to **DD**, **D**ivide by 60.

Exercise 7-1B

Perform the following conversions.

	DD (to four decimal places)	DMS
1.	83.5616°	
2.	58.4291°	
3.		14° 25′ 39″
4.		62° 18′ 41″
5.	78.0039°	
6.		73° 0′ 15″
7.	121.4100°	
8.		21° 1′ 1″
9.	10.3260°	
10.		14° 48′
11.	42.5000°	
12.	142.3875°	
13.		52° 0′ 20″
14.	89.9950°	
15.	20.9998°	
16.		179° 59′ 60″

When multiplying or dividing angles, it is best to use decimal degrees. Sometimes this involves converting DMS to DD and then converting back to DMS after performing the multiplication or division.

Examples

J. Multiply 38° 19′ 27″ by 5.

38° 19′ 27″ = 38.3242°	1. Convert from DMS to DD.
38.3242° × 5 = 191.6208°	2. Multiply the decimal degrees by 5.
191.6208° = 191° 37′ 15″	3. Convert back to DMS.

K. Divide 13° 41′ 18″ by 2.

13° 41′ 18″ = 13.6883°	1. Convert from DMS to DD.
13.6883° ÷ 2 = 6.8442°	2. Divide by 2.
6.8442° = 6° 50′ 39″	3. Convert back to DMS.

Exercise 7-1C

Multiply and divide the following angles using the methods of Examples J and K. Round to the nearest second, if necessary.

1. $42° 19' 34'' \div 2$
2. $18° 41' 55'' \times 4$
3. $106° 4' 45'' \div 7$
4. $38° 15' \times 4$
5. $41° 26' \times 3$
6. $18° 14' 9'' \div 3$
7. $64° 8' 30'' \div 5$
8. $0° 29' 14'' \times 7$
9. $0° 54' 12'' \times 5$
10. $22° \div 36$

7-2 CONGRUENT AND SIMILAR TRIANGLES

OBJECTIVES

Upon completing this section, the student will be able to:

1. Identify the different types of triangles.
2. Identify congruent triangles.
3. Identify similar triangles.
4. Solve problems involving congruent triangles.
5. Solve problems involving similar triangles.

The name **triangle** means "three angles." A triangle is a three-sided, closed figure (Figure 7-7). Where two sides of a triangle meet, the *vertex* of the angle is formed. *The sum of the angles of every triangle equals 180°.*

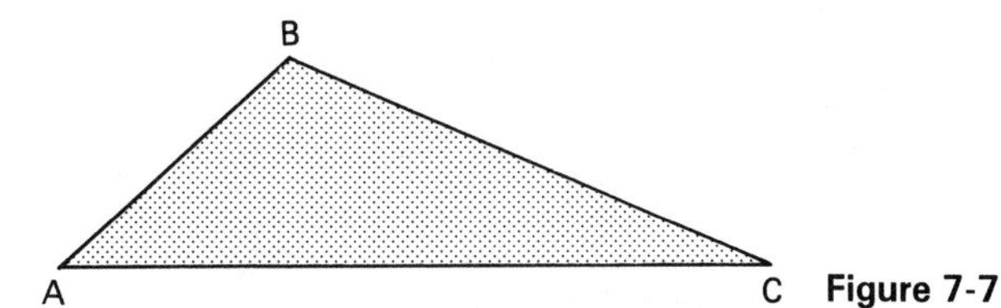

Figure 7-7

Triangles are generally specified by capital letters at their vertices. The triangle shown in Figure 7-7 would be named $\triangle ABC$. The symbol $\triangle$ means triangle. It would be equally correct to name the triangle $\triangle BCA$, $\triangle BAC$, $\triangle CAB$, and so on. The sides of a triangle are identified with the lowercase letter corresponding to the capital letter associated with the opposite angle. Note that $\angle A$ in Figure 7-8 is formed by sides b and c. Therefore, side a is the *opposite* side.

Triangles can be classified by angles and by sides.

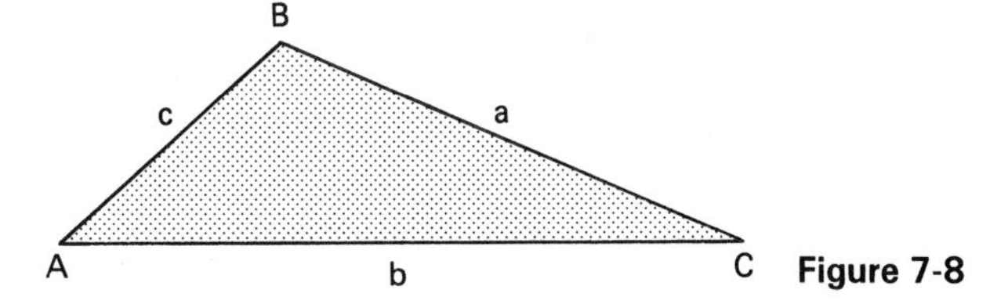

Figure 7-8

Classification by Sides

1. *Scalene triangle.* All three sides are unequal (Figure 7-9a).

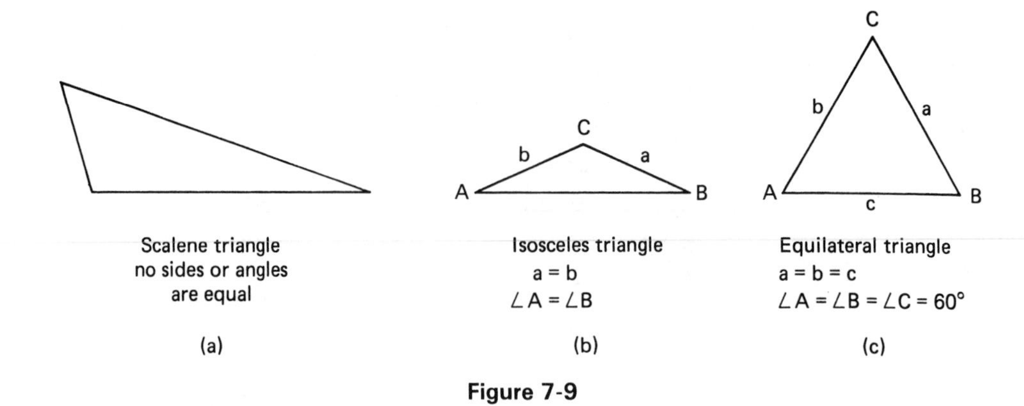

Figure 7-9

2. *Isosceles triangle.* Two sides are equal (Figure 7-9b). The unequal side is called the *base*. The two angles opposite the equal sides are also equal.
3. *Equilateral triangle.* All three sides are equal (Figure 7-9c). All three angles are also equal. Since the sum of the three angles is 180°, each angle in an equilateral triangle is 60°.

In any triangle, the largest angle will be opposite the longest side, the smallest angle will be opposite the shortest side, and the middle angle will be opposite the middle side. If two angles are equal, the sides opposite them will also be equal.

Classification by Angles

1. *Right triangle.* Any triangle containing a right (90°) angle (Figure 7-10a).
2. *Oblique triangle.* Any triangle that does not contain a right angle (Figure 7-10b).
3. *Obtuse triangle.* Any triangle that contains an obtuse angle (Figure 7-10c).
4. *Acute triangle.* Any triangle with three acute angles (Figure 7-10d). (*Every* triangle has at least two acute angles.)

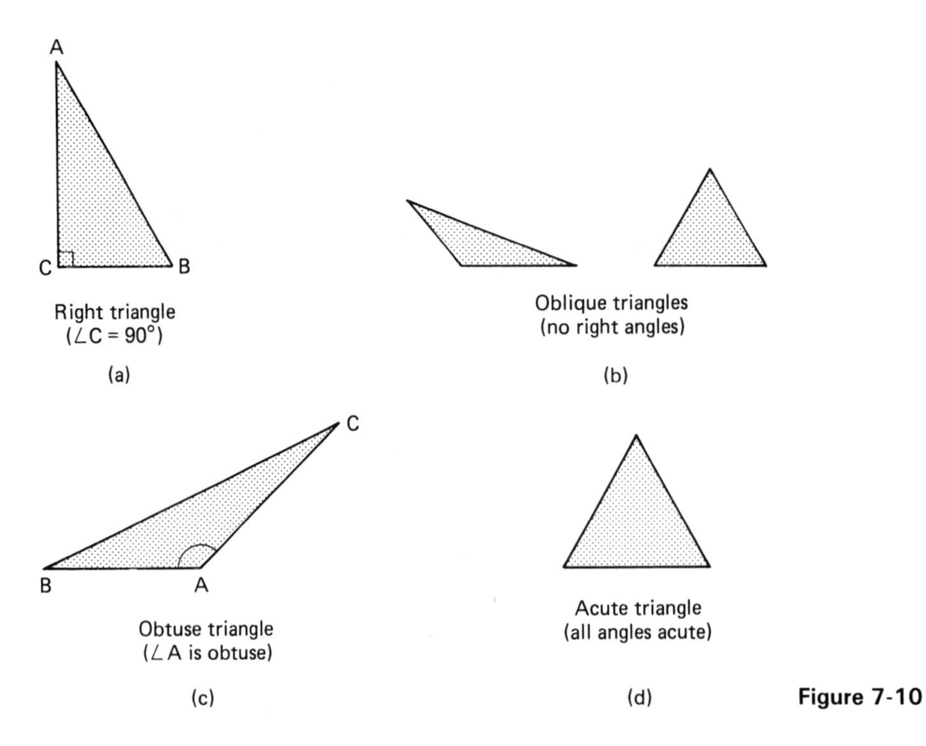

Figure 7-10

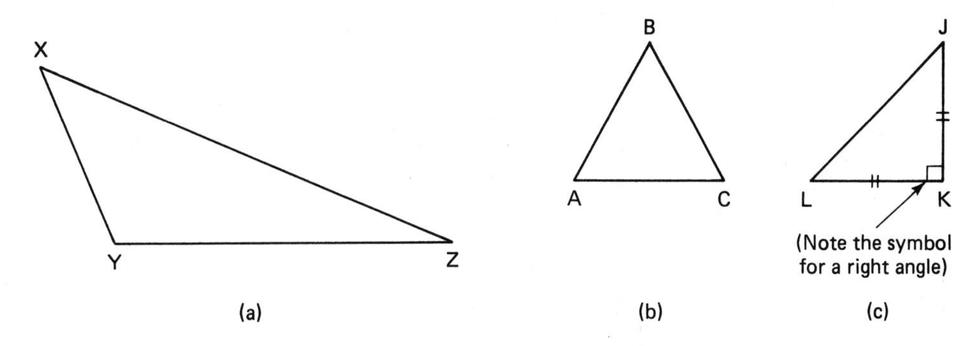

Figure 7-11

A triangle may have more than one classification. Referring to Figure 7-11a, $\triangle XYZ$ can be classified as:

1. Scalene (no sides equal)
2. Oblique (no right angle)
3. Obtuse (one obtuse angle)

$\triangle ABC$ in Figure 7-11b can be classified as:

1. Equilateral (all sides equal)
2. Oblique (no right angle)
3. Acute (all angles acute)

$\triangle JKL$ in Figure 7-11c can be classified as:

1. Isosceles (two sides equal)
2. Right ($\angle K$ is a right angle)

If two triangles are exactly identical in shape and size they are said to be **congruent.** $\triangle ABC$ in Figure 7-12 could fit exactly on top of $\triangle DEF$. Therefore, $\triangle ABC \cong \triangle DEF$. The symbol $\cong$ means "is congruent to." In Figure 7-12,

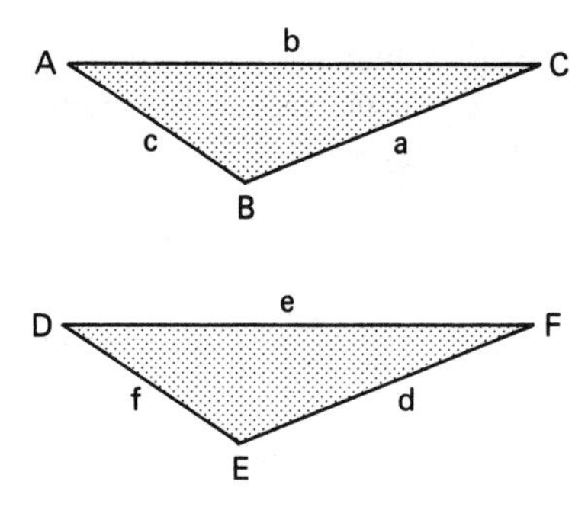

Figure 7-12

$$\angle A = \angle D \qquad a = d$$

$$\angle B = \angle E \qquad b = e$$

$$\angle C = \angle F \qquad c = f$$

In congruent triangles, all corresponding sides and corresponding angles are equal.

If two triangles have the same shape but not the same size, they are said to be **similar** (see Figure 7-13). In order to have the same shape, the angles in one triangle

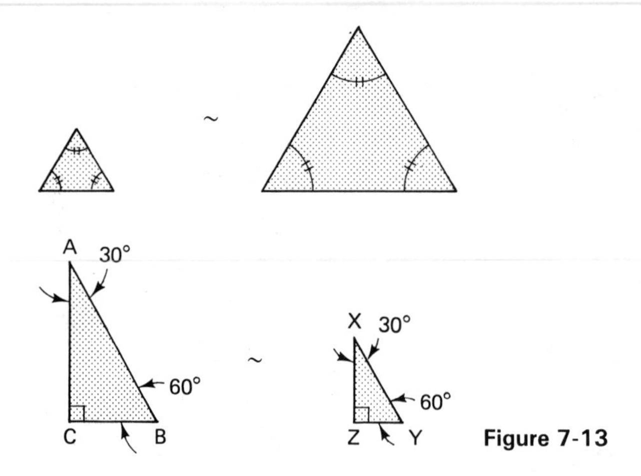

Figure 7-13

must equal the corresponding angles in the other triangle. All equilateral triangles are similar. In Figure 7-13, $\triangle ABC \sim \triangle XYZ$. The symbol for similar triangles is $\sim$.

In similar triangles, the ratios of the corresponding sides are equal. This property makes it possible to solve for missing sides of similar triangles.

Examples

A. Given that $\triangle ABC \sim \triangle XYZ$. What is the length of side y?

$$\frac{4}{10} = \frac{y}{16}$$

$$y = 6.4$$

Since the ratios of the corresponding sides are equal, y can be solved as a proportion.

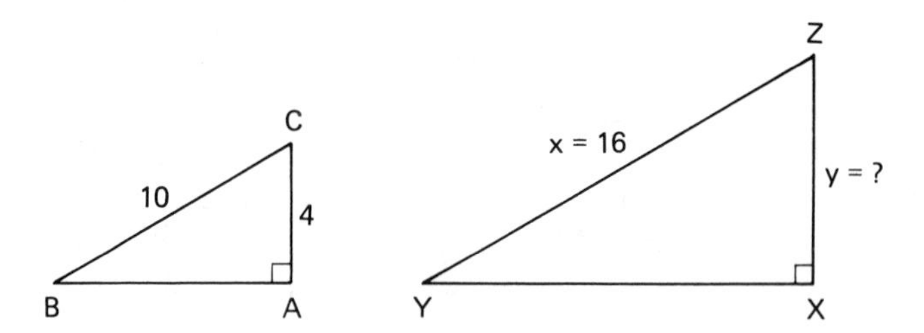

B. A flagpole casts a shadow that is 37 ft long at a certain time of the day. A stake measured to be 2 ft 6 in. casts a shadow 4 ft long at the same time. What is the height of the flagpole?

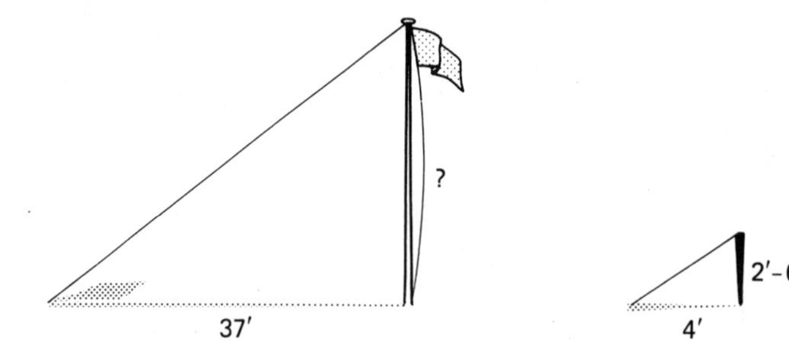

$$\frac{2.5 \text{ ft}}{4 \text{ ft}} = \frac{X}{37 \text{ ft}}$$

1. Since the triangles are similar, the ratio of the height of the flagpole to its shadow is the same as the ratio of the height of the stake to its shadow.

$$\frac{\text{flagpole}}{\text{flagpole shadow}} = \frac{\text{stake}}{\text{stake shadow}}$$

$4X = 2.5(37)$ 2. Cross-multiply and solve for X.

$X = 23.125$ ft

$= 23$ ft $1\frac{1}{2}$ in. 3. Convert to feet and inches.

C. $\triangle ADE \sim \triangle ABC$. Find the distance X.

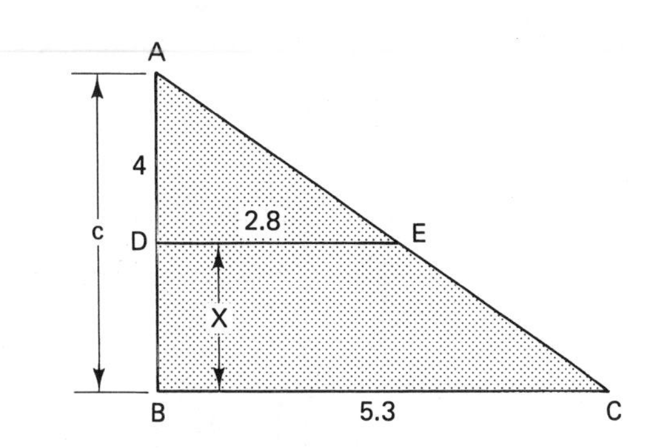

$$\frac{4}{2.8} = \frac{c}{5.3}$$ 1. Set up the proportion.

$c = 7.57$ 2. Cross-multiply and solve for c.

$X = 7.57 - 4$ 3. X is the difference between c and 4.

$= 3.57$

Exercise 7-2

1. $\triangle ABC \cong \triangle XYZ$. Find x, y, and z; and $\angle X$, $\angle Y$, and $\angle Z$.

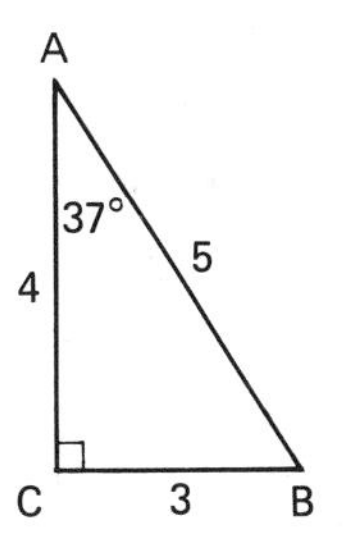

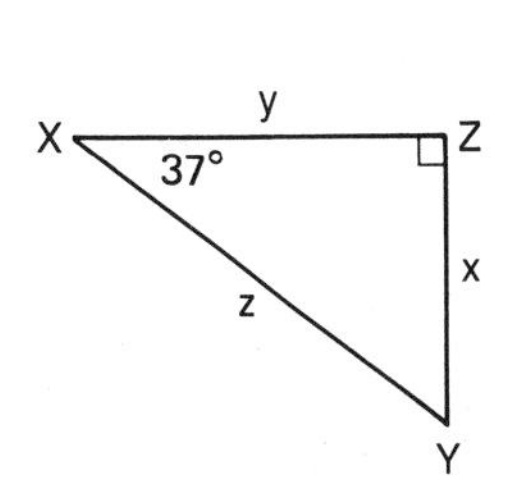

2. $\triangle ABC \sim \triangle JKL$. Find the lengths of sides j and l. Round to one decimal place.

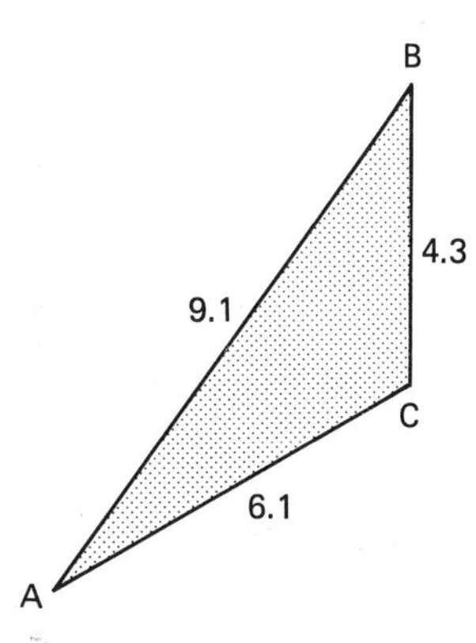

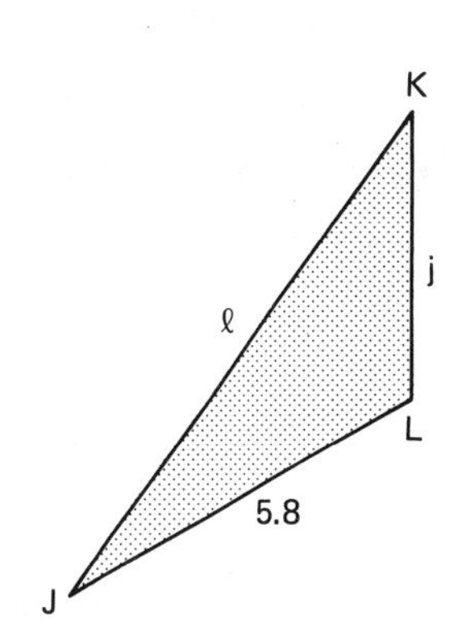

3. $\triangle ABC \sim \triangle ADE$. Find the length of x. Round to one decimal place.

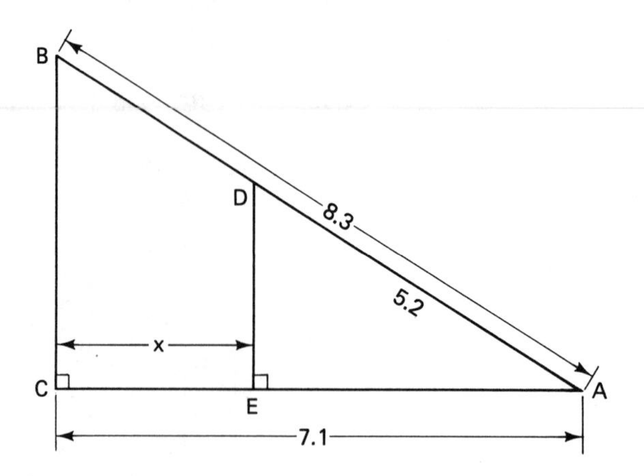

4. What is the height of the tree? Round to the nearest foot.

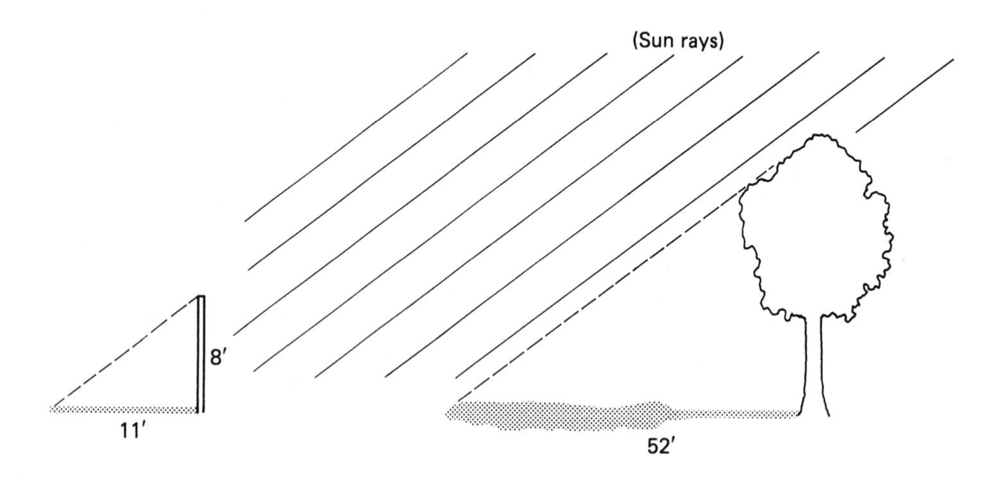

5. Find the length of the unknown side.

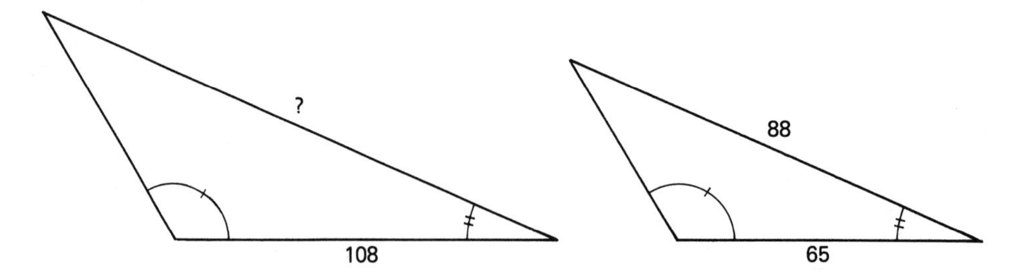

7-3 THE PYTHAGOREAN THEOREM

OBJECTIVES

Upon completing this section, the student will be able to:

1. Solve right triangles using the Pythagorean theorem.
2. Solve word problems involving right triangles.

Before introducing the Pythagorean theorem, a brief discussion of squares and square roots of numbers is important. 5^2 is read "5 squared" and means 5 times 5. Similarly, 4^2 means 4×4 and 18.72^2 means 18.72×18.72. Any number squared is multiplied by itself. $\sqrt{16}$ is read "the square root of 16" and indicates the number that when multiplied by itself equals 16. $\sqrt{16} = 4$ since 4×4 or $4^2 = 16$. $\sqrt{25} = 5$ since 5×5 or $5^2 = 25$. $\sqrt{2} \doteq 1.414$ because $1.414^2 \doteq 2$. $\sqrt{3} \doteq 1.73$ because $1.73^2 \doteq 3$. Although the student is expected to use a calculator to find square roots, it is useful to be able to recognize the first 10 common squares and square roots.

$1 = \sqrt{1}$	$6 = \sqrt{36}$	$1^2 = 1$	$6^2 = 36$
$2 = \sqrt{4}$	$7 = \sqrt{49}$	$2^2 = 4$	$7^2 = 49$
$3 = \sqrt{9}$	$8 = \sqrt{64}$	$3^2 = 9$	$8^2 = 64$
$4 = \sqrt{16}$	$9 = \sqrt{81}$	$4^2 = 16$	$9^2 = 81$
$5 = \sqrt{25}$	$10 = \sqrt{100}$	$5^2 = 25$	$10^2 = 100$

Exercise 7-3A

Using the square and square root keys on a calculator, determine the following answers, rounding to three decimal places where necessary.

1. 8.24^2
2. 15^2
3. 47.81^2
4. 0.564^2
5. 9.3^2
6. 16.25^2
7. 42^2
8. 10^2
9. 5.6^2
10. 1^2
11. $\sqrt{25}$
12. $\sqrt{16}$
13. $\sqrt{1}$
14. $\sqrt{4}$
15. $\sqrt{81}$
16. $\sqrt{9.36}$
17. $\sqrt{17.5}$
18. $\sqrt{10}$
19. $\sqrt{600}$
20. $\sqrt{851}$

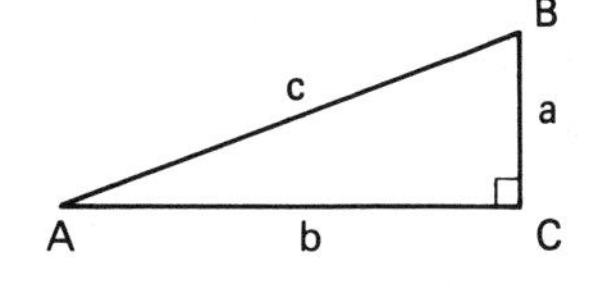

Figure 7-14

In a right triangle, the three sides have specific names. The **hypotenuse** is the side opposite the right angle. The other two sides are called **legs.** The **Pythagorean theorem** states a relationship among the three sides of a right triangle. In Figure 7-14,

$$c^2 = a^2 + b^2$$

where c is the length of the hypotenuse, and a and b are the lengths of the other two sides. If any two sides of a right triangle are known, the other side can be found using the Pythagorean theorem. *Remember:* The Pythagorean theorem applies only to *right triangles*.

Examples

A. Find the missing side of $\triangle ABC$.

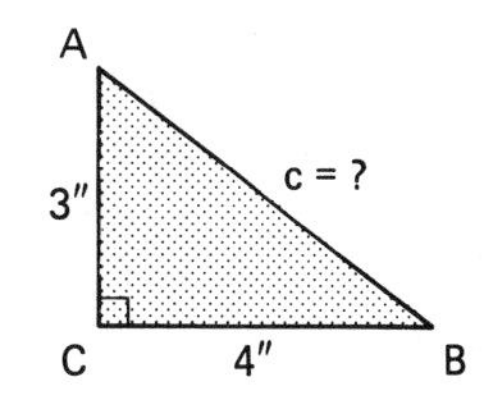

$c^2 = a^2 + b^2$	1. The hypotenuse c is the unknown.
$= 4^2 + 3^2$	2. Substitute the known values.
$= 16 + 9$	3. a and b must be squared before adding them together.
$= 25$ $c = \sqrt{25}$	4. Take the square root of 25 to find the value of c.
$= 5$	5. Note that c is the longest side. The hypotenuse is *always* the longest side of a right triangle.

B. Find the length of side a in $\triangle ABC$.

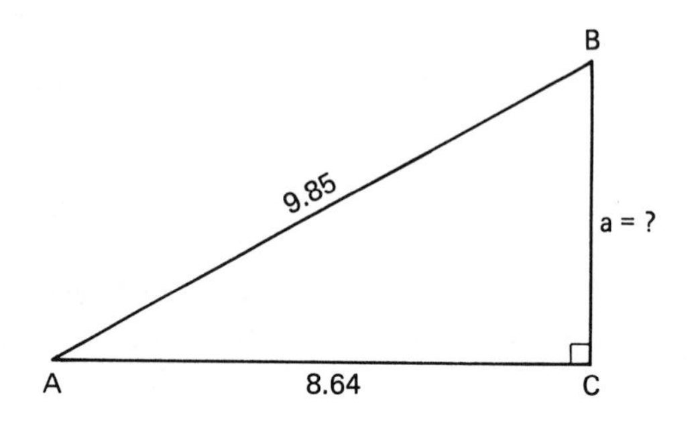

$c^2 = a^2 + b^2$	1. In this example the hypotenuse is given.
$c^2 - b^2 = a^2$	2. Whenever the hypotenuse is given, the square of the known side is *subtracted* from the square of the hypotenuse.
$9.85^2 - 8.64^2 = a^2$ $97.02 - 74.65 = a^2$	3. Solve for a. Remember to square c and b before subtracting.
$22.37 = a^2$ $a = \sqrt{22.37}$	4. Take the square root to find a.
$= 4.73$	5. Note that both legs are smaller than the hypotenuse.

C. A rectangular foundation has sides that measure 36 ft long and 28 ft wide. To check that the foundation is a true rectangle, what would the diagonals measure?

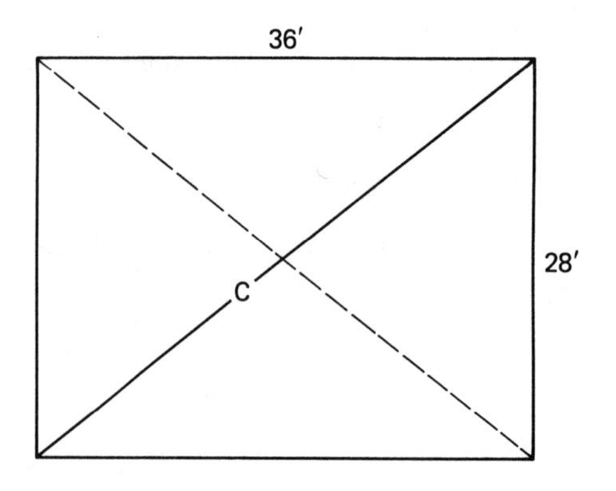

	1. The diagonal, labeled c, is the hypotenuse of the two identical right triangles.
$c^2 = 36^2 + 28^2$	2. Find c^2 and take the square root.
$c = 45.607$ ft	
$45.607 \text{ ft} = 45 \text{ ft } 7\frac{1}{4} \text{ in.}$	3. Convert to feet and inches to the nearest $\frac{1}{4}$ in.
	4. The diagonals should be equal and should measure 45 ft $7\frac{1}{4}$ in. if the foundation is truly rectangular.

Exercise 7-3B

Complete the table for △ ABC. ∠ C is the right angle.

	Side a	Side b	Side c
1.		16.14	18.25
2.	2.23	1.65	
3.	8.92	5.36	
4.	62.17	14.59	
5.	8.35	6.23	
6.	19.35		27.33

7. A rectangular foundation measures 25 ft × 40 ft. What should the diagonals measure, to the nearest $\frac{1}{4}$ in.?

8. A square deck is 8 ft on a side. What is the length of the diagonal of the deck? Give answer in feet and inches to the nearest $\frac{1}{16}$ in.

9. A central vacuum system has an outlet installed in one corner of a 12 ft × 18 ft room. To the nearest inch, what is the longest distance the vacuum hose must be able to reach in that room?

10. A roof has a rise of 6 ft 5 in. and a run of 12 ft 8 in. To the nearest $\frac{1}{4}$ in., find the length of the roof rafters, as shown. Disregard any overhang.

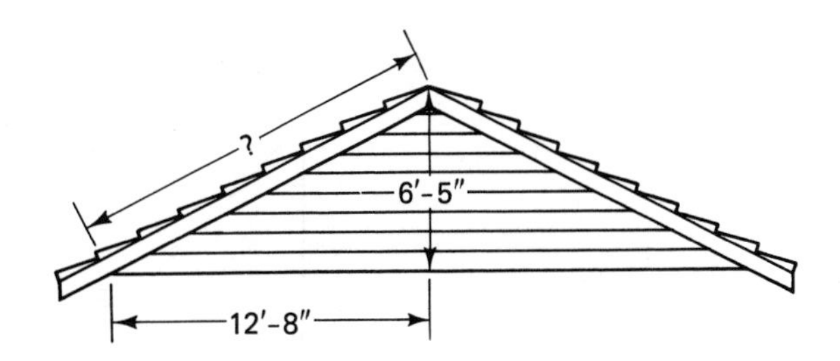

11. The diagonal of a rectangular foundation measures 51 ft $1\frac{1}{4}$ in. If the length of the foundation is 44 ft, what is the width to the nearest $\frac{1}{4}$ in.?

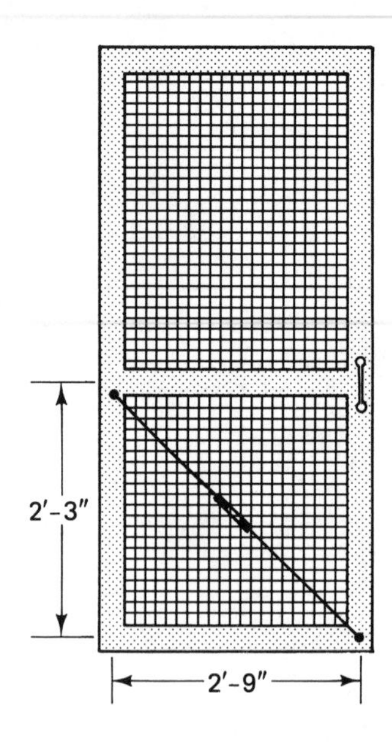

12. A screen door is to have a diagonal brace connected as shown. What should the length of the brace be to the nearest $\frac{1}{8}$ in.?

7-4 SPECIAL RIGHT TRIANGLES

> OBJECTIVES
>
> Upon completing this section, the student will be able to:
>
> 1. Solve 45°–45°–90° triangles (isosceles right triangles).
> 2. Solve 30°–60°–90° triangles.
> 3. Solve 3–4–5 and 5–12–13 right triangles.
> 4. Solve word problems involving special right triangles.

There are two right triangles that are classified as "special right triangles." If the ratios of their sides are known, only one side needs to be given in order to find the remaining two sides. (To solve by the Pythagorean theorem, two sides must be known.)

45°–45°–90° Triangle or Isosceles Right Triangle

This triangle has two equal sides or legs as shown in Figure 7-15. If the sides (legs) are each 1 unit long, the hypotenuse is $\sqrt{2}$ or 1.414 units long. If the legs are 8 in. long, the hypotenuse is $8 \cdot \sqrt{2}$ in. or 11.31 in. long. If the legs are 19.65 ft long, the hypotenuse

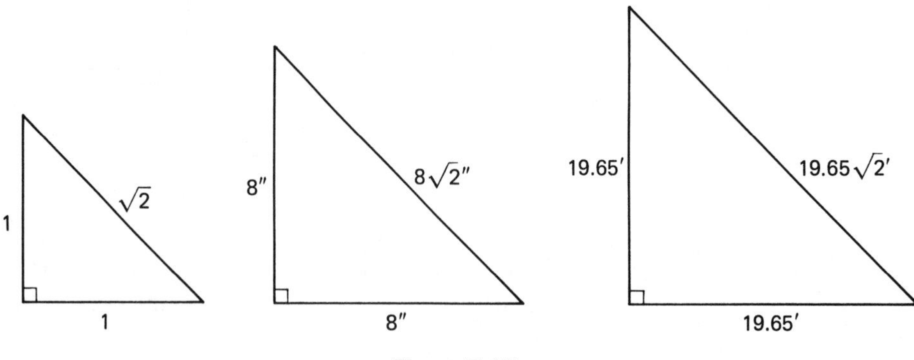

Figure 7-15

is $19.65 \cdot \sqrt{2}$ ft or 27.79 ft long. In other words, the sides are always in the ratio $1 : 1 : \sqrt{2}$, where $\sqrt{2}$ represents the hypotenuse.

Examples

A. Given $\triangle ABC$ with side $b = 4.26$. Find sides a and c.

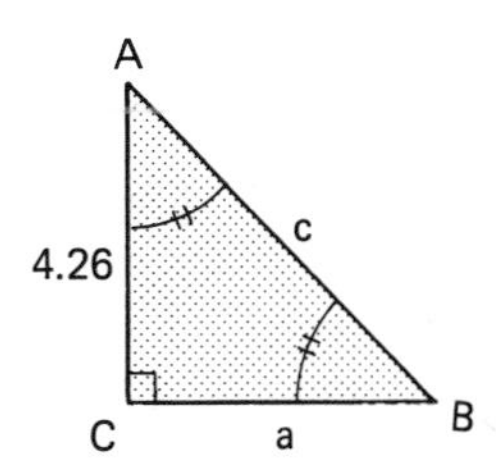

Since $b = 4.26$, $a = 4.26$

1. This is a 45°–45°–90° triangle because $\angle A = \angle B$. Therefore, side a = side b.

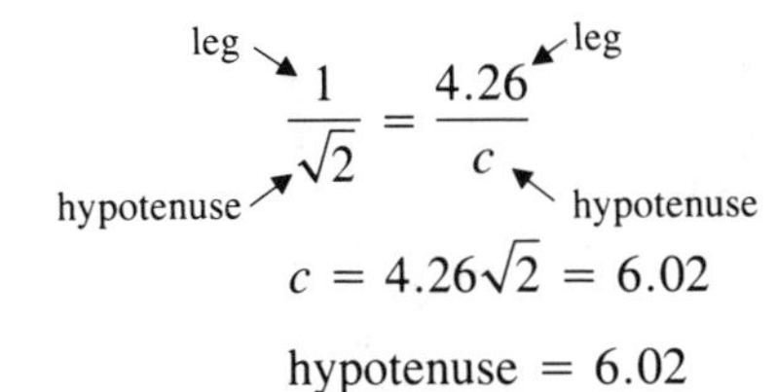

$$\frac{1}{\sqrt{2}} = \frac{4.26}{c}$$

2. Set up the ratio of leg/hypotenuse = leg/hypotenuse.

$c = 4.26\sqrt{2} = 6.02$

3. Cross-multiply and solve for c.

hypotenuse = 6.02

4. In a 45°–45°–90° triangle, if the length of the leg is given, the hypotenuse will always be that length times $\sqrt{2}$.

B. Given $\triangle XYZ$ with side $z = 19.23$ in. Find sides x and y.

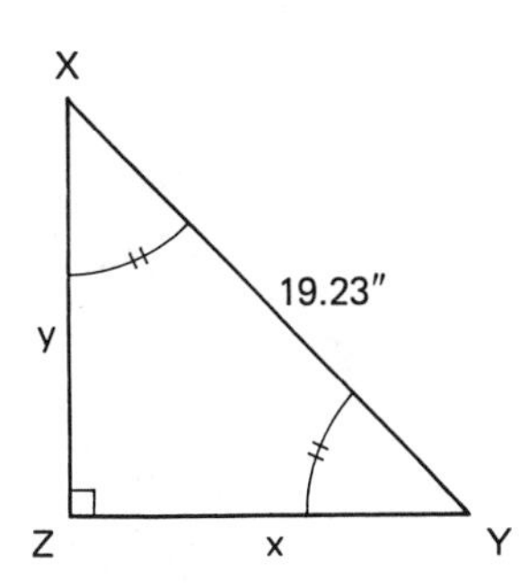

$z = 19.23$ in.

1. This time the hypotenuse is the given value in the 45°–45°–90° triangle.

$$\frac{1}{\sqrt{2}} = \frac{x}{19.23 \text{ in.}}$$

2. Set up the proportion and cross-multiply.

$\sqrt{2} \cdot x = 19.23$ in.

3. Solve for x by dividing 19.23 by $\sqrt{2}$.

$x = 19.23 \div \sqrt{2} = 13.60$ in.

4. Whenever the hypotenuse is known, divide by $\sqrt{2}$ to find the length of the legs in a 45°–45°–90° triangle.

$x = 13.60$ in.; $y = 13.60$ in.

5. The legs x and y are equal.

30°–60°–90° Triangle

This is another very useful right triangle. The ratio of its sides are $1 : \sqrt{3} : 2$ (Figure 7-16). The smallest side, opposite the 30° angle, is 1. The side opposite the 60° angle

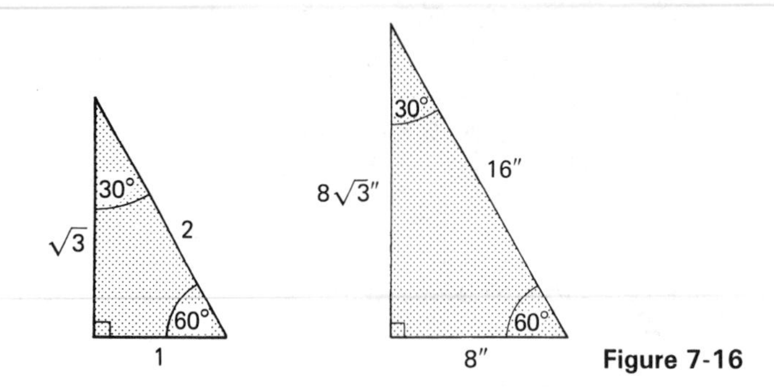

Figure 7-16

is $\sqrt{3}$ or 1.73. The longest side, the hypotenuse, is 2. Any 30°–60°–90° triangle will have its sides in this same ratio. For instance, if the shortest side of a 30°–60°–90° triangle is 8 in., the longer leg will be $8 \cdot \sqrt{3}$ or 13.86 in., and the hypotenuse will be $8 \cdot 2$ or 16 in.

Example

C. Given the 30°–60°–90° triangle shown with side $a = 18.35$. Find sides b and c.

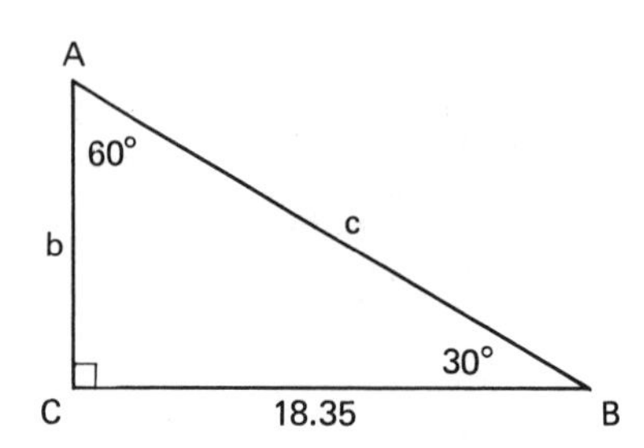

$$\frac{1}{\sqrt{3}} = \frac{b}{18.35}$$

1. The ratio of $b : a$ is $1 : \sqrt{3}$.

$$\sqrt{3} \cdot b = 18.35$$
$$b = 10.59$$

2. Cross-multiply and divide by $\sqrt{3}$ to solve for b.

$$\frac{1}{2} = \frac{10.59}{c}$$

3. When the smallest side is found, use it to find the remaining side. The hypotenuse is always twice the smallest side.

$$c = 21.19$$

4. Cross-multiply to solve for c.

The ratios of the sides of these triangles are equal (because they are similar triangles).

3–4–5 and 5–12–13 Triangles

Two other right triangles used commonly in building construction are the 3–4–5 triangle and the 5–12–13 triangle. The **3–4–5 triangle** has legs of 3 and 4 units and the hypotenuse is 5 units. This triangle is used commonly to check for right angles.

Example

D. To lay out a wall perpendicular to an existing wall, measure 3 ft on the existing wall, measure 4 ft on the floor where the new wall will be, and the distance between these two points should be 5 ft apart (the hypotenuse of right triangle thus formed). If the distance does not measure 5 ft, the angle between the existing wall and the wall to be constructed is not 90°, and the location of the new wall must be adjusted. If greater accuracy is desired, multiples of the

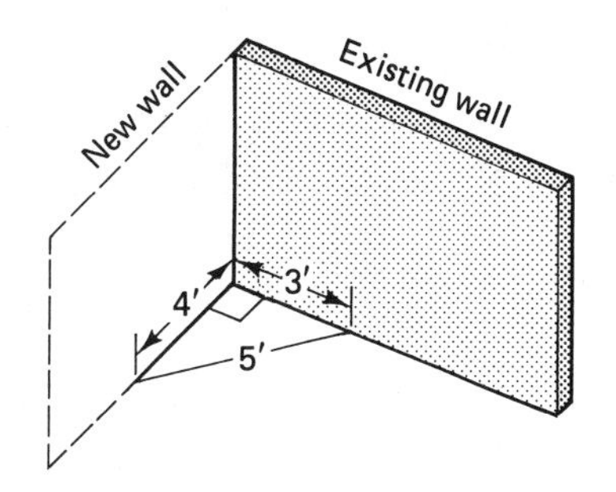

3–4–5 triangle can be used. The existing wall could be measured at 6 ft, the location of the new wall at 8 ft, and the distance between the two points should be 10 ft.

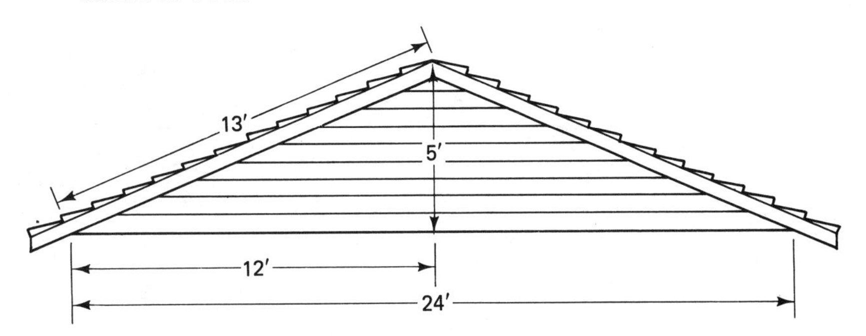

Figure 7-17

The **5–12–13 triangle** shown in Figure 7-17 is useful in building construction because it corresponds to the 5/12 roof used commonly on ranch houses. (5/12 refers to the unit rise, not the pitch. See Chapter 15 for more detailed information on unit rise and pitch.) The run of 12 ft and the rise of 5 ft require a roof rafter 13 ft long (disregarding overhang).

Exercise 7-4

1. $\triangle ABC$: $A = 60°$, $B = 30°$, $C = 90°$; $a =$, $b =$, $c = 17.83$

2. $\triangle XYZ$: $X = 45°$, $Y = 45°$, $Z = 90°$; $x =$, $y = 36.64$, $z =$

3. $\triangle JKL$: $J = 30°$, $K = 60°$, $L = 90°$; $j =$, $k = 46.23$, $l =$

4. $\triangle RST$: $R = 30°$, $S = 60°$, $T = 90°$; $r = 41.75$, $s =$, $t =$

5. $\triangle XYZ$: $X = 45°$, $Y = 45°$, $Z = 90°$; $x =$, $y =$, $z = 17.85$

6. Right triangle JKL with $\angle L = 90°$: $j = 3$, $k = 4$, $l =$

7. Right triangle ABC with $\angle C = 90°$: $a = 9$ ft, $b = 12$ ft, $c =$

8. Right triangle LMN with $\angle N = 90°$: $l = 5$ in., $m =$, $n = 13$ in.

9. Right triangle NOP with $\angle P = 90°$: $n = 10$ ft $o = 24$ ft
$p =$

10. Right triangle XYZ with $\angle Z = 90°$: $x = 6$ ft $y = 8$ ft
$z =$

11. A shed roof has a slope of 30°. If the run is 12 ft, what is the length of the rafters? Ignore overhang. Give the answer to nearest $\frac{1}{8}$ in.

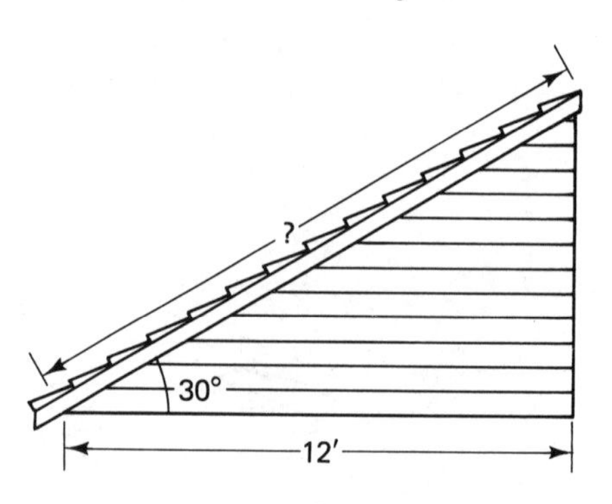

12. A square room measures 15 ft 3 in. by 15 ft 3 in. What is the length in feet and inches of the diagonal? (*Hint:* Change feet and inches to decimal feet before finding the diagonal, then change back to feet and inches.) Give answer in feet and inches to the nearest $\frac{1}{16}$ in.

13. Stairs sloped at an angle of 30° must reach 10 ft 8 in. to the next floor. What is the total run of the stairs? Give answer in feet and inches to the nearest $\frac{1}{16}$ in.

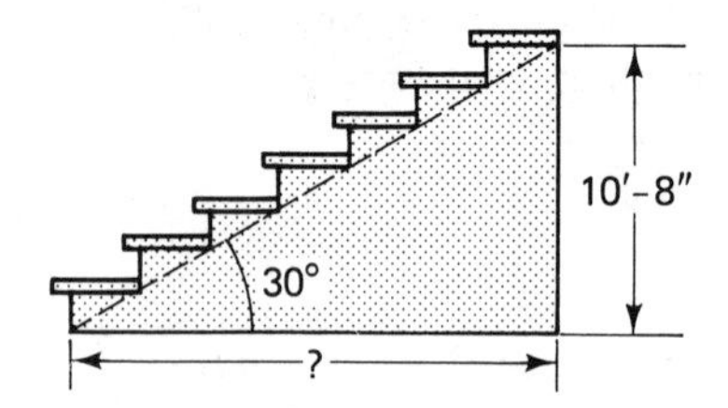

14. The roof shown has a 30° slope and a height of 5 ft $9\frac{1}{4}$ in. Determine the run.

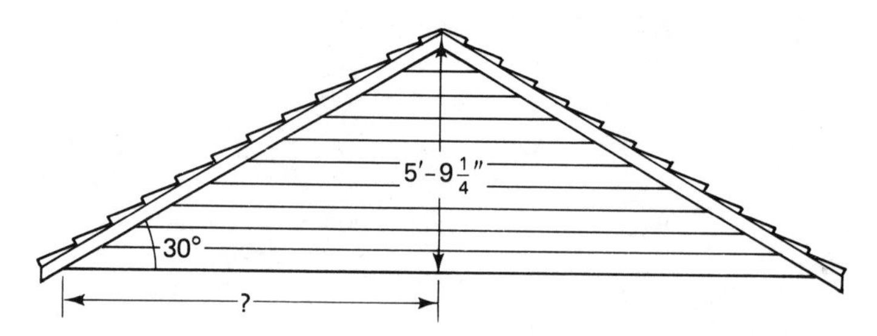

15. A square foundation has diagonals of 33 ft $11\frac{1}{4}$ in. What is the length of the sides? Round to the nearest whole inch.

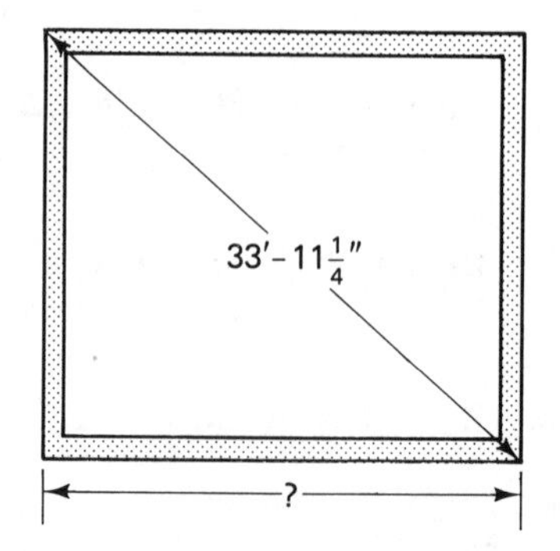

OBJECTIVES

Upon completing this section, the student will be able to:

1. Determine the perimeter of triangles.
2. Solve for the area of right triangles.
3. Solve for the area of oblique triangles.
4. Solve word problems involving perimeter and area of triangles.

The **perimeter** of a triangle is the distance around the figure. Hence for $\triangle ABC$ shown in Figure 7-18, the perimeter

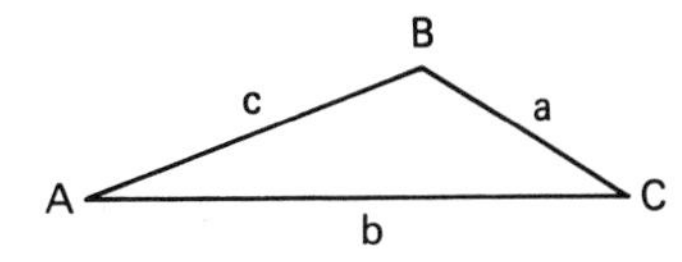

Figure 7-18

$$P = a + b + c$$

Examples

A. Find the perimeter of a triangle with sides of 8, 5, and 7 in.

$P = a + b + c$ — 1. The perimeter is the sum of the three sides.

$= 8 \text{ in.} + 5 \text{ in.} + 7 \text{ in.}$ — 2. The perimeter of the triangle = 20 in.

$= 20 \text{ in.}$

B. Find the perimeter of $\triangle ABC$.

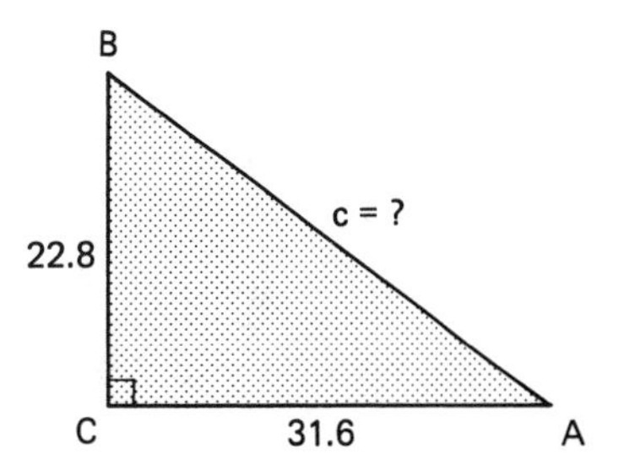

$c^2 = 22.8^2 + 31.6^2$

$c = 39.0$

1. Since $\triangle ABC$ is a right triangle, side c can be found by the Pythagorean theorem.

$P = 22.8 + 31.6 + 39.0$

$= 93.4$

2. The perimeter is the sum of the three sides.

The area of a triangle is given by

$$A = \tfrac{1}{2}bh$$

This is the formula for the area of a triangle where A is the area, b the base, and h the height of the triangle.

Any of the three sides can be considered the base, but the height must always be perpendicular to the base, as shown in Figure 7-19. In a right triangle, the base and height are the legs of the right triangle. In an oblique triangle, the base is one of the sides, but the height is not.

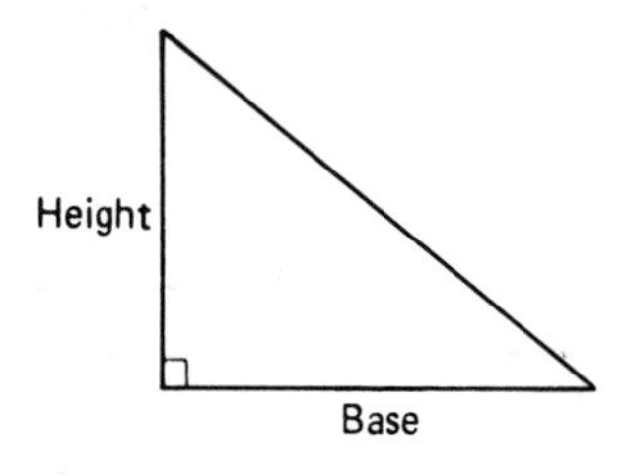

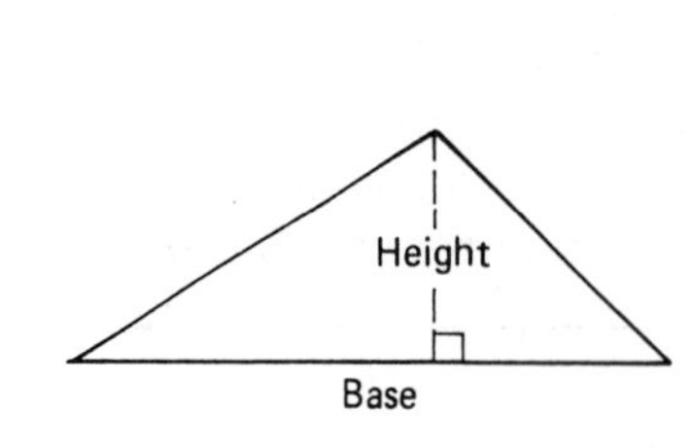

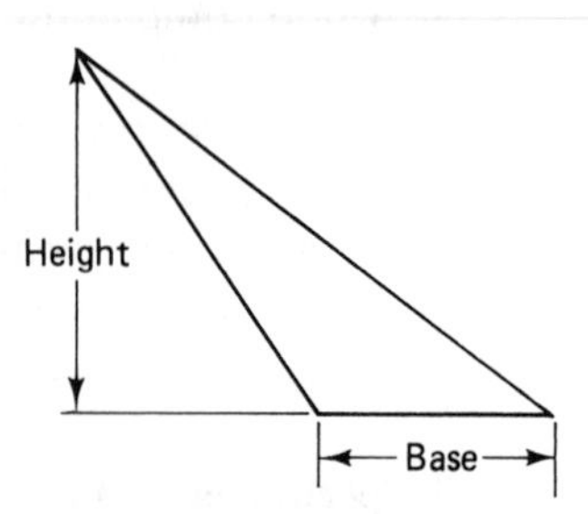

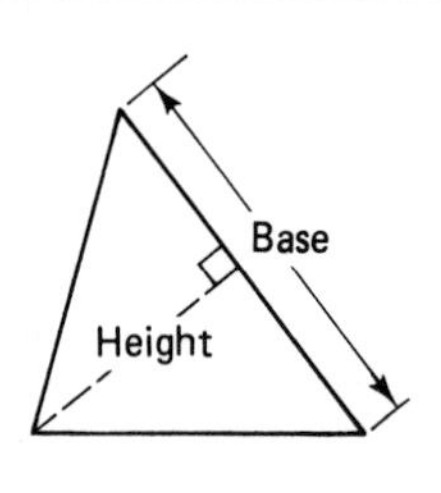

Figure 7-19

Examples

C. Find the area of $\triangle ABC$.

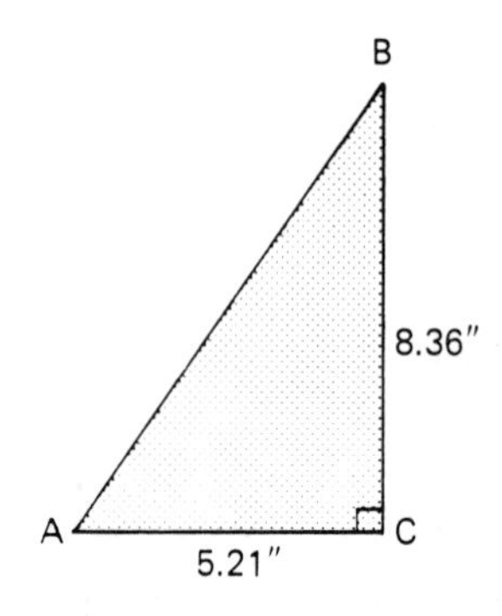

1. Since $\triangle ABC$ is a right triangle, the two legs of the triangle can be considered the height and base.

$A = \frac{1}{2}bh$

2. Substitute the given values for b and h and calculate.

$= \frac{1}{2}(5.21 \text{ in.})(8.36 \text{ in.})$

$= 21.78 \text{ in.}^2$

3. Area is measured in square units.

D. Find the area of $\triangle XYZ$.

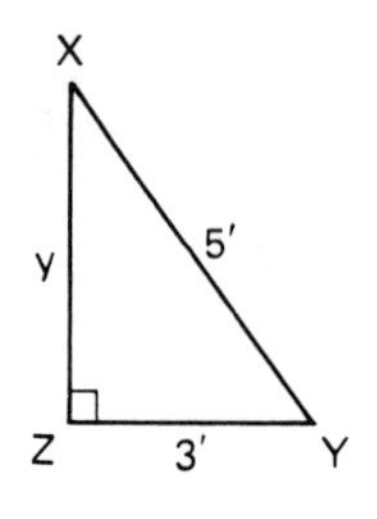

1. Since this is a right triangle, sides x and y can be considered to be the base and height.

$y^2 = 5^2 - 3^2 = 16 \text{ ft}^2$

$y = 4 \text{ ft}$

2. Using the Pythagorean theorem, solve for y.

$A = \frac{1}{2}bh$

$= \frac{1}{2}(3 \text{ ft})(4 \text{ ft})$

3. Solve for the area using the values 3 ft and 4 ft for the base and height.

$= 6 \text{ ft}^2$

4. Area is measured in square feet.

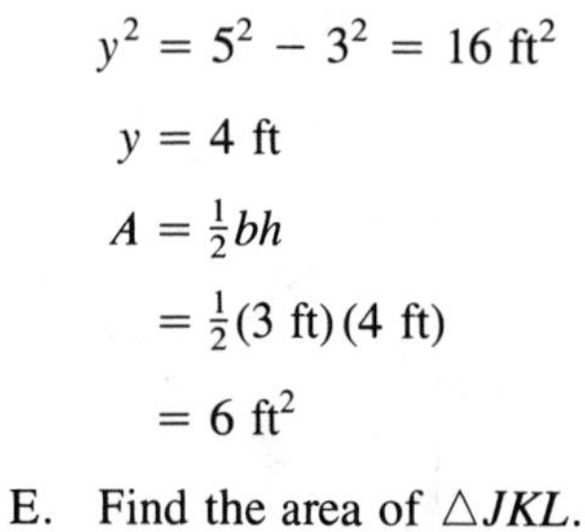

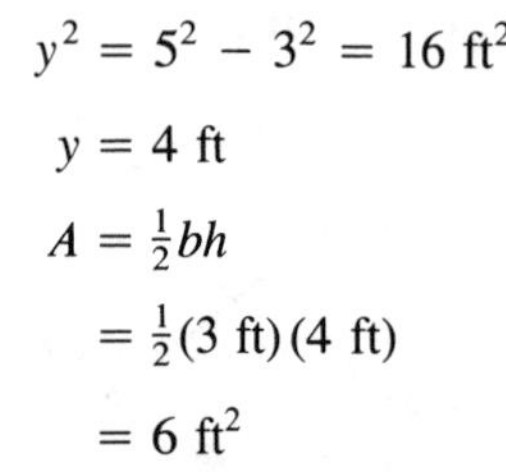

E. Find the area of $\triangle JKL$.

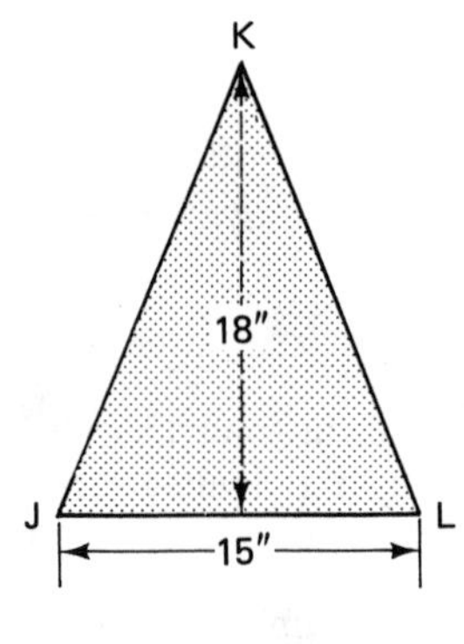

1. Since this is not a right triangle, the altitude is not one of the sides.

$A = \frac{1}{2}(15 \text{ in.})(18 \text{ in.})$

$= 135 \text{ in.}^2$

2. Calculate the area using the height given.

F. Find the perimeter and area of $\triangle ABC$.

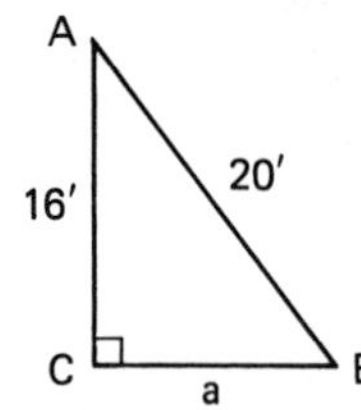

1. Since this is a right triangle, the base can be found by the Pythagorean theorem.

$a^2 = 20^2 - 16^2$

$a = 12 \text{ ft}$

2. The base is 12 ft.

$P = 16 \text{ ft} + 20 \text{ ft} + 12 \text{ ft}$

$= 48 \text{ ft}$

3. The perimeter is the sum of the three sides.

$A = \frac{1}{2}(12 \text{ ft})(16 \text{ ft})$

$= 96 \text{ ft}^2$

4. Find the area.

$P = 48 \text{ ft}; \quad A = 96 \text{ ft}^2$

5. Note that perimeter is in linear feet and area is in square feet.

Area of a Triangle by Hero's Formula

If not enough information is given to find the height of an oblique triangle but all three sides are known, the area can be computed by means of **Hero's formula:**

$$A = \sqrt{s(s - a)(s - b)(s - c)}$$

where s is $\frac{1}{2}$ the perimeter.

Example

G. Find the area of $\triangle ABC$ if $a = 8$ ft, $b = 9$ ft, and $c = 15$ ft.

$P = 8 \text{ ft} + 9 \text{ ft} + 15 \text{ ft} = 32 \text{ ft}$

1. Find the perimeter of the triangle.

$s = \frac{1}{2}(32) = 16 \text{ ft}$

2. s equals one-half the perimeter.

$A = \sqrt{s(s - a)(s - b)(s - c)}$

3. $(s - a)$ is the difference between s and side a.

$= \sqrt{16(16 - 8)(16 - 9)(16 - 15)}$

$= \sqrt{16(8)(7)(1)}$

4. Multiply and take the square root.

$= \sqrt{869}$

$= 29.9 \text{ ft}^2$

Since the formula $A = \frac{1}{2}bh$ is simpler, it is generally used whenever possible. In this case there was not enough information to use the simpler formula.

Exercise 7-5

Given right triangle ABC, find the missing side and the perimeter. Assume that $\angle C$ is the right angle.

	Side a	Side b	Side c	Perimeter
1.	8.26	4.11		
2.	3.58	6.24		
3.	4.12	6.85		
4.	3.00		5.00	
5.		5.00	13.00	

Given right triangle ABC, complete the table. $\angle C$ is the right angle.

	Side a	Side b	Side c	Area
6.	8.21	4.15		
7.		4	5	
8.	6.62	5.17		
9.	3.2	6.4		
10.		12	13	

Given the oblique triangle ABC, use Hero's formula to find the area.

	Side a	Side b	Side c	Area
11.	8.26	4.17	9.23	
12.	27.0	35.6	31.8	
13.	8.3	8.3	8.3	
14.	7.4	5.2	7.4	
15.	5	12	15	

16. Find the area of the triangle.

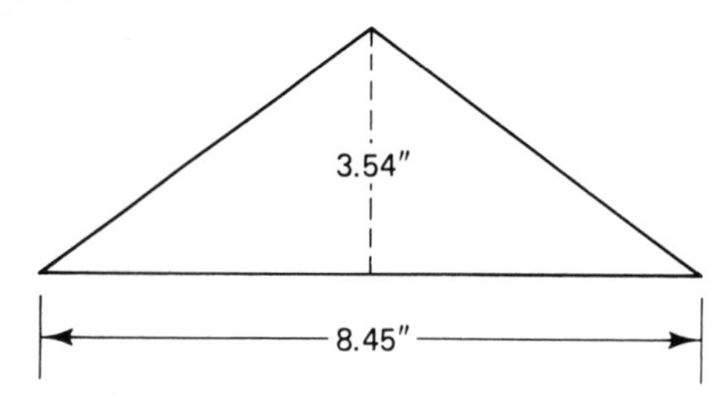

17. Find the area of the triangle.

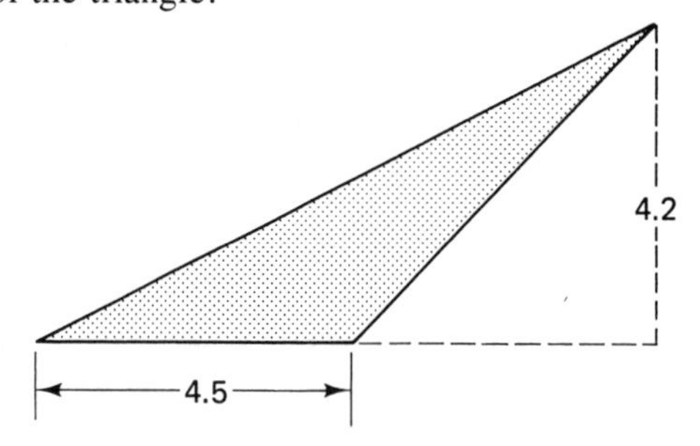

18. Given the 45°–45°–90° triangle shown, find the area of the triangle.

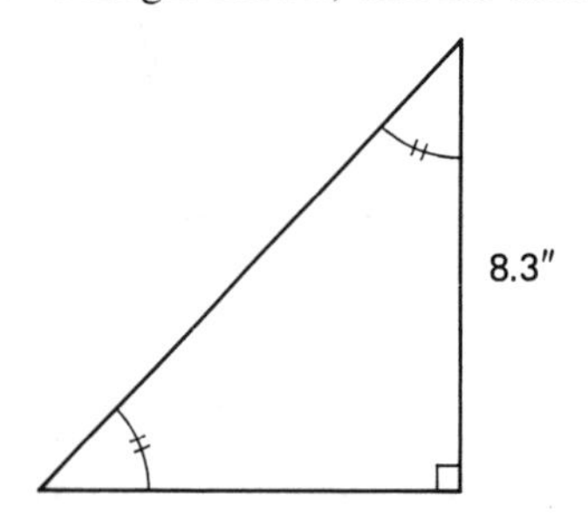

19. A triangular piece of land is bordered by a road on one side, a stone wall on the second side, and a barbed wire fence on the third side. If the lot borders the road for 428 ft, the stone wall for 1000 ft, and the barbed wire fence for 862 ft, what size is the lot in acres? (1 acre = 43,560 ft^2)

20. A house with cathedral ceilings has windows in the gable ends that are right triangles. Find the area in square feet of the glass in each window.

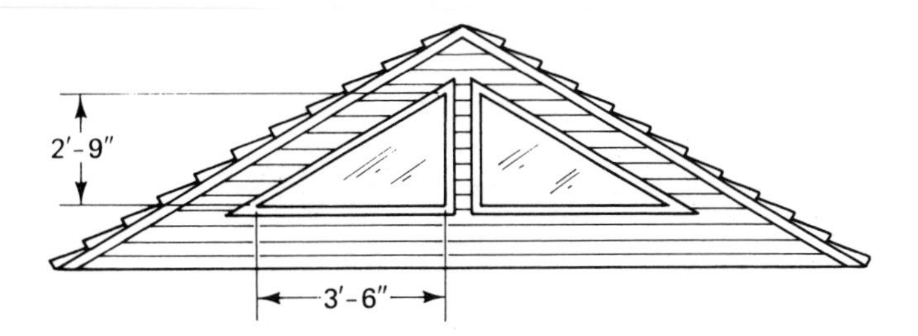

21. In Problem 20, find the amount of trim needed to go around each window. Add 2 ft of trim per window for cutting waste. Give the answer to the nearest foot.

22. A passive solar collector is attached to the side of a house as shown. The collector is 4 ft deep and 5 ft 3 in. high. To the nearest $\frac{1}{16}$ in., how long must the glass front in the collector be? The frame is 1″ wide.

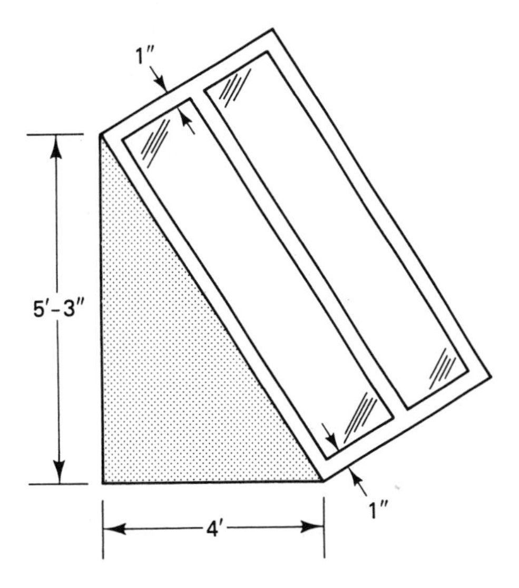

23. The triangular supports in a geodesic dome house are equilateral triangles 4 ft on a side. What is the area of each triangular segment of the roof?

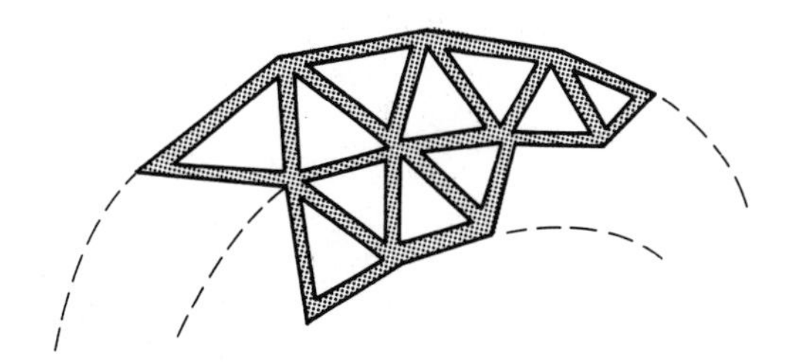

24. The roof shown is designed for a solar hot-water collector. What is the area of end shown?

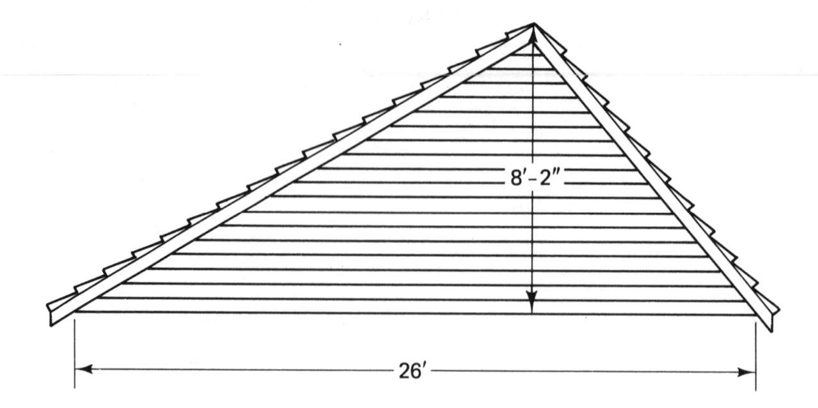

25. A triangular building on a city lot has the dimensions shown. What is the area in square feet of each floor of the building?

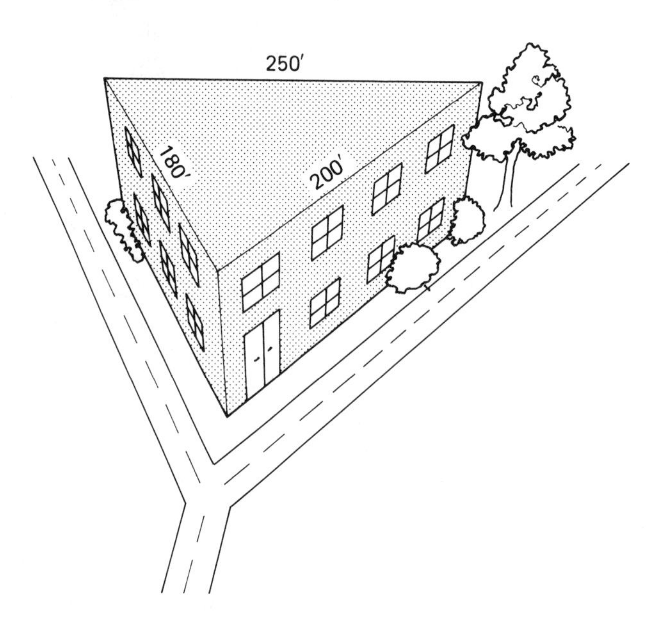

26. An A-frame cottage is 32 ft 8 in. tall and 20 ft wide. If the ends are to be sided with cedar shingles, how many square feet, to the nearest whole foot, are to be sided on each end? Ignore openings for doors and windows.

chapter 8

Areas and Perimeters

8-1 QUADRILATERALS

OBJECTIVES

Upon completing this section, the student will be able to:

1. Find the perimeter and area of a rectangle.
2. Find the perimeter and area of a square.
3. Find the perimeter and area of a parallelogram.
4. Find the perimeter and area of a trapezoid.
5. Solve word problems involving perimeters and areas of the figures noted above.

A **quadrilateral** is a closed, four-sided figure. *Quad* means "four," and *lateral* means "side."

Parallel lines are lines that are the same distance apart the entire length of the lines, as shown in Figure 8-1. No matter how far they are extended, they will never meet or cross. Parallel lines are in the same **plane.** Think of a plane as a two-dimensional flat surface such as a floor or a sheet of paper.

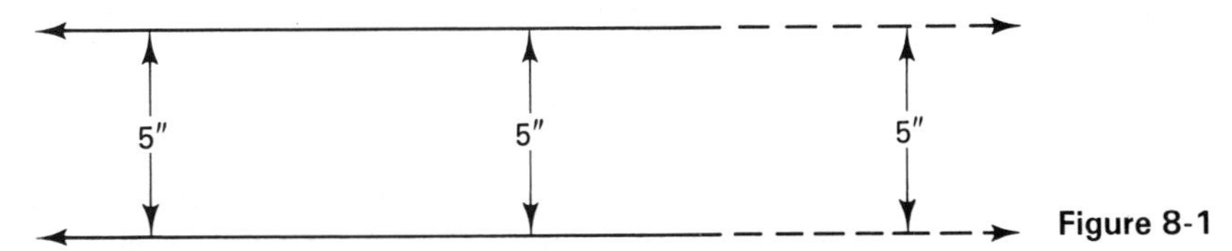

Figure 8-1

A **parallelogram** is a quadrilateral with opposite sides equal and parallel. In parallelogram *ABCD* in Figure 8-2, sides *AB* and *DC* are opposite sides. $AB = DC$ and *AB*

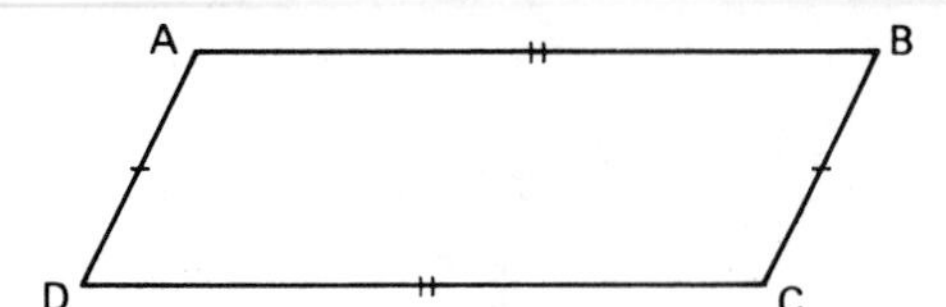

Figure 8-2

$\parallel DC$ (the symbol $\parallel$ means "parallel to"). Opposite sides AD and BC are also equal and parallel to each other. Opposite angles are also equal. Therefore, $\angle A = \angle C$ and $\angle B = \angle D$. The sum of $\angle A + \angle B + \angle C + \angle D = 360°$. The four angles of any four-sided figure total 360°.

A **rectangle** is a parallelogram with four right angles (Figure 8-3). Any parallelogram that has at least one right angle must be a rectangle. Why?

Figure 8-3

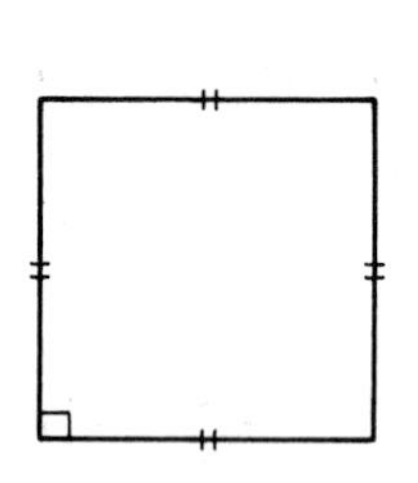

Figure 8-4

A **square** is a rectangle with four equal sides (Figure 8-4). Because squares and rectangles are special cases of parallelograms, every square is also a rectangle, and every rectangle is also a parallelogram.

Figure	*Area*	*Perimeter*	*Diagram*
Rectangle	$A = L \times W$	$P = 2(L + W)$ or $P = 2L + 2W$	
Square	$A = S^2$	$A = 4S$	
Parallelogram	$A = b \times h$	$P = 2a + 2b$	
Trapezoid	$A = \frac{1}{2}h\,(B + b)$	$P = a + b + c + B$	

Perimeter and Area of Rectangles

The **perimeter** of a rectangle is the distance around it. Therefore, the perimeter is the sum of all four sides. The formula for perimeter is usually written in either of two forms:

$$P = 2(L + W) \quad \text{or} \quad P = 2L + 2W$$

where P is the perimeter, L the length, and W the width.

Examples

A. Before ordering baseboard trim, the perimeter of a 12 ft × 18 ft room must be found. What is the perimeter of the room?

$P = 2(L + W)$

1. This is frequently the simpler formula to use.

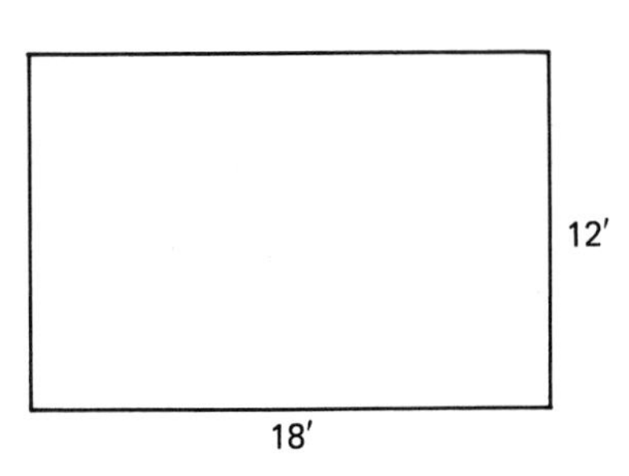

$= 2(12 \text{ ft} + 18 \text{ ft})$

2. Add the length and width in the parentheses. The length plus the width represent halfway around the room.

$= 2(30 \text{ ft})$

$= 60 \text{ ft}$

3. Multiply the sum of the length and width by 2 to find the entire distance around.

B. The perimeter of a deck is 48 ft. If the length is 16 ft, what is the width?

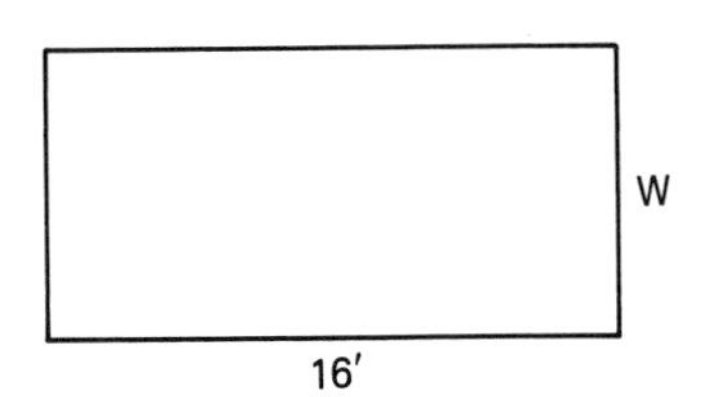

$P = 2(L + W)$

$48 \text{ ft} = 2(16 \text{ ft} + W)$

1. If the entire distance around is $2(L + W)$, then halfway around the deck will be $(L + W)$.

$24 \text{ ft} = 16 \text{ ft} + W$

2. Halfway around (half the perimeter) is half of 48 ft.

$24 \text{ ft} - 16 \text{ ft} = W$

3. Subtract the length from half the perimeter.

$W = 8 \text{ ft}$

4. The width is 8 ft.

Area is measured in square units. The area of a house is the amount of floor space it has. The area of a rectangle is found by the formula

$$A = L \times W$$

where A is the area, L the length, and W the width.

Examples

C. Find the floor space of a rectangular house that is 26 ft wide by 30 ft long.

$A = L \times W$ — 1. Area of a rectangle equals length times width.

$= 26 \text{ ft} \times 30 \text{ ft}$ — 2. Substitute known values into the formula and solve for area.

$= 780 \text{ ft}^2$ — 3. Area is in square units.

D. Find the floor space in a two-story rectangular house that measures 24 ft 9 in. by 30 ft 6 in.

24 ft 9 in. = 24.75 ft

30 ft 6 in. = 30.5 ft — 1. To find area the length and width must be in the same units.

$A = 24.75 \text{ ft} \times 30.5 \text{ ft}$ — 2. Find the area of each floor.

$= 754.875 \text{ ft}^2$

Total area $= 2 \times 754.875 \text{ ft}^2$ — 3. Multiply by 2 to find the floor space on both floors.

$= 1510 \text{ ft}^2$ — 4. Round to the nearest square foot.

E. A rectangle has a length of 10 in. and an area of 50 in.^{2}. What is its perimeter?

$A = L \times W$

$50 \text{ in.}^2 = 10 \text{ in.} \cdot W$ — 1. Substitute the given values in the formula for area of a rectangle.

$\frac{50 \text{ in.}^2}{10 \text{ in.}} = W$ — 2. Rearrange and solve for W.

$W = 5$ in.

$P = 2L + 2W$

$= 2(10 \text{ in.}) + 2(5 \text{ in.})$ — 3. Substitute length and width into the formula for perimeter.

$= 30$ in. — 4. Solve for P.

F. Find the perimeter of a square if one side is 8 in.

$P = 4S$ — 1. Since the length and width are equal on a square, the perimeter is four times the length of one side.

$P = 4(8 \text{ in.})$ — 2. Substitute the known side and multiply.

$= 32$ in.

G. Find the area of a square with sides of 5 in.

$A = S^2$ — 1. Since the length and width are the same on a square, the area is any side squared.

$A = 5^2$

$= 25 \text{ in.}^2$ — 2. A square 5 in. on a side has an area of 25 in.^{2}.

H. A guest house has a square floor with 576 ft^2 of area. What is the linear distance around the house?

$A = S^2$

$576 \text{ ft}^2 = S^2$

$\sqrt{576} = S$

$S = 24 \text{ ft}$

$P = 4S$

$= 4(24 \text{ ft})$

$= 96 \text{ ft}$

1. If the area of a square is given, the length of any side can be found.
2. The length of the side of a square is the square root of its area.
3. Knowing the length of a side, the perimeter (distance around) can be found.
4. The distance around the house is 96 ft.

I. Find the area of the parallelogram shown.

$A\square = b \times h$

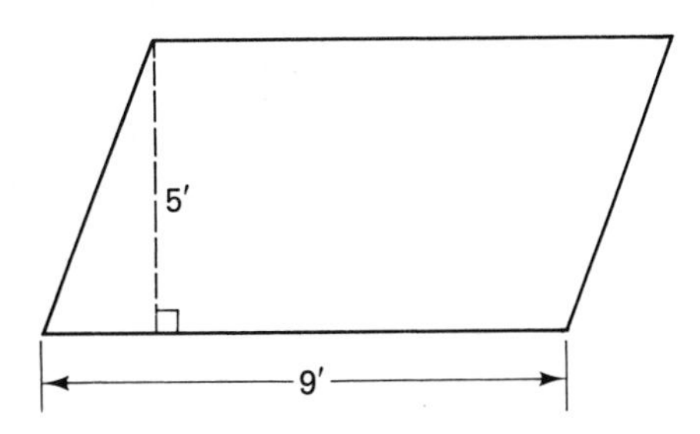

The area of a parallelogram is the height times the base. *Remember:* The height is always perpendicular to the base.

$A = b \times h$

$= 9 \text{ ft} \times 5 \text{ ft}$

$= 45 \text{ ft}^2$

1. Substitute given quantities into the formula.
2. Area is 45 ft^2.

J. Find the area of the parallelogram shown.

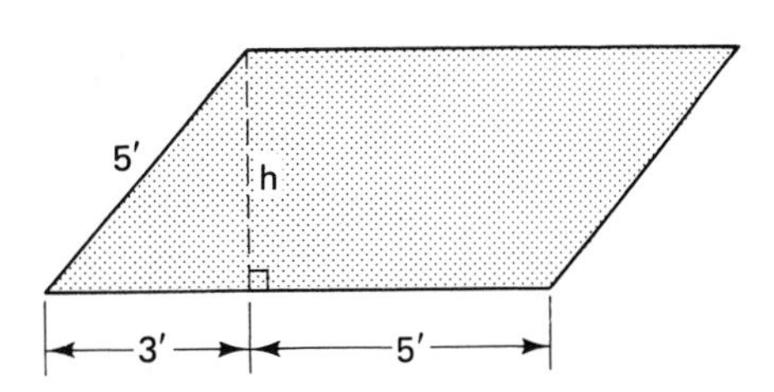

$h^2 = 5^2 - 3^2$

$h = 4 \text{ ft}$

$b = 3 \text{ ft} + 5 \text{ ft} = 8 \text{ ft}$

$A = 4 \text{ ft} \times 8 \text{ ft} = 32 \text{ ft}^2$

1. Find the height by using the Pythagorean theorem.
2. Determine the base.
3. Find the area by the formula.

Perimeter and Area of Trapezoids

A **trapezoid** is a four-sided figure with *two sides parallel* and *unequal*. Several types of trapezoids are shown in Figure 8-5. The two sides that are parallel and unequal are called the **bases.** The longer base is usually represented by B and the shorter base by b. The

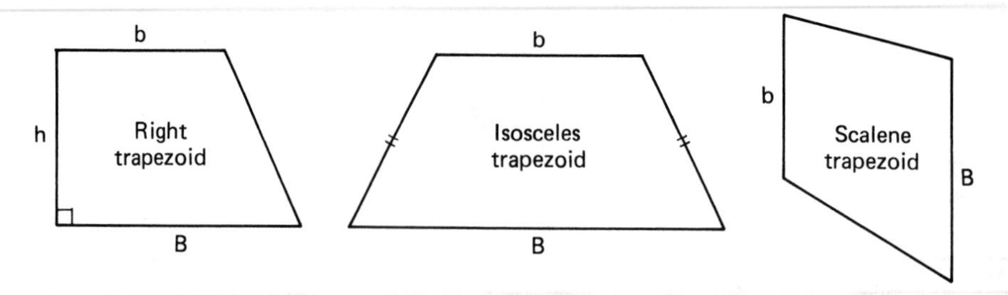

Figure 8-5

perimeter of a trapezoid is the sum of its four sides. The area of a trapezoid is given by the formula

$$A = \tfrac{1}{2}h(B + b)$$

where A is the area, h the height, B the longer base, and b the shorter base.

Examples

K. Find the area of the trapezoid shown.

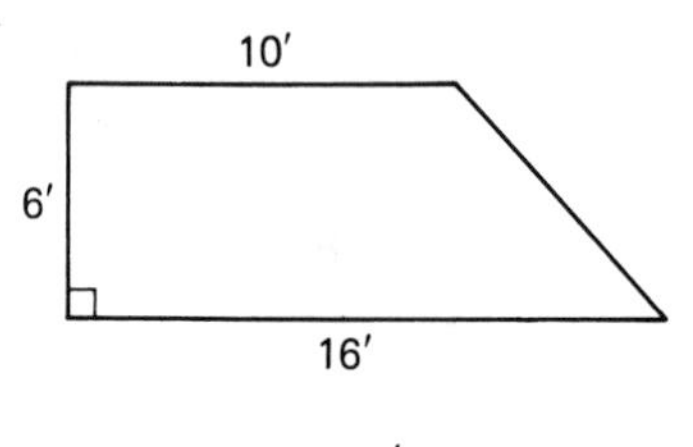

$$A = \tfrac{1}{2}h(B + b)$$

In a right trapezoid the height h is the side perpendicular to the bases.

$A = \frac{1}{2}(6 \text{ ft})(10 \text{ ft} + 16 \text{ ft})$ — 1. Substitute the known values into the formula.

$= \frac{1}{2}(6 \text{ ft})(26 \text{ ft})$ — 2. Perform the addition inside the parentheses first.

$= 78 \text{ ft}^2$ — 3. Multiply to determine the area in square feet.

L. Find the perimeter of the following trapezoid.

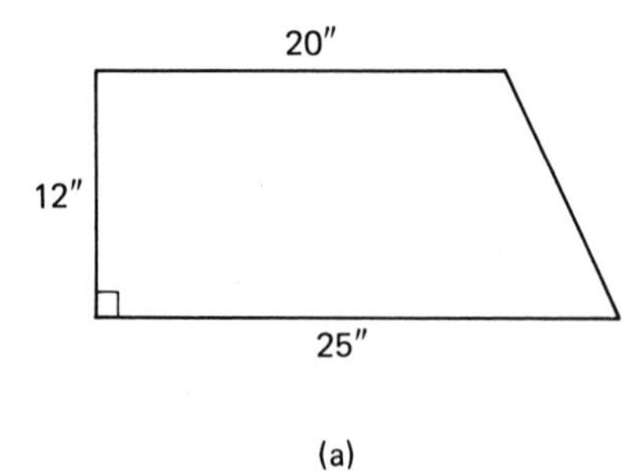

(a)

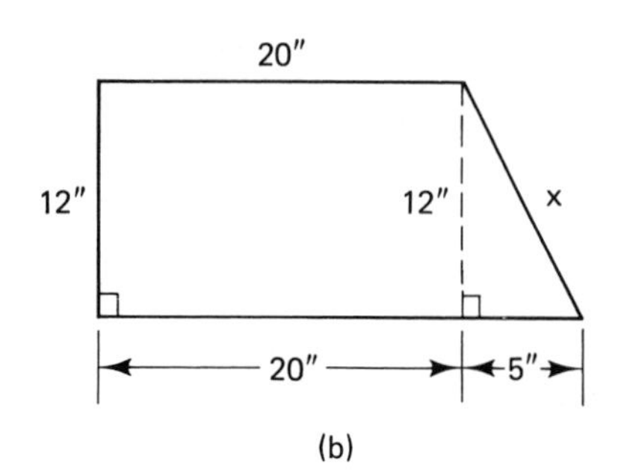

(b)

1. Draw in the height, forming a right triangle and a rectangle. Label the sides with the information given.

$X^2 = 12^2 + 5^2$

$X = 13$ in.

2. Find side X using the Pythagorean theorem.

$P = 25 \text{ in.} + 12 \text{ in.} + 20 \text{ in.} + 13 \text{ in.}$

$= 70$ in.

3. The perimeter is the sum of the sides.

Exercise 8-1

The student may find it useful to draw and label a diagram when one is not given.

1. Find the perimeter of a rectangle that is 8 in. long and 5 in. wide.

2. Find the area of the rectangle in Problem 1.

3. Find the perimeter of ▱$ABCD$.

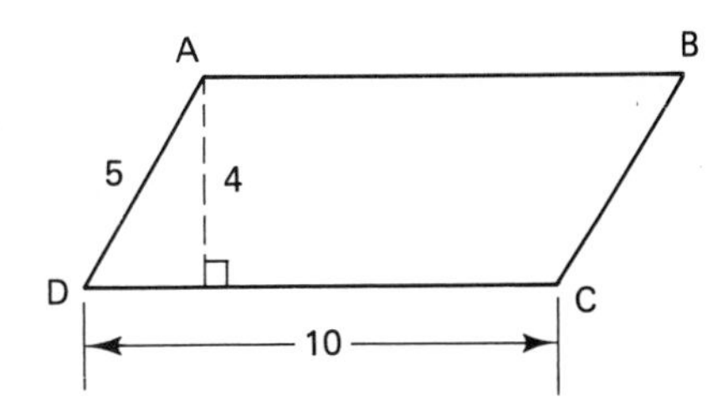

4. Find the area of ▱$ABCD$ in Problem 3.

5. Find the area of a square which is 3 ft on a side.

6. Find the perimeter of the square in Problem 5.

7. A square has a perimeter of 20 ft. What is its area?

8. A rectangle has an area of 200 ft^2 and a width of 8 ft. What is its length?

9. What is the perimeter of the rectangle in Problem 8?

10. Find the perimeter of ▱$ABCD$.

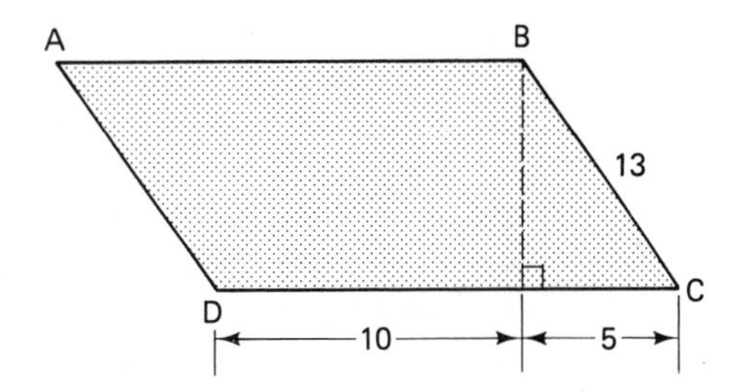

11. Find the area of ▱$ABCD$ in Problem 10.

12. A trapezoid has bases of 10 and 12 ft and a height of 14 ft. Find the area of the trapezoid.

Complete the table for the **rectangle** described.

	Length	Width	Perimeter	Area
13.	10 in.		28 in.	
14.	16 ft			192 ft^2
15.	15 ft 3 in.	8 ft 6 in.		
16.		26 ft	112 ft	

17. A rectangular window calls for a rough opening of 2 ft 6 in. by 1 ft 8 in. What is the area, rounded to the nearest square foot, of the R.O. (rough opening)? (*Hint:* Do not round until the final answer is obtained.)
18. A door is 6 ft 6in. high by 2 ft 4 in. wide. What is the area, in square inches, of the door?
19. A square table is to be covered with plastic laminate costing $1.78 per square foot. If the perimeter of the table is 12 ft, what is the cost of the laminate to cover it?
20. A room 12 ft × 16 ft 6 in. is to be carpeted. Assuming no waste, how many square feet of carpeting is needed?
21. If carpeting costs $27.95 per square yard installed, what will it cost to carpet the room in Problem 20?
22. Floor tiles 1 ft by 1 ft are to be installed on a kitchen floor measuring 8 ft 8 in. × 4 ft 8 in. How many tiles are needed? Add six tiles for waste and proper fitting.
23. A sill is to be placed on top of a rectangular foundation that is 26 ft wide and has an area of 780 ft^2. Ignoring waste, how many lineal feet of 2 × 6 lumber will be needed for the sill?
24. A straight driveway 15 ft wide and 93 ft long is to be hot-topped at a cost of $12.50 per square yard. What will it cost to hot-top the driveway?
25. A square gazebo is 48 ft around. What is the area of the gazebo?

8-2 CIRCLES

OBJECTIVES

Upon completing this section, the student will be able to:

1. Find the circumference of a circle.
2. Find the area of a circle.
3. Solve problems involving circles, semicircles, and quarter circles.

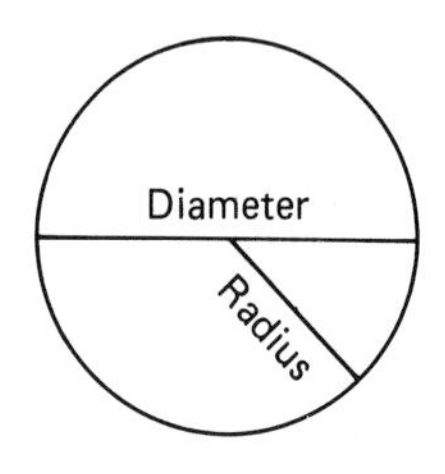

Figure 8-6

The perimeter of a circle is called the **circumference.** The distance to the center of the circle is called the **radius** (see Figure 8-6). The distance across a circle through the center is the **diameter.** The diameter is twice the length of the radius.

Formulas for the circumference and area of a circle involve π (pi). π represents the ratio of the circumference of any circle to its diameter, and it is always the same regardless of the size of the circle. π is approximately 22/7 or 3.14. Many calculators have a π key because it is a value that is used so frequently. The area of a circle is found by the formula

$$A = \pi r^2$$

where A is the area and r is the radius (π is always approximately 3.14).

Example

A. Find the area of a circle with a radius of 10 in.

$A = \pi r^2$	1. Formula for the area of a circle.
$= \pi(10)^2$	2. Be sure to square the radius before multiplying by π.
$= \pi(100)$	
$= 314 \text{ in.}^2$	3. The area is in square inches.

The circumference of a circle is found by the formula

$$C = \pi d \qquad \text{or} \qquad C = 2\pi r$$

where C is the circumference, d the diameter, and r the radius.

Examples

B. Find the circumference of a circle if the diameter is 20 ft.

$C = \pi d$

1. Substitute the given values into the formula.

$= \pi(20 \text{ ft})$

$= 62.8 \text{ ft}$

2. The circumference is in linear feet.

C. Find the radius of a circle if its circumference is 80 ft.

$C = \pi d$

1. Substitute the known values into the formula.

$80 \text{ ft} = \pi d$

$\dfrac{80 \text{ ft}}{\pi} = d$

2. Rearrange and solve for d. Use $\pi = 3.14$ if a calculator with a π key is not available.

$d = 25.5 \text{ ft}$

3. Since the diameter is twice the radius, divide by 2 to find r.

$r = 25.5 \text{ ft} \div 2$

$= 12.7 \text{ ft}$

D. Find the area of a circle if its circumference is 100 in.

$C = \pi d$

1. Substitute into the formula for circumference.

$100 \text{ in.} = \pi d$

2. Divide 100 by π to solve for d.

$31.8 \text{ in.} = d$

$r = \dfrac{31.8 \text{ in.}}{2} = 15.9 \text{ in.}$

3. Solve for the radius ($\frac{1}{2}$ the diameter).

$A = \pi r^2$

4. Solve for the area by substituting into the area formula.

$= \pi(15.9)^2$

$= 795.8 \text{ in.}^2$

E. Find the area of the ring shown (the shaded area).

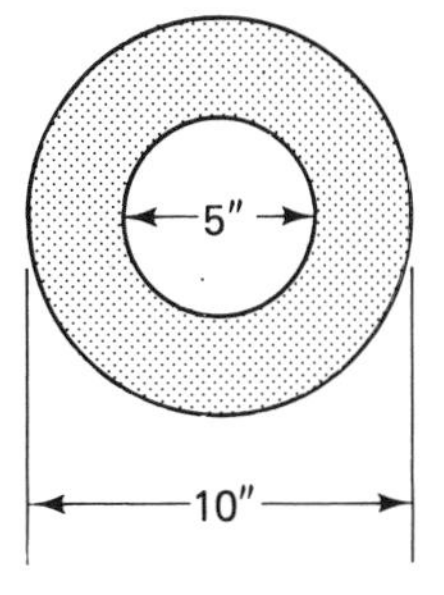

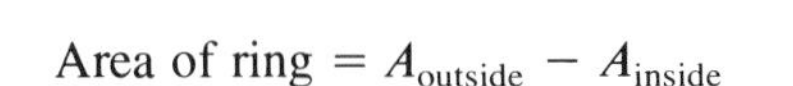

Area of ring $= A_{\text{outside}} - A_{\text{inside}}$

1. To find the area of the ring, find the area of the inside circle and subtract it from the area of the outside circle.

$A_{\text{outside}} = \pi r^2$

2. Find the area of the outside circle. Remember to square the radius, not the diameter.

$= \pi(5)^2$

$= 78.5 \text{ in.}^2$

$A_{\text{inside}} = \pi r^2$ — 3. Find the area of the inside circle.

$= \pi(2.5)^2$

$= 19.6 \text{ in.}^2$

$A_{\text{ring}} = 78.5 \text{ in.}^2 - 19.6 \text{ in.}^2$

4. The area of the ring is the difference between the areas of the outside and inside circles.

$= 58.9 \text{ in.}^2$

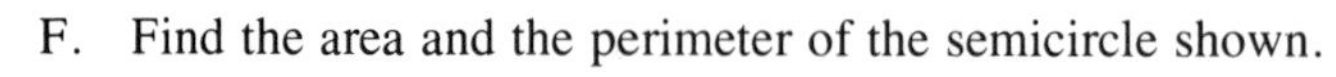
F. Find the area and the perimeter of the semicircle shown.

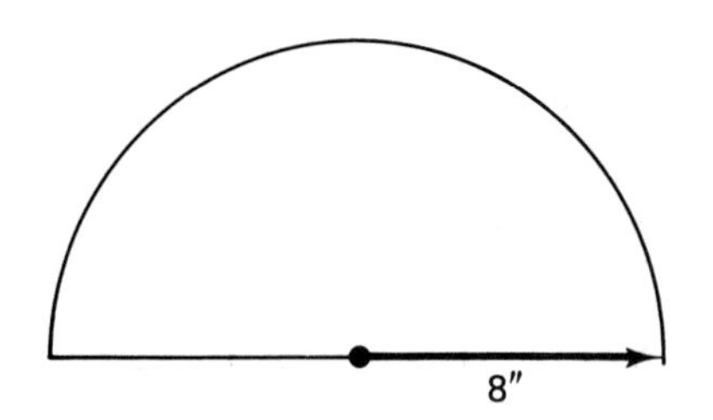

$A = \pi(8)^2$ — 1. This is the area of a circle with radius = 8 in.

$= 201 \text{ in.}^2$

Area of semicircle $= 100.5 \text{ in.}^2$ — 2. The area of the semicircle will be $\frac{1}{2}$ the area of the circle.

$C = \pi d$ — 3. Find the circumference and divide by 2.

$= \pi(16) = 50.3 \text{ in.}$

$\frac{1}{2} \cdot$ circumference $= 25.1$ in.

$P = \frac{1}{2} C + d$ — 4. The perimeter equals half the circumference plus the diameter of the circle.

$= 25.1 \text{ in.} + 16 \text{ in.}$

$= 41.1 \text{ in.}$

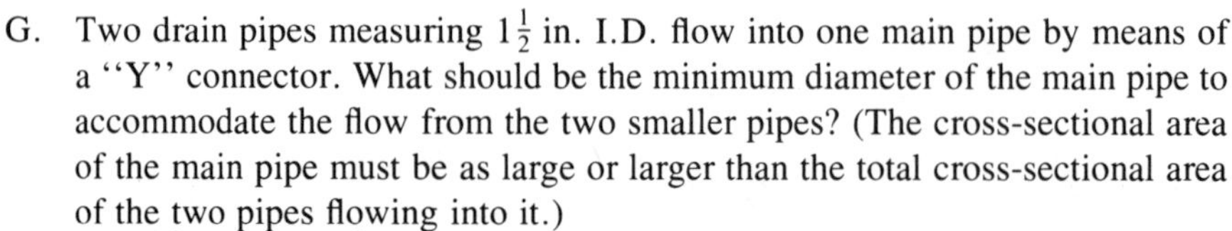
G. Two drain pipes measuring $1\frac{1}{2}$ in. I.D. flow into one main pipe by means of a "Y" connector. What should be the minimum diameter of the main pipe to accommodate the flow from the two smaller pipes? (The cross-sectional area of the main pipe must be as large or larger than the total cross-sectional area of the two pipes flowing into it.)

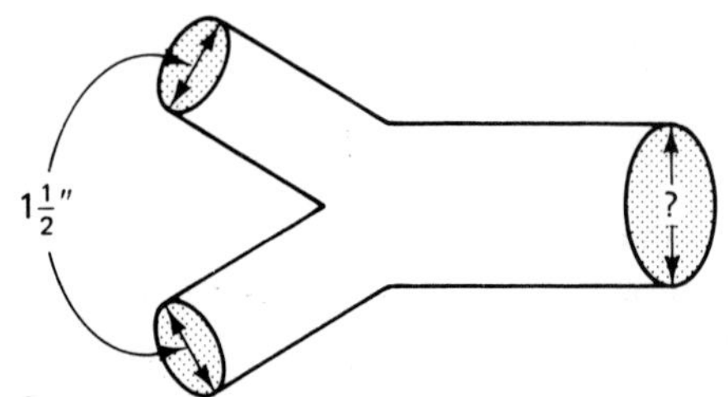

$A = \pi r^2$ — 1. Find the cross-sectional area of each of the smaller pipes.

$= \pi(0.75 \text{ in.})^2$

$= 1.77 \text{ in.}^2$

$A_{\text{TOTAL}} = 2 \times 1.77 \text{ in.}^2$ — 2. This is the minimum cross-sectional area the main pipe can have.

$= 3.53 \text{ in.}^2$

$A = \pi r^2$ — 3. Substitute the area into the formula.

$3.53 = \pi r^2$

$r^2 = \dfrac{3.53}{\pi} = 1.125 \text{ in.}^2$ — 4. Rearrange and solve for r^2.

$r = \sqrt{1.125} = 1.06 \text{ in.}$ — 5. Take the square root to solve for r.

$d = 2.12$ in. — 6. Find the diameter and convert to the nearest $\frac{1}{8}$ in. Round up.

$= 2\frac{1}{8}$ in. I.D. — 7. Note that doubling the diameter does *not* double the area.

Exercise 8-2

Complete the table for the circles described.

	Radius	Diameter	Circumference	Area
1.	5 in.			
2.		12 ft		
3.			82.6 ft	
4.		10.35 in.		
5.	8 in.			
6.		2.6 in.		
7.			31.4 in.	
8.	$6\frac{1}{4}$ in.			
9.				50.265 in.^{2}

10. A concrete storm pipe has an inside diameter of 5 ft and an outside diameter of 7 ft. What is the cross-sectional area of the concrete walls?

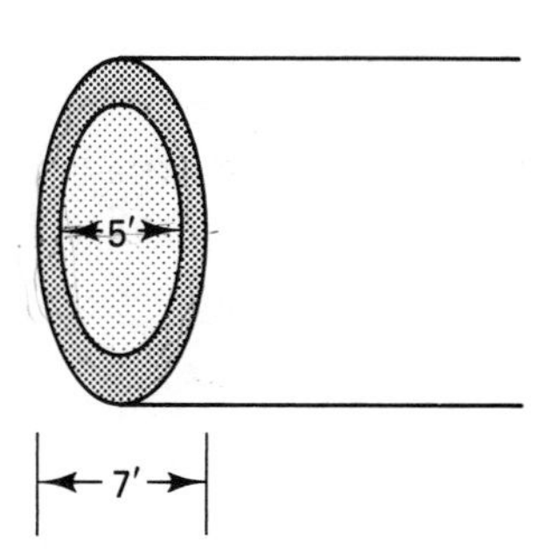

11. A window is a quarter circle with a radius of 1 ft 4 in. How many square inches of glass are in the window?

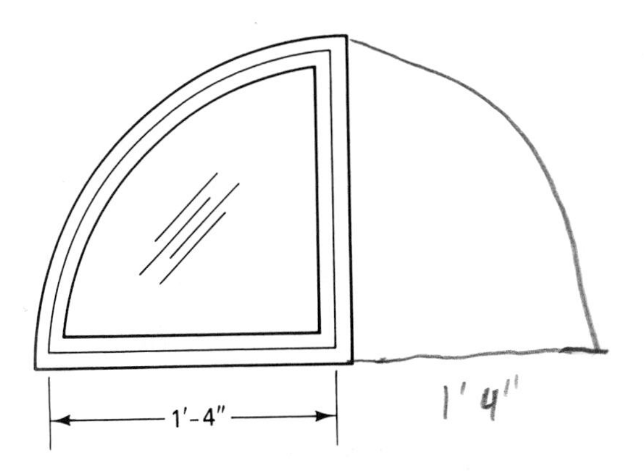

12. To the nearest $\frac{1}{8}$ in., what is the perimeter of the window in Problem 11?

13. A walkway is made of circular concrete slabs 18 in. in diameter. What is the area of each slab?

14. A circular dining room table with a diameter of 40 in. is to be covered with woodgrain plastic laminate. How many square feet of laminate are needed? Round up to the next whole number of square feet.

15. The table in Problem 14 is to have the edge trimmed with a thin strip of laminate. How long should the laminate strip be? Round up to the next whole number of feet.

16. A tree being sawed into lumber measures 32 in. around. To the nearest inch, what is the diameter of the tree?

17. A tapered newel post has a diameter of $3\frac{1}{8}$ in. at the top and $4\frac{1}{2}$ in. at the bottom. To the nearest square inch, what is the difference in the cross-sectional area at the top and the bottom of the post?

18. A wooden dowel has a diameter of $\frac{3}{4}$ in. What is its cross-sectional area?

19. If a 10-in. pizza feeds one medium-hungry carpenter, what size of pizza, to the nearest inch, would be needed to split between two medium-hungry carpenters? (10 in. refers to the diameter.)

20. Two $1\frac{1}{2}$-in. and one 2-in. drain pipes merge into a main drain pipe. What is the minimum size the main pipe can be to have a cross-sectional area that is the same as or greater than the three pipes draining into it? Round up to the nearest $\frac{1}{4}$ in. The sizes given are diameters. (Refer to Example G.)

8-3 COMPOSITE FIGURES

OBJECTIVES

Upon completing this section, the student will be able to:

1. Determine the area of composite figures.
2. Determine the perimeter of composite figures.

To determine the areas and perimeters of composite shapes, divide them into sections that can be determined individually.

Examples

A. Here are three different methods to find the area of the house plan shown.

40′

30′

24′

12′

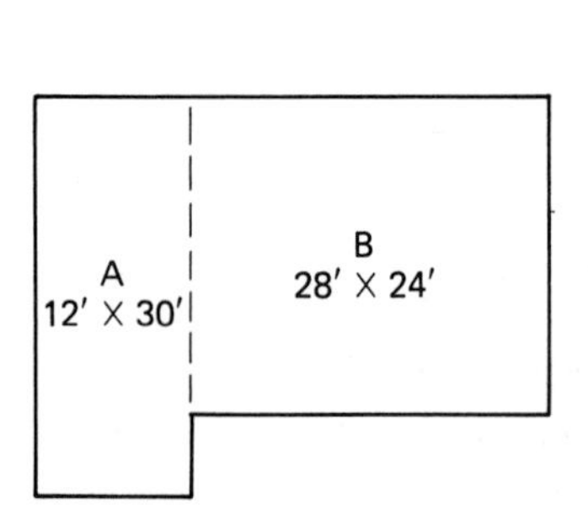

$A = 30 \text{ ft} \times 12 \text{ ft} = 360 \text{ ft}^2$

$B = 24 \text{ ft} \times 28 \text{ ft} = \underline{672 \text{ ft}^2}$

Area of house: 1032 ft^2

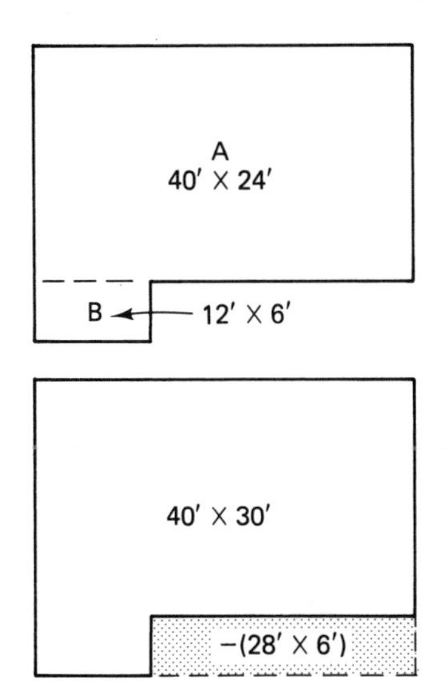

$$A = 24 \text{ ft} \times 40 \text{ ft} = 960 \text{ ft}^2$$

$$B = 6 \text{ ft} \times 12 \text{ ft} = 72 \text{ ft}^2$$

Area of house: 1032 ft^2

$$40 \text{ ft} \times 30 \text{ ft} = 1200 \text{ ft}^2$$

$$-(28 \text{ ft} \times 6 \text{ ft}) = -168 \text{ ft}^2$$

Area of house: 1032 ft^2

In the third case, find the area of the expanded rectangle and subtract the portion that is not on the blueprint.

B. The 5-ft-wide track shown below is to be paved. How many square feet are to be paved? Assume that the ends are semicircles.

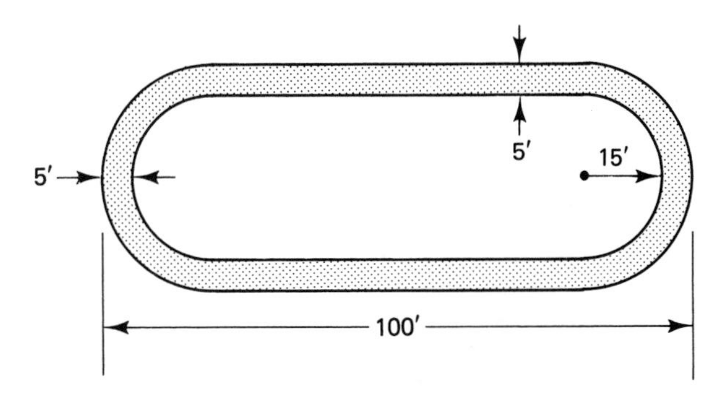

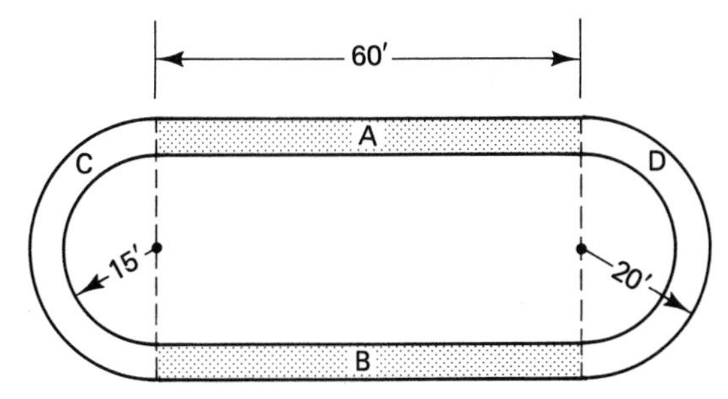

1. Areas A and B are rectangles 60 ft × 5 ft. Study how the length of 60 ft was determined.

$$A + B = 2(60 \times 5) = 600 \text{ ft}^2$$

$$\text{Area}_{C+D} = \pi(20)^2 - \pi(15)^2 = 550 \text{ ft}^2$$

2. The two ends combined form a ring with outside radius = 20 ft and inside radius = 15 ft.

$$\text{Total area} = (A + B) + (C + D)$$

3. Find the sum of all four areas.

$$= 600 \text{ ft}^2 + 550 \text{ ft}^2$$

$$= 1150 \text{ ft}^2$$

C. Find the perimeter of the figure shown.

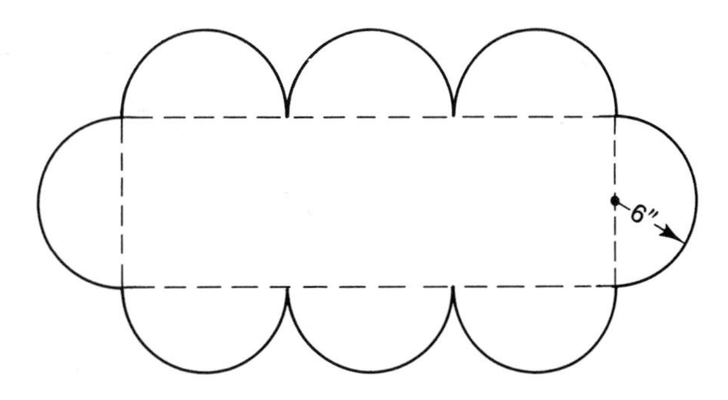

Perimeter $= 4(\pi d)$	1. The perimeter is the equivalent of the circumferences of four circles.
$P = 4(\pi \cdot 12 \text{ in.})$	2. The radius is shown as 6 in.; hence the diameter $d = 12$ in.
$= 150.8$ in.	3. Solve for P.

D. Find the perimeter of the figure shown.

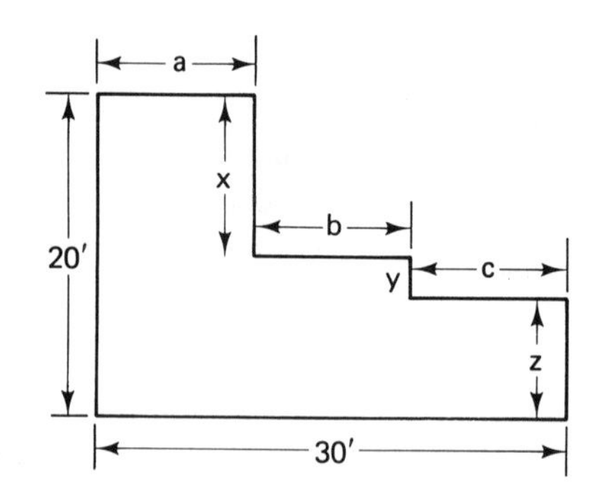

$a + b + c = 30$ ft $x + y + z = 20$ ft	1. The three horizontal segments a, b, and c total 30 ft. The vertical segments x, y, and z total 20 ft.
$P = 2(30 \text{ ft}) + 2(20 \text{ ft})$	2. The perimeter will be twice the horizontal distance of 30 ft and twice the vertical distance of 20 ft.
$= 100$ ft	3. Note that it was unnecessary to know the individual values of a, b, c, x, y, and z in this example.

Exercise 8-3

1. Find the area.

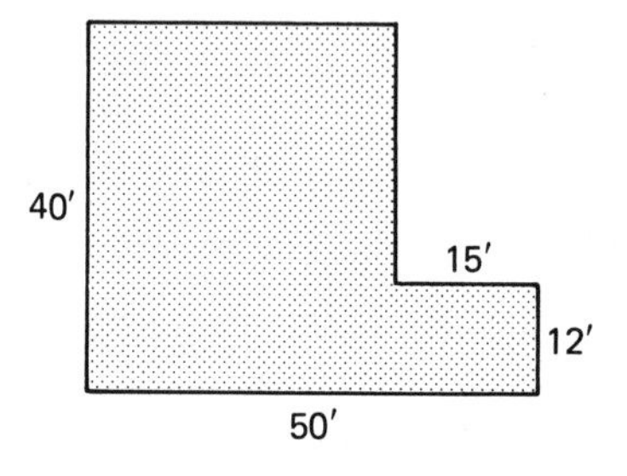

2. Find the perimeter of the figure in Problem 1.

3. Find the area of the track.

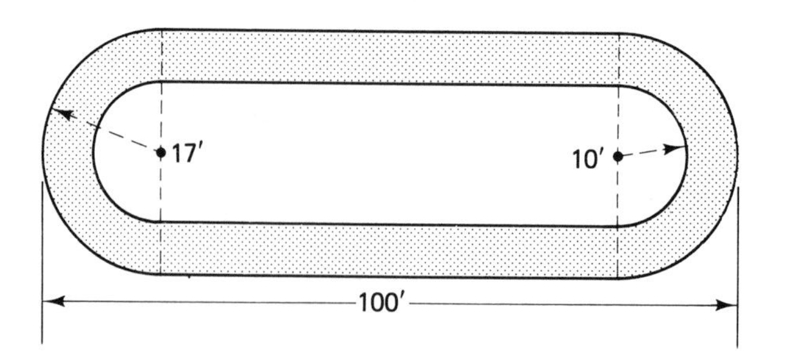

4. Find the outside perimeter of the track in Problem 3.

5. Find the inside perimeter of the track in Problem 3.

6. The square in the center has an area of 100 in.^{2}. Find the area of the entire figure. (*Hint:* The sides of the square are the diameters of the semicircles.)
7. Find the perimeter of the figure shown in Problem 6.

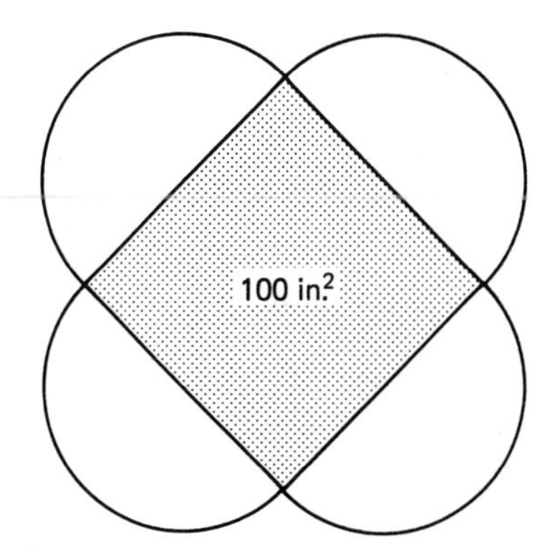

8. Find the shaded area and the perimeter of the square.

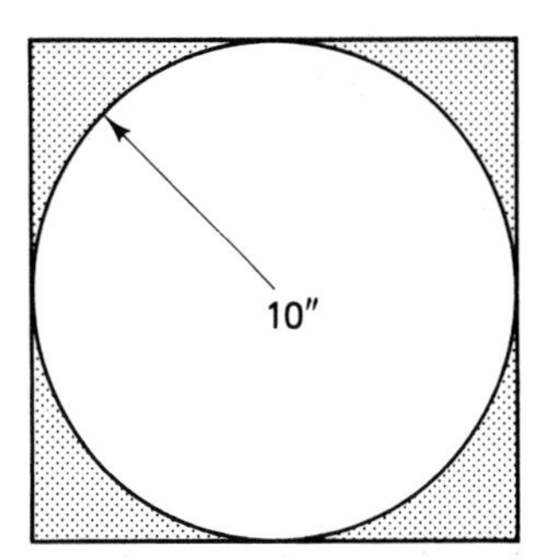

Find the perimeter and area of the following figures.

9.

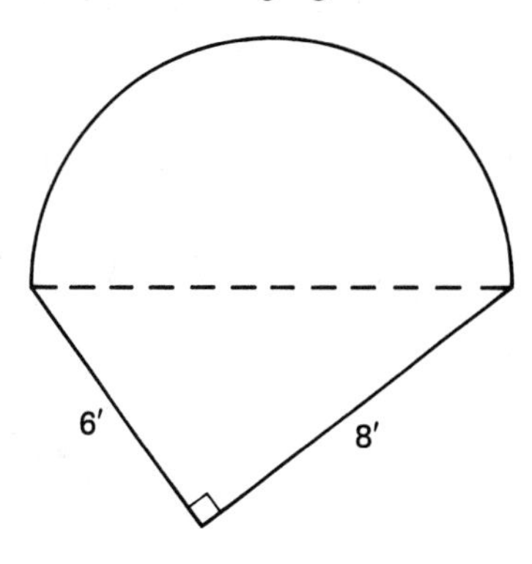

10.

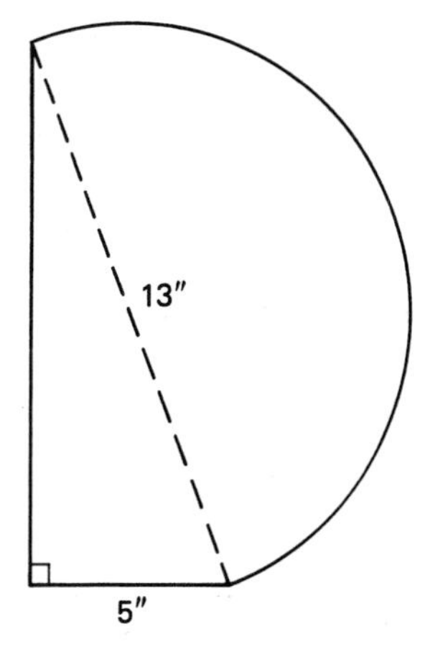

11.

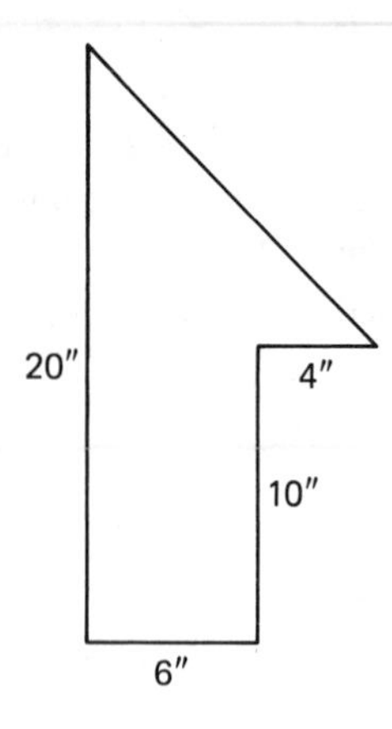

12. A semicircular window is positioned atop a rectangular window as shown. Determine the total window area.

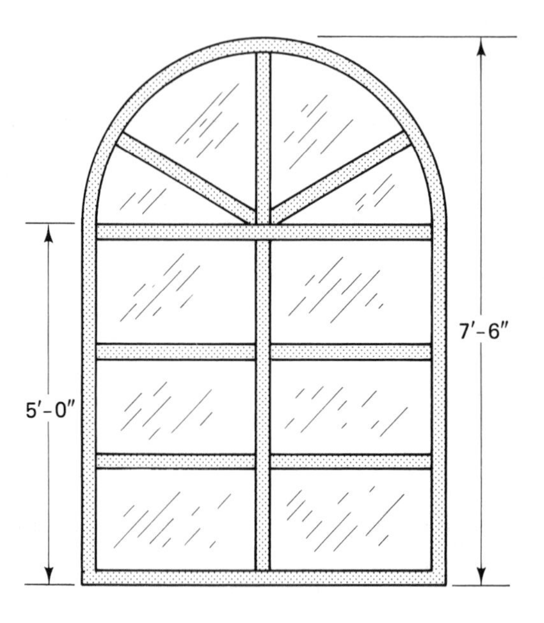

13. Determine the perimeter of the window in Problem 12.

14. A new house is to be sided with clapboards. What is the area in square feet of the end shown? Do not deduct for windows and doors. (*Hint:* Divide into two trapezoids as shown.)

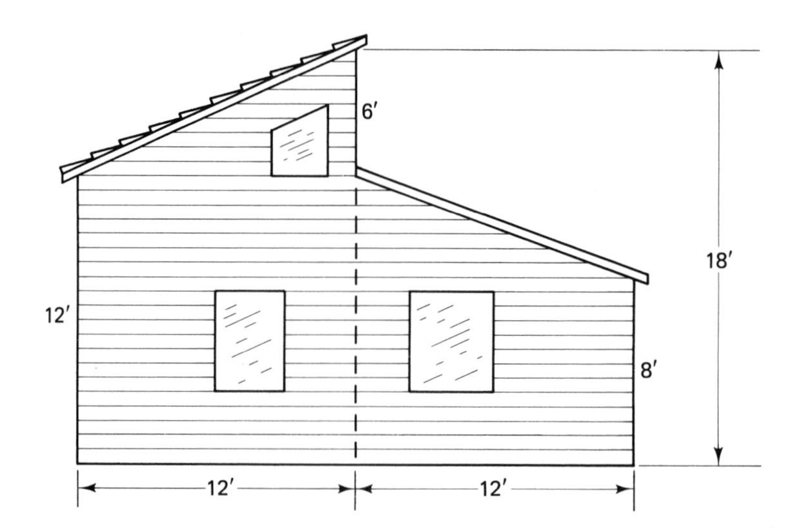

15. A tabletop for a corporation conference room is shown. Find the area of the tabletop.

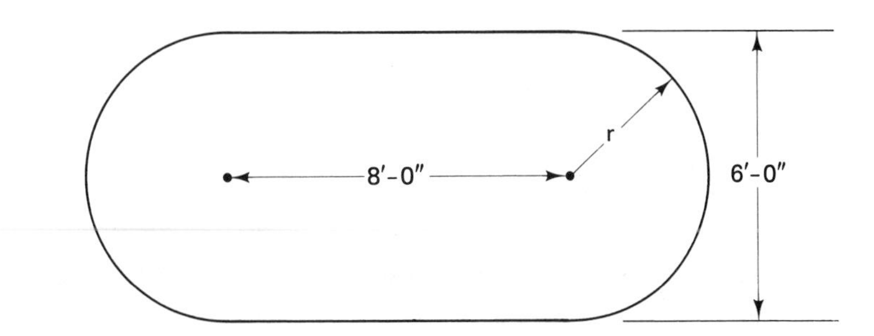

16. Find the perimeter of the tabletop in Problem 15.

17. Find the area of the cross section of a steel I-beam as shown.

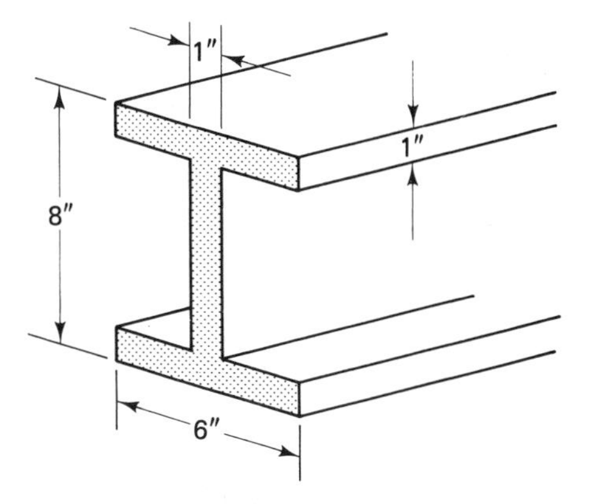

18. Find the perimeter of the cross section of the I-beam in Problem 17.

19. A rectangular pool has a semicircular spa attached to one side. Find the total area of the pool and spa.

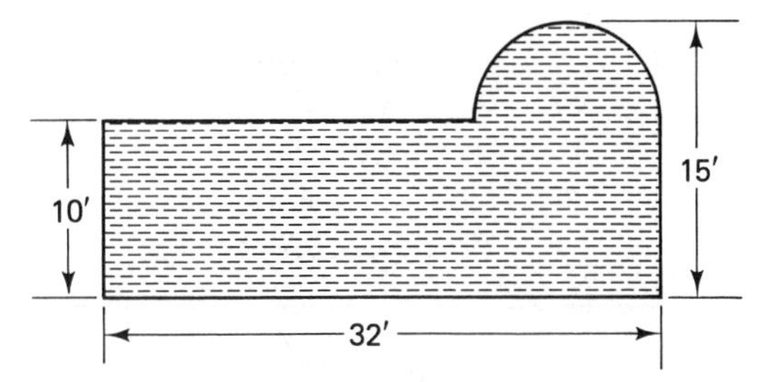

20. Find the perimeter of the pool and spa in Problem 19.

chapter 9

Volume and Surface Area of Solids

9-1 RECTANGULAR SOLIDS AND PRISMS

OBJECTIVES

Upon completing this section, the student will be able to:

1. Find the volume of rectangular solids.
2. Find the LSA (lateral surface area) and TSA (total surface area) of rectangular solids.
3. Find the volume of right prisms.
4. Find the LSA and TSA of right prisms.

Rectangular solids are three-dimensional figures with six sides, all of which are rectangles (or squares). The adjoining sides of a rectangular solid are perpendicular to each other. Figure 9-1 shows examples of rectangular solids.

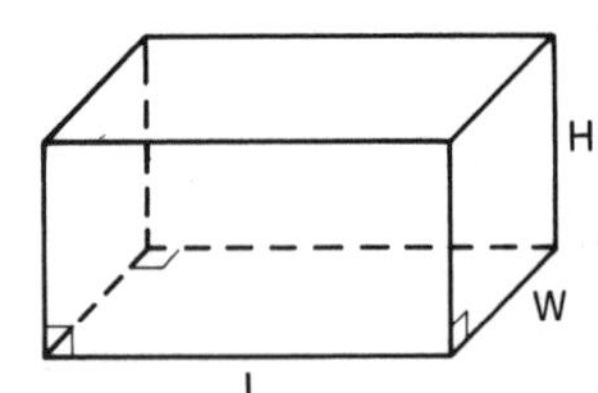

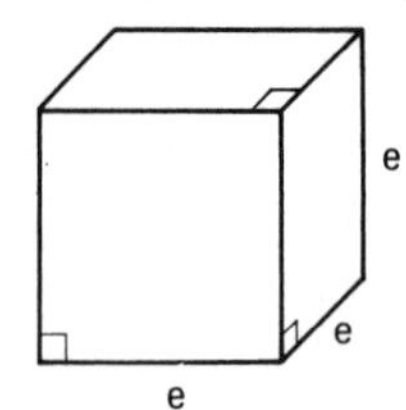

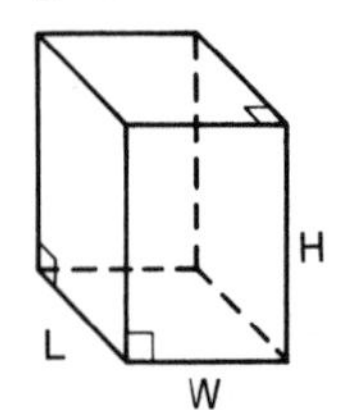

Figure 9-1

The volume of a rectangular solid is found by the formula

$$V = LWH$$

where V is the volume, L the length, W the width, and H the height. When no operation symbol is shown, multiplication is implied. Hence L, W, and H are multiplied together.

Examples

A. Find the volume of air to be heated in a room that measures 12 ft × 18 ft with 8-ft-high ceilings.

$V = LWH$	1. The area of the room ($L \times W$) times the height equals the volume.
$= 18 \text{ ft} \times 12 \text{ ft} \times 8 \text{ ft}$	2. Substitute given values into the formula.
$= 1728 \text{ ft}^3$	3. Volume is measured in cubic units.

B. What is the volume of a cube that is 10 in. on an edge?

$V = e^3$	1. A cube is a rectangular solid with length, width, and height equal. Therefore, $V = LWH = e^3$, where e stands for any edge.
$= (10 \text{ in.})^3$	2. Substitute known values into the formula.
$= 1000 \text{ in.}^3$	3. in.^3 is the abbreviation for cubic inches.

C. The volume of a rectangular watering trough is 75 ft³. If the trough is 2 ft 6 in. deep and 10 ft long, how wide is it?

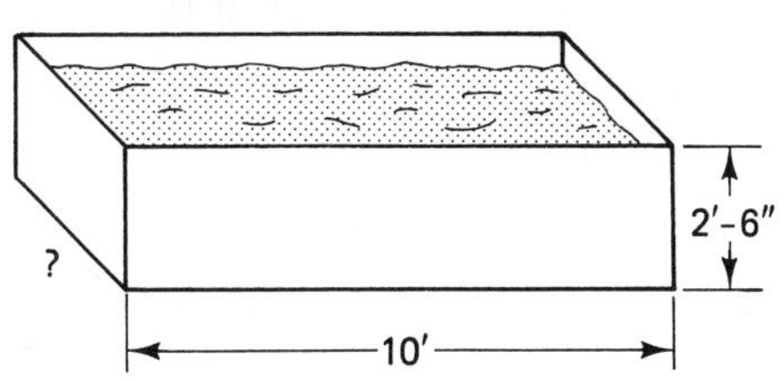

$V = LWH$ $75 \text{ ft}^3 = 10 \text{ ft} \times W \times 2.5 \text{ ft}$	1. Change 2 ft 6 in. to 2.5 ft and substitute known values into the formula.
$= 25 \cdot W$ $\frac{75 \text{ ft}^3}{25 \text{ ft}^2} = W$	2. Simplify and rearrange formula.
$W = 3 \text{ ft}$	3. Divide to obtain the width. Note that cubic feet ÷ square feet yield (linear) feet.

D. A concrete foundation 8 ft deep measures 24 ft × 40 ft (outside dimensions) and has walls 9 in. thick. How many cubic yards of concrete is there in the foundation walls?

$V_{\text{outside}} = (40 \text{ ft})(24 \text{ ft})(8 \text{ ft})$ $= 7680 \text{ ft}^3$	1. Find the volume of the outside dimensions of the walls.

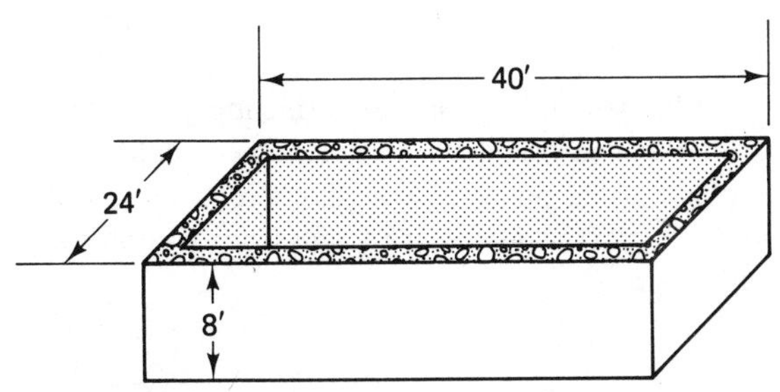

$40 \text{ ft} - 1.5 \text{ ft} = 38.5 \text{ ft}$ $24 \text{ ft} - 1.5 \text{ ft} = 22.5 \text{ ft}$	2. Find the inside dimensions of the foundation by subtracting twice the thickness of the walls from the outside dimensions. (9-in. walls on both sides = 18 in. = 1.5 ft)
$V_{\text{inside}} = (38.5 \text{ ft})(22.5 \text{ ft})(8 \text{ ft}) = 6930 \text{ ft}^3$	3. Find the volume of the inside dimensions of the walls.
$V_{\text{walls}} = 7680 \text{ ft}^3 - 6930 \text{ ft}^3$	4. The difference in the volume of the outside and inside dimensions is the volume of the walls in cubic feet.
$= \frac{750 \text{ ft}^3}{1} \times \frac{1 \text{ yd}^3}{27 \text{ ft}^3}$	5. Convert cubic feet to cubic yards.
$= 27.8 \text{ yd}^3$	6. Approximately 28 yd^3 of concrete are required for the foundation walls. For another approach, see Chapter 12.

The **total surface area** (TSA) is the sum of the areas of all the sides of a rectangular solid.

Example

E. Find the total surface area of the rectangular solid shown.

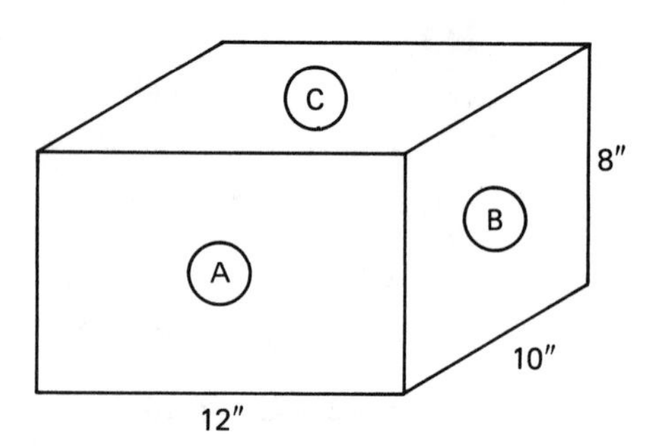

$\text{TSA} = 2A + 2B + 2C$	1. The total surface area (TSA) is twice the area of sides A, B, and C.
$A = 12 \text{ in.} \times 8 \text{ in.} = 96 \text{ in.}^2$ $B = 10 \text{ in.} \times 8 \text{ in.} = 80 \text{ in.}^2$ $C = 10 \text{ in.} \times 12 \text{ in.} = 120 \text{ in.}^2$	2. Find the area of each side A, B, and C.
$\text{TSA} = 2(96) + 2(80) + 2(120)$	3. There are 2 sides with the dimensions of A, 2 with the dimensions of B, and 2 sides with the dimensions of C.
$= 592 \text{ in.}^2$	4. Surface area is in square units.

A **prism** is a solid figure with two bases that are congruent (identical) polygons, and faces that are rectangles. The faces of a right prism are always perpendicular to the bases. Figure 9-2 shows several examples of right prisms. (Note that a rectangular solid is a special case of a prism. In this book, only right prisms are discussed; therefore, any reference to ''prism'' will mean ''right prism.'')

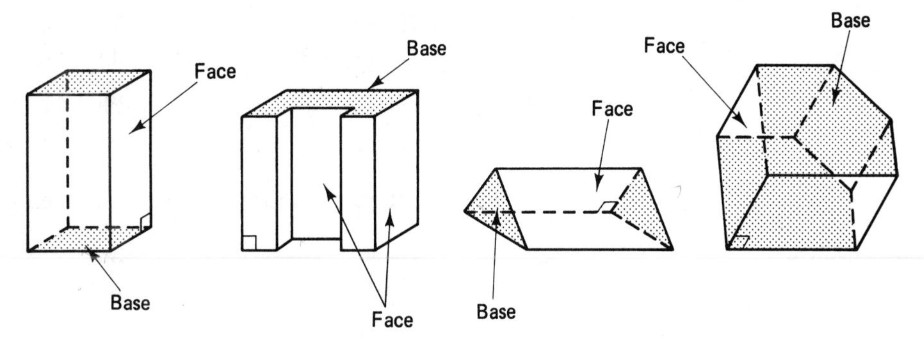

Figure 9-2

The volume of any right prism is found by the same formula, regardless of its shape:

$$V = B \times h$$

where V is the volume of prism, B the area of the base, and h the height of the prism. (Height is the dimension perpendicular to the base.)

Example

F. Find the volume of the prism shown. The area of each base is 42.8 in.^{2}, and its height is 10 in.

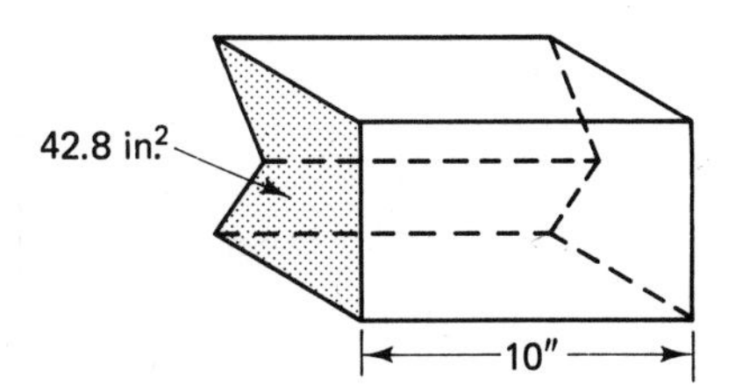

$V = B \times h$ — 1. Substitute given values into the formula.

$= 42.8 \text{ in.}^2 \times 10 \text{ in.}$

$= 428 \text{ in.}^3$ — 2. Volume is in cubic inches.

The **lateral surface area** (LSA) of a prism is the area of all the *faces*. The **total surface area** (TSA) is the area of all the *faces* plus the *two bases*.

Example

G. Find the LSA and TSA of the prism shown.

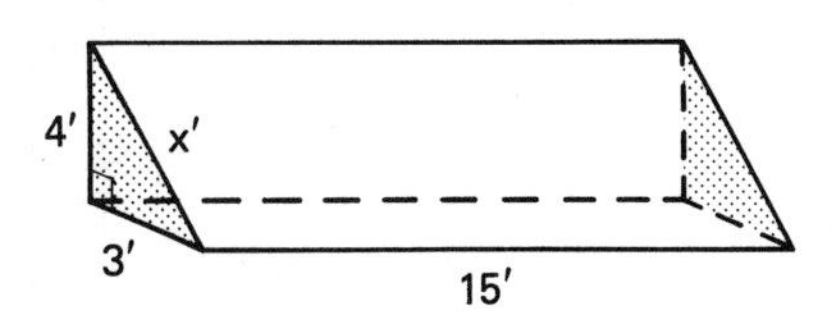

$x^2 = 3^2 + 4^2$ — 1. Find the unknown side of the base by the Pythagorean theorem.

$= 5 \text{ ft}$

$\text{LSA} = 3(15) + 4(15) + 5(15)$ — 2. Find the area of each rectangular face.

$= 180\ \text{ft}^2$ — 3. The LSA is the sum of the area of the three faces.

$\text{Area}_{\text{base}} = \frac{1}{2}bh$ — 4. Find the area of each base.

$= \frac{1}{2}(4)(3) = 6\ \text{ft}^2$

$\text{TSA} = 2(6\ \text{ft}^2) + 180\ \text{ft}^2$ — 5. The area of the two bases plus the LSA equals the total surface area.

$= 192\ \text{ft}^2$

Exercise 9-1

Find the LSA, TSA, and volume of the following figures.

1.

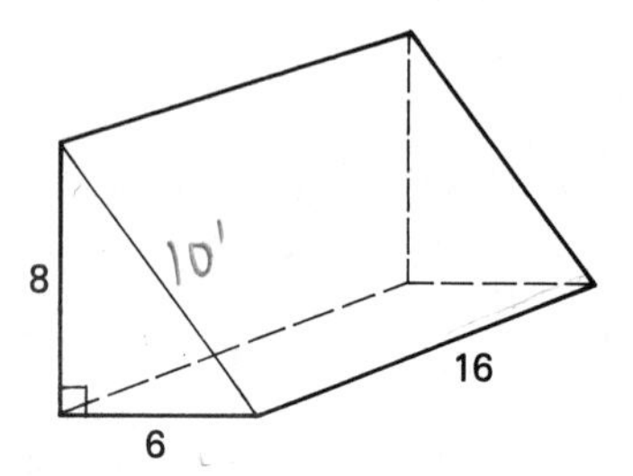

2.

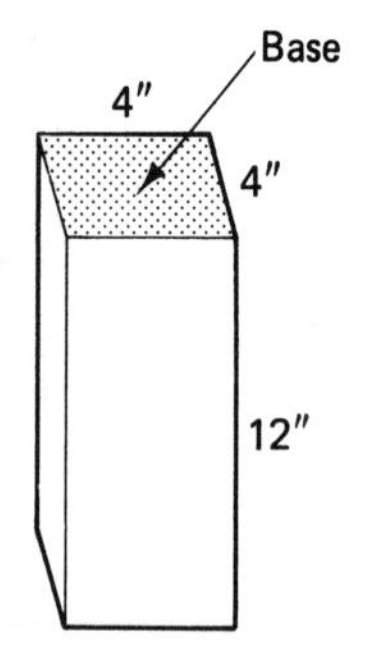

3.

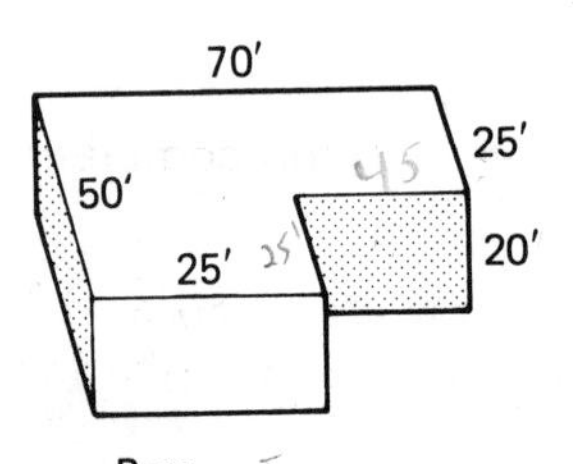

4.

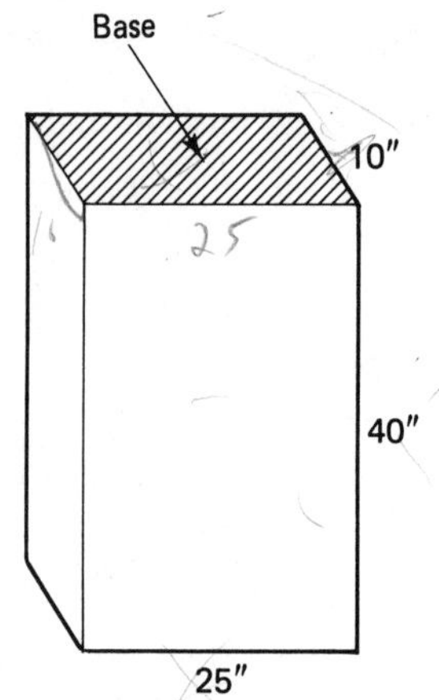

5. A foundation is to have outside measurements of 28 ft × 36 ft. If the foundation is to be 7 ft high and the walls 9 in. thick, how many cubic yards of concrete is needed for the foundation walls? Round up to a whole number of cubic yards. (*Hint:* Refer to Example D.)

6. The outside of the foundation walls in Problem 5 is to be waterproofed. What is the surface area of the exterior walls?

7. A roof has the dimensions shown. How many cubic feet of air space is there in the attic?

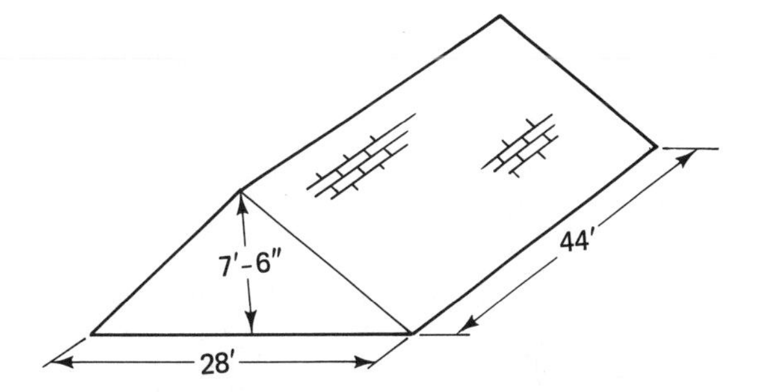

8. A small shed, open in the front, is to be completely shingled on the outside with cedar shingles. Assuming minimal waste, how many bundles of shingles are required to shingle the three sides and roof of the shed? One bundle covers 25 ft^2.

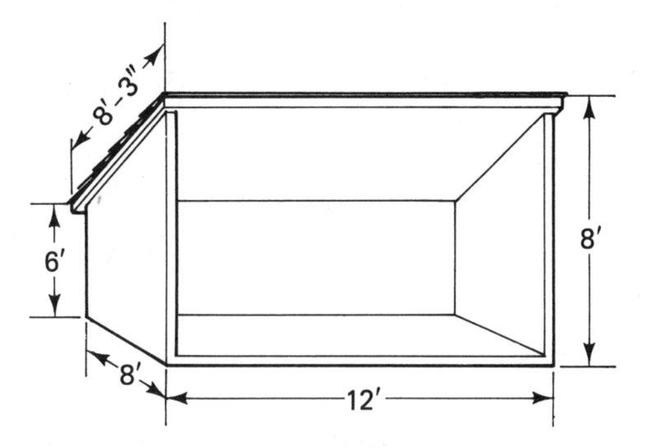

9-2 CYLINDERS AND CONES

> **OBJECTIVES**
>
> Upon completing this section, the student will be able to:
>
> 1. Find the volume of right cylinders.
> 2. Find the LSA and TSA of right cylinders.
> 3. Find the volume of cones.
> 4. Find the LSA and TSA of cones.

The formula for the volume of a cylinder (Figure 9-3) is

$$V = \pi r^2 h$$

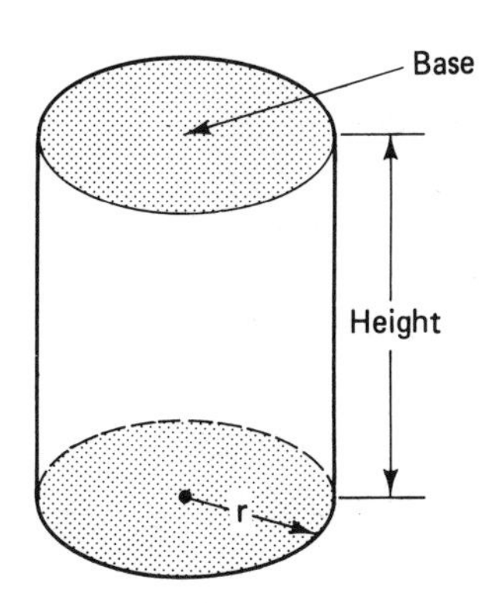

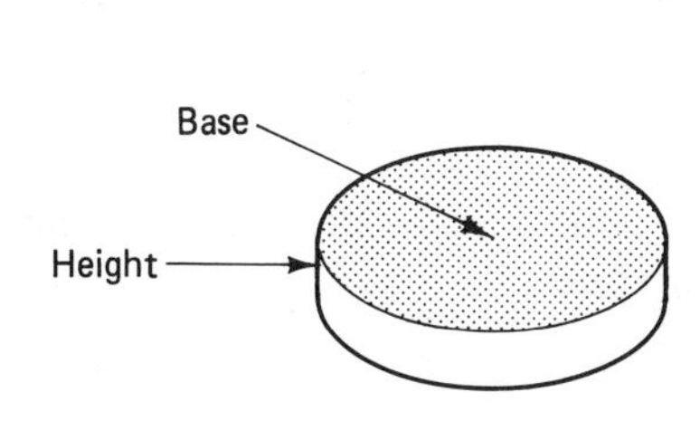

Figure 9-3

where V is the volume, r the radius of the base, and h the height. Since πr^2 is the area of the circular base, the volume of a cylinder, like that of a prism, is actually the area of the base × height.

Examples

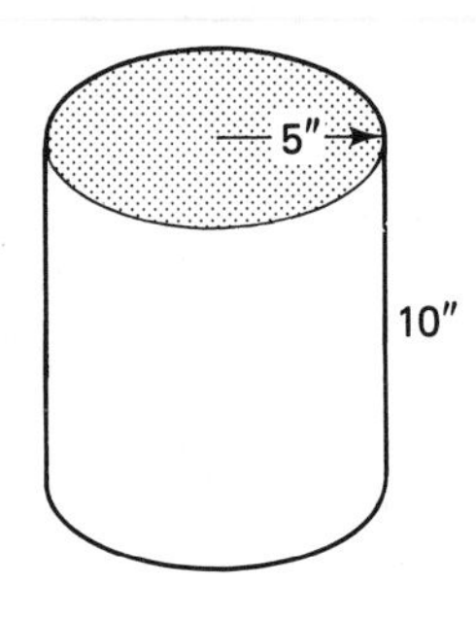

A. Find the volume of a cylinder with a radius of 5 in. and a height of 10 in.

$V = \pi r^2 h$ — 1. Use the formula for the volume of a cylinder.

$= \pi(5 \text{ in.})^2(10 \text{ in.})$ — 2. Substitute the given values into the formula. Be sure to square the radius before multiplying by the height.

$= 785.4 \text{ in.}^3$ — 3. The volume is in cubic inches.

B. A copper water pipe with $\frac{3}{4}$ in. I.D. is 87 ft long. How many gallons of water are there in the pipe? (1 gal = 231 in.^{3})

$87 \text{ ft} \times 12 \text{ in./ft} = 1044 \text{ in.}$ — 1. Change the length to inches.

$V = \pi r^2 h$ — 2. A pipe is a cylinder. Be sure to square the radius, not the diameter.

$= \pi(0.375 \text{ in.})^2(1044 \text{ in.})$

$= 461 \text{ in.}^3$

$= \dfrac{461 \text{ in.}^3}{1} \times \dfrac{1 \text{ gal}}{231 \text{ in.}^3}$ — 3. Convert cubic inches to gallons.

$= 2 \text{ gal}$ — 4. The pipe contains approximately 2 gal of water.

The **lateral surface area** of a cylinder (think of the label on a tin can) can be found by the formula

$$\text{LSA} = 2\pi rh \quad \text{or} \quad \text{LSA} = \pi dh$$

If the label is cut and removed as shown in Figure 9-4, the rectangle thus formed has a length of $2\pi r$ or πd (the circumference of the cylinder), and a width of h (the height of the cylinder).

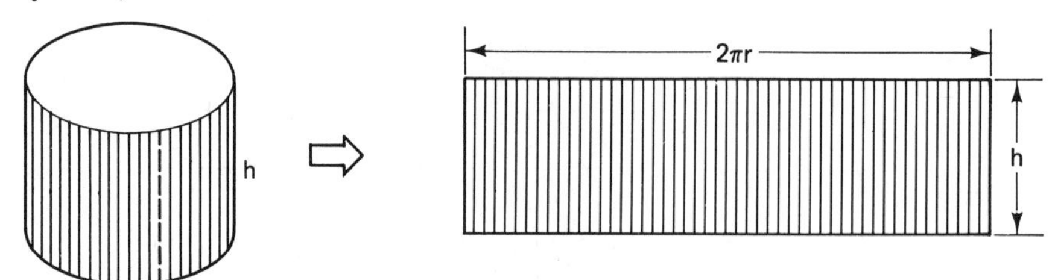

Figure 9-4

The **total surface area** (TSA) is the entire surface area of the cylinder.

$$\text{TSA} = 2\pi rh + 2\pi r^2$$

If the formula looks complicated, just remember it as the lateral surface area plus the area of the two bases.

$$\text{TSA} = \underbrace{2\pi rh}_{\text{LSA}} + \underbrace{2\pi r^2}_{\text{bases}}$$

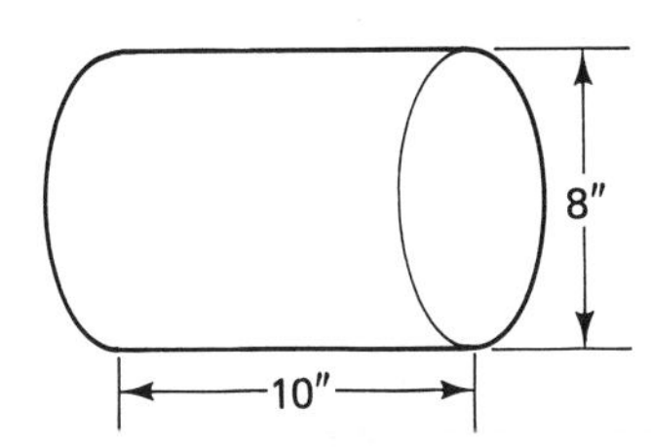

Examples

C. Find the LSA and the TSA of a cylinder with a height of 10 in. and a diameter of 8 in.

$$\begin{aligned} \text{LSA} &= 2\pi rh \\ &= 2\pi(4 \text{ in.})(10 \text{ in.}) \\ &= 251 \text{ in.}^2 \end{aligned}$$

1. Substitute known values into the formula for LSA.

$$\begin{aligned} \text{TSA} &= \text{LSA} + 2\pi r^2 \\ &= 251 \text{ in.}^2 + 2\pi(4)^2 \\ &= 352 \text{ in.}^2 \end{aligned}$$

2. Find the area of the bases and add to the LSA.

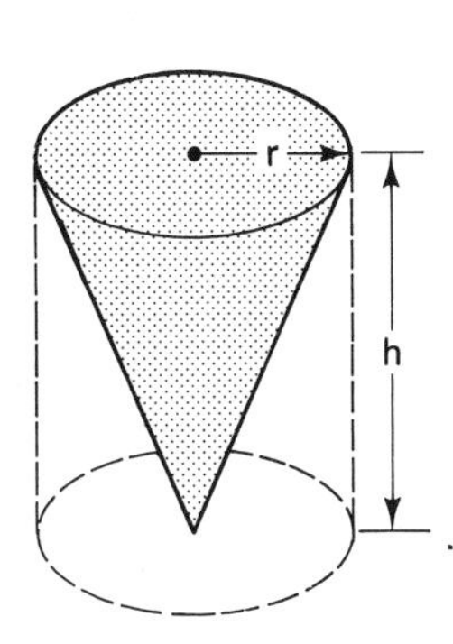

Figure 9-5

The volume of a cone (Figure 9-5) is $\frac{1}{3}$ that of a cylinder with the same base and height.

$$V = \tfrac{1}{3}\pi r^2 h$$

Therefore, the formula for the volume of a cone is identical to the formula for a cylinder except that it is multiplied by $\frac{1}{3}$.

Example

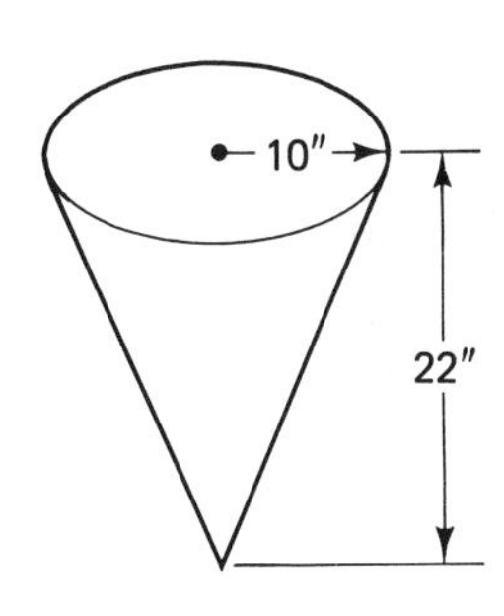

D. A cone has a 10-in. radius and is 22 in. high. How many gallons will it hold? (1 gal = 231 in.^{3})

$$\begin{aligned} V &= \tfrac{1}{3}\pi r^2 h \\ &= \tfrac{1}{3}\pi(10 \text{ in.})^2(22 \text{ in.}) \\ &= 2304 \text{ in.}^3 \end{aligned}$$

1. Formula for volume of a cone.
2. Substitute and solve for volume.

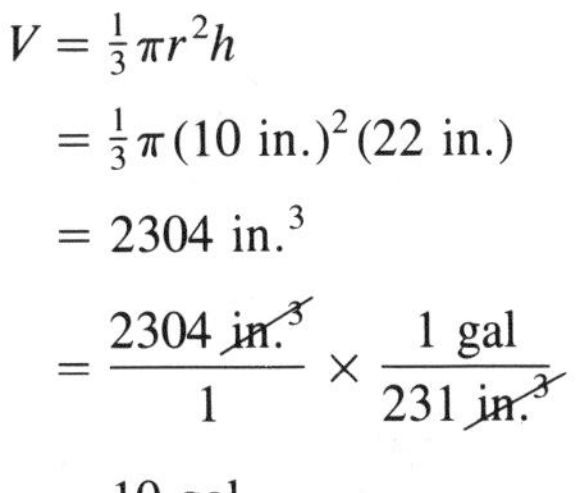

$$= \frac{2304 \cancel{\text{in.}^3}}{1} \times \frac{1 \text{ gal}}{231 \cancel{\text{in.}^3}}$$

$$= 10 \text{ gal}$$

3. Convert to gallons.

The LSA and TSA of a cone are found by the formulas

$$\text{LSA} = \pi rs$$

$$\text{TSA} = \pi rs + \pi r^2$$

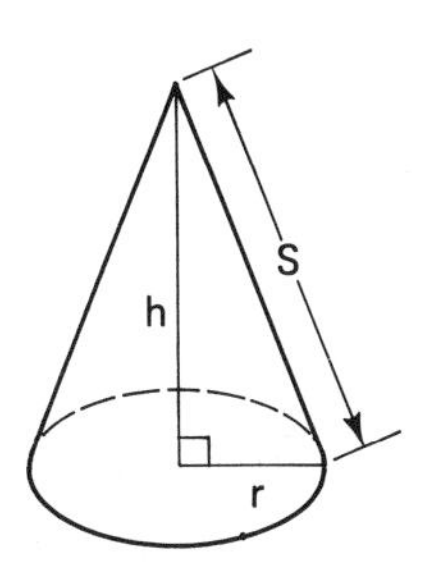

Figure 9-6

where r is the radius and s is the slant height of the cone. If the height h and the radius r are given, the slant height s can be found by using the Pythagorean theorem, where the slant height is the hypotenuse (see Figure 9-6).

Example

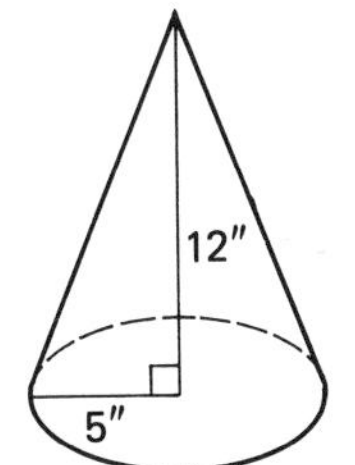

E. Find the TSA of a cone with the dimensions shown.

$$\begin{aligned} s^2 &= 5^2 + 12^2 \\ s &= 13 \text{ in.} \end{aligned}$$

1. Use the Pythagorean theorem to find the slant height.

$$\begin{aligned} \text{TSA} &= \pi rs + \pi r^2 \\ &= \pi(5 \text{ in.})(13 \text{ in.}) + \pi(5^2) \\ &= 283 \text{ in.}^2 \end{aligned}$$

2. Substitute into the formula to solve for TSA.
3. Note that the TSA is found by adding the LSA with the area of the circular base.

Exercise 9-2

Complete the table for the cylinders described.

	Radius	Diameter	Circum-ference	Area of base	Height	Volume	LSA	TSA
1.	8 in.				5 in.			
2.		12 in.			8 in.			
3.			15.7 in.		5 in.			
4.	$3\frac{1}{4}$ in.				8 in.			
5.		1 ft 10 in.			1 ft 3 in.			

6. A cone has a radius of 5 in. and a height of 10 in. What is its volume?

7. A cone and a cylinder have the same height and the same size base. If the cone has a volume of 125 in.^{3}, what is the volume of the cylinder?

8. A cone has a base with an area of 300 in.^{2}. How high must it be to have a volume of 500 in.^{3}?

9. A drilled well has a diameter of 6 in. The water standing in the well is 200 ft deep. How many gallons of water is standing in the well? (A drilled well is a cylinder.) (1 ft^3 = 7.5 gal)

10. A conical (cone-shaped) pile of sand is being mixed with cement and gravel to make concrete. If the pile of sand is 5 ft high and 7 ft across at its base, how many cubic feet of sand are in the pile?

11. What is the lateral surface area of the pile of sand in Problem 10.

12. A 1-gal paint can has a radius of approximately $3\frac{1}{4}$ in. What is the height of the can? Round to the nearest inch. (1 gal = 231 in.^{3})

13. Water pipes with an outside diameter of 1 in. are to be insulated with a thin sheet of foam. Assuming no waste or overlap, how many square inches of foam is needed to insulate pipes with a total length of 40 ft? Round to the nearest square inch.

14. A cylindrical drum used for storing kerosene has a diameter of 2 ft 6 in. and is 4 ft long. What is the total surface area of the drum?

15. A tent is to be constructed in the form of a tepee (cone shaped) 12 ft high and 10 ft wide at the base. Ignoring openings, determine the number of square feet of canvas needed to construct the tent. The floor is also canvas. Add 20% for waste and round to the nearest whole square foot.

9-3 SPHERES

OBJECTIVES

Upon completing this section, the student will be able to:

1. Find the volume of spheres.
2. Find the surface area of spheres.
3. Solve word problems involving volume and surface area of spheres.

The formula for the volume of a sphere is

$$V = \frac{4}{3}\pi r^3$$

The surface area of a sphere is found by the formula

$$SA = 4\pi r^2$$

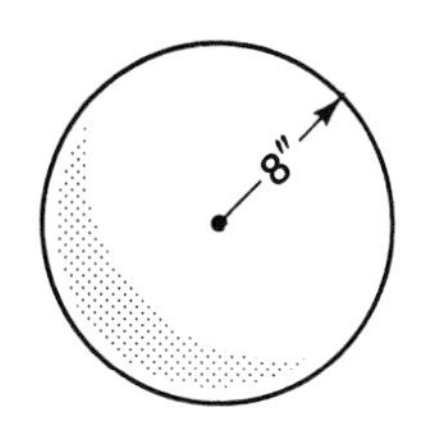

Examples

A. Find the volume of a sphere that has a radius of 8 in.

$V = \frac{4}{3}\pi r^3$	1. Formula for the volume of a sphere.
$= \frac{4}{3}\pi(8 \text{ in.})^3$	2. Find 8^3 before multiplying by $\frac{4}{3}\pi$.
$= 2145 \text{ in.}^3$	3. Volume is in cubic units.

B. Find the surface area of a sphere with a diameter of 22 ft.

$SA = 4\pi r^2$	1. Formula for surface area of a sphere.
$= 4\pi(11 \text{ ft})^2$	2. Be sure to square radius, not diameter.
$= 1521 \text{ ft}^2$	3. Surface area is in square units.

C. A spherical water tank has a circumference of 62 ft. (Circumference of a sphere is the distance around at its widest point.) Find the number of gallons of water it can hold. ($1 \text{ ft}^3 = 7.5$ gal.)

$C = \pi d$	1. Use the formula for the circumference of a circle or a sphere.
$62 = \pi d$	2. Find the diameter of the sphere.
$d = 19.74$ ft	
$r = 9.87$ ft	3. The values for diameter and radius shown here are rounded. When using a calculator it is best not to round until the final answer is obtained.
$V = \frac{4}{3}\pi(9.87 \text{ ft})^3$	4. Substitute the radius into the formula for volume.
$= 4024.6 \text{ ft}^3$	5. To find $(9.87)^3$ on a calculator with a Y^x key: 9.87 Y^x 3 =.
$= \frac{4024.6 \cancel{\text{ft}^3}}{1} \cdot \frac{7.5 \text{ gal}}{1 \cancel{\text{ft}^3}}$	6. Convert cubic feet to gallons.
$= 30{,}185$ gal	

Exercise 9-3

Complete the table for the spheres described.

	Radius	Diameter	Circumference	Volume	Surface area
1.	8 in.				
2.	5 ft				
3.		16 ft			
4.	3 ft 4 in.				
5.			22 ft		
6.	15 in.				
7.	8.6 in.				
8.		12 ft 6 in.			

OBJECTIVES

Upon completing this section, the student will be able to:

1. Find the volume of composite shapes.
2. Find the surface area of composite shapes.
3. Solve problems involving volume and surface area of composite shapes.

To find the volume and surface area of composite shapes, divide the form into parts that can be determined individually.

Examples

A. A propane gas storage tank is a cylinder with hemispheric ends. (A hemisphere is half a sphere.) Find the volume of the tank. The tank is 20 ft long and has a diameter of 5 ft.

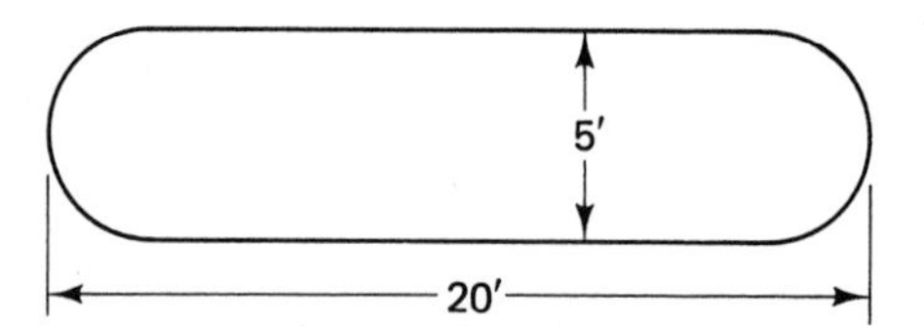

$V_{sphere} = \frac{4}{3}\pi r^3$ — 1. The volume of the two hemispheres = the volume of a sphere with radius 2.5 ft.

$= \frac{4}{3}\pi(2.5 \text{ ft})^3$

$= 65.4 \text{ ft}^3$ — 2. Substitute and solve for volume.

$V_{cylinder} = \pi r^2 h$ — 3. Solve for the volume of the cylinder. Why is the height of the cylinder 15 ft?

$= \pi(2.5 \text{ ft})^2(15 \text{ ft})$

$= 294.5 \text{ ft}^3$ — 4. The volume of the cylindrical portion of the tank.

$V_{tank} = 65.4 \text{ ft}^3 + 294.5 \text{ ft}^3$ — 5. Find the total volume of the tank.

$= 360 \text{ ft}^3$ — 6. Round to the nearest cubic foot.

B. The gas tank in Example A is to be painted with rust-inhibiting paint. How many square feet is to be painted?

$\text{SA}_{sphere} = 4\pi r^2$ — 1. Find the surface area for the two hemispheres.

$= 4\pi(2.5 \text{ ft})^2$

$= 78.5 \text{ ft}^2$

$\text{LSA}_{cylinder} = 2\pi rh$ — 2. Find the LSA of the cylindrical middle part of the tank.

$= 2\pi(2.5 \text{ ft})(15 \text{ ft})$

$= 235.6 \text{ ft}^2$

$$TSA_{tank} = 78.5 \text{ ft}^2 + 235.6 \text{ ft}^2$$

3. Find the surface area of the entire figure.

$$= 314 \text{ ft}^2$$

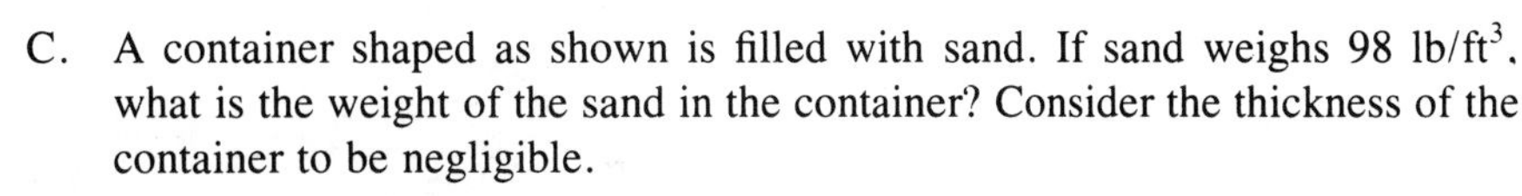

C. A container shaped as shown is filled with sand. If sand weighs 98 lb/ft³, what is the weight of the sand in the container? Consider the thickness of the container to be negligible.

$$V_{cone} = \tfrac{1}{3}\pi r^2 h$$

1. Find the volume of the conical (cone-shaped) top.

$$= \tfrac{1}{3}\pi(1 \text{ ft})^2 \cdot (4 \text{ ft})$$

$$= 4.2 \text{ ft}^3$$

$$V_{cylinder} = \pi r^2 h$$

2. Find the volume of the cylindrical bottom.

$$= \pi(1 \text{ ft})^2 \cdot (6 \text{ ft})$$

$$= 18.8 \text{ ft}^3$$

$$V_{total} = 4.2^3 + 18.8^3 = 23 \text{ ft}^3$$

3. Find the total volume.

$$\text{Weight} = \frac{23 \cancel{\text{ft}^3}}{1} \times \frac{98 \text{ lb}}{1 \cancel{\text{ft}^3}}$$

4. Use unit conversion to find weight.

$$= 2250 \text{ lb}$$

5. Approximate weight of sand.

Exercise 9-4

Find the total surface area and the volume of the following figures.

1.

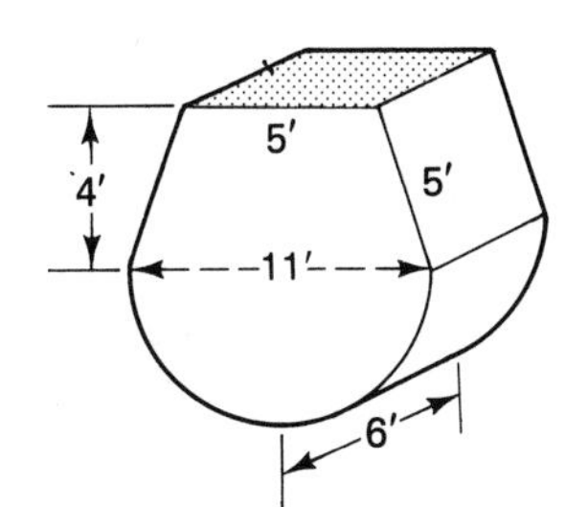

2.

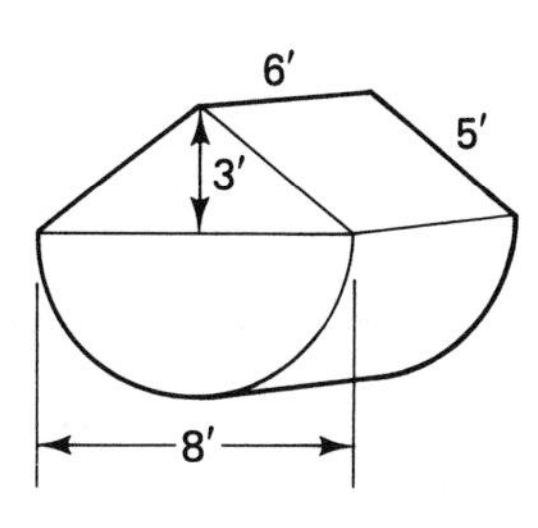

3.

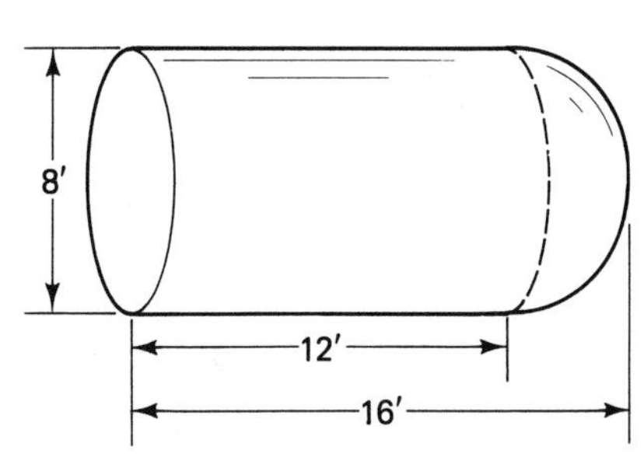

4. A small house is to be sided with clapboards. How many square feet is to be sided? (Don't put clapboards on the roof!) Ignore openings for doors and windows.

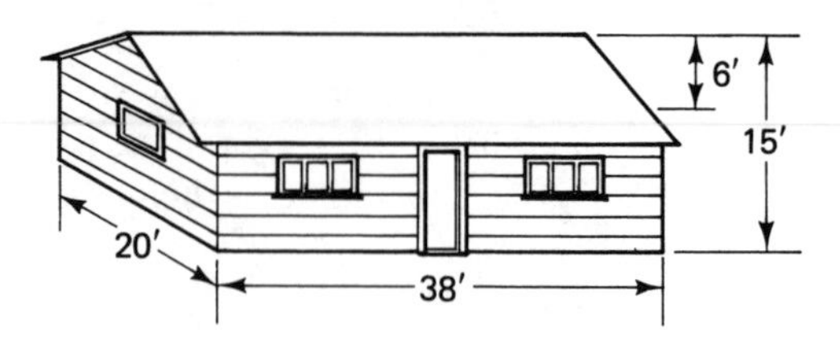

5. A metal silo has the dimensions shown. The silo is to be painted with two coats of metal paint. How many gallons is needed if each gallon of paint covers 350 ft^2? Round up.

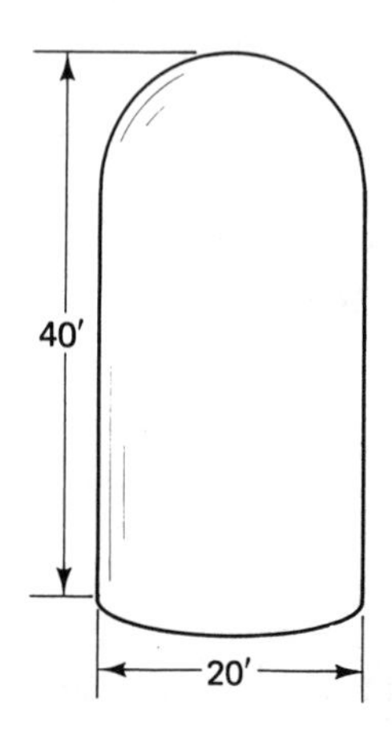

chapter 10

Board Measure

OBJECTIVES

Upon completing this chapter, the student will be able to:

1. Calculate the number of board feet for any size and quantity of lumber.
2. Determine the number of linear feet of lumber needed for a certain job, given the number of board feet.

The basic unit of measure for lumber is board feet. One **board foot** equals 144 in.^{3} of wood.

A common example of 1 board foot is a board that is 1 in. thick and 12 in. square. Use the volume formula for a rectangular solid:

$$V = \text{length} \times \text{width} \times \text{depth}$$
$$= 1 \text{ in.} \times 12 \text{ in.} \times 12 \text{ in.} = 144 \text{ in.}^3$$

That is, 1 board foot equals 144 in.^{3} of wood. This leads to the conclusion that the number of board feet in a piece of wood can be calculated by finding its volume in cubic inches and dividing by 144. That is,

$$\text{board feet} = \frac{\text{volume}}{144}$$

However, a more convenient formula is

$$BF = \frac{T \times W \times L}{12}$$

where T is the thickness in inches, W is the width in inches, and L is the length in feet.

Take note that the length is not converted to inches (by multiplying by 12). Not multiplying by 12 in the numerator means that one factor of 12 in 144 may be eliminated.

EXAMPLE 10-1

Find the number of board feet in a 2 × 10 − 16 ft long.

***Solution*:** $BF = \frac{2 \times 10 \times 16}{12} = 26\frac{2}{3}$ board feet

---

This is a convenient formula to use since the variables are defined in the same way that dimensions of lumber are usually given. For purposes of calculating board feet, the dimensions of lumber as used in the formula are the rough-sawn dimensions. Framing lumber (2 × 4's, 2 × 6's, 2 × 8's, etc.) is rough cut 2 in. thick; but after being dressed and dried, it is closer to $1\frac{1}{2}$ in. thick. Nevertheless, its thickness is 2 in. for purposes of calculating board feet. In a similar manner, 1-in. boards actually measure $\frac{3}{4}$ in. thick, but 1 in. should still be used when calculating board feet. A product sometimes used for porch decking is called *five-quarter decking*. This material has a finished thickness of 1 in., but its thickness in the board-foot formula will be $\frac{5}{4}$ or $1\frac{1}{4}$ in. A similar relation exists with the width of lumber. The rough-sawn width should be used for the board-foot calculation. Thus, while a 2 × 4 is actually $3\frac{1}{2}$ in. wide, the width is 4 when calculating board feet.

EXAMPLE 10-2

Find the number of board feet in 200 pieces of 2 × 6 each 8 ft long.

***Solution*:** $BF = 200 \times \frac{2 \times 6 \times 8}{12} = 1600$

Notice that a factor for the number of pieces has been built into the formula, rather than making a separate calculation.

A second use of the board-foot formula is to calculate the number of linear feet of lumber needed to cover a certain area—such as a deck, a patio, or a floor. Lumber is often sold by the linear foot and priced by the board foot. Thus familiarization with the conversion from board feet to linear feet is important.

When an area is covered with 1-in. boards, each square foot is equal to 1 board foot. When an area is covered with 2-in. lumber, each square foot is equal to 2 board feet. (This can be verified by calculating the number of board feet in a piece of lumber 12 in. square by 1 in. thick—or 2 in. thick.)

EXAMPLE 10-3

Find the number of linear feet of 1 × 8 boards needed to cover a deck measuring 14 ft × 20 ft.

***Solution*:** First, calculate the area to be covered.

$$A = 20 \times 14 = 280 \text{ ft}^2$$

Now use the board-foot formula with BF = 280 and solve for the length, L. (*Remember:* Square feet is the same as board feet when 1-in.-thick boards are used.)

$$280 = \frac{1 \times 8 \times L}{12}$$

$$280 \times 12 = 8 \times L$$

$$L = \frac{280 \times 12}{8}$$

$$= 420 \text{ ft}$$

That is, it takes 420 linear feet of 1 × 8 boards to equal 280 board feet. (Since a 1 × 8 board is actually closer to $7\frac{1}{2}$ in. wide, some error is present in this calculation. If the boards are to be spaced, some of this error will be eliminated. The estimator will generally add 10 to 15% more material to the order to cover error and waste.)

Now the carpenter can decide what lengths would provide the most efficient use of the lumber and order accordingly. These lengths will be determined by the spacing of the floor joists and whether the 1 × 8's are to be laid with the 20-ft length or the 14-ft width.

REVIEW EXERCISES

10-1. The most commonly used sizes in framing lumber are 2 × 4's, 2 × 6's, 2 × 8's, 2 × 10's, and 2 × 12's. For each size, calculate the number of board feet in 1 linear foot of material. Then notice the sequence formed by the answers. This is an easy-to-remember sequence and will be useful in estimating prices of lumber.

10-2. Use your results from Problem 10-1 to calculate the number of board feet in each of the following quantities of lumber.

(a) 120 pieces of 2 × 4 − 8 ft long (*Hint:* Find the total length; then multiply by $\frac{2}{3}$.)

(b) 84 pieces of 2 × 10 − 14 ft long

(c) 26 pieces of 2 × 6 − 12 ft long

(d) 33 pieces of 2 × 8 − 10 ft long

(e) 15 pieces of 2 × 12 − 16 ft long

10-3. Find the number of board feet in the following quantities of lumber.

(a) 535 pieces of 2 × 4 − 8 ft long

(b) 96 pieces of 2 × 10 − 14 ft long

(c) 14 pieces of 2 × 6 − 12 ft long

(d) 76 pieces of 2 × 8 − 10 ft long

(e) 18 pieces of 2 × 12 − 14 ft long

10-4. How many linear feet of 1 × 8 boards is needed to cover an area of subflooring 26 ft wide by 42 ft long? (*Hint:* Use the board-foot formula with L as the unknown.)

10-5. A deck measuring 8 ft × 14 ft is to be covered with 2 × 6's. Find the number of linear feet needed.

10-6. Five-quarter decking, 4 in. wide, will be used in covering a porch 4 ft wide and 26 ft long. How many board feet is needed?

10-7. Find the number of linear feet of five-quarter decking needed to cover the deck described in Problem 10-6.

10-8. The roof of a house contains 1120 ft^2. How many linear feet of 1 × 8 roofers would be needed to cover this roof? (Roofers are a tongue-and-grooved board sometimes used for boarding in an area.)

10-9. How many sheets of $\frac{5}{8}$-in. CDX plywood would be needed to cover the roof described in Problem 10-8? (Plywood measures 4 ft × 8 ft.)

10-10. A deck measuring 10 ft × 16 ft is to be covered with 2 × 6's. How many linear feet of 2 × 6's are needed?

10-11. It has been estimated that the corner boards and fascia trim for a particular house require the following amounts of No. 2 pine boards:

1 × 4: 8/10's ("8/10's" means 8 pieces, 10 ft long each)
1 × 6: 6/8's; 10/12's; 2/14's
1 × 8: 8/10's

Find the total number of board feet in the materials list above.

chapter 11

Lumber Pricing

OBJECTIVES

Upon completing this chapter, the student will be able to:

1. Calculate the cost of lumber based on a price quoted per board foot.
2. Calculate the cost of lumber based on a price quoted per linear foot.
3. Convert a price per board foot to either a price per piece or to a price per linear foot.
4. Convert a price per linear foot to either a price per board foot or to a price per piece.
5. Convert a price per piece to either a price per board foot or to a price per linear foot.

In the building trade, there is little consistency used in methods for pricing lumber. One building supply firm may price lumber by the board foot, while another prices the same material by the linear foot, and still another by the piece. In fact, some suppliers will use all three methods for different types of materials. Because of this lack of consistency, it is important that a buyer know whether a quoted price is "per piece," "per board foot," "per thousand board feet" (mbf), or "per linear foot." It is also important that a buyer know how to convert from one method of pricing to another.

EXAMPLE 11-1

If 2×4's are quoted at 24 cents per board foot, find the cost of a piece 16 ft long.

***Solution*:** First, find the number of board feet.

$$\frac{2 \times 4 \times 16}{12} = 10.667$$

Second, multiply by the price.

$$0.24 \times 10.667 = \$2.56$$

When using a calculator to solve any problem, it is easiest to set up the entire problem before using the calculator. Then the calculations can be done in a continuous sequence. The setup for Example 11-1 would look like this:

$$0.24 \times \frac{2 \times 4 \times 16}{12} = \$2.56$$

EXAMPLE 11-2

Find the cost of 20 pieces of 2 × 8, each 12 ft long, if the price is quoted \$325 mbf.

Solution: First, convert the quoted price to a price per board foot. Since "mbf" means per 1000 board feet, divide the quote by 1000, getting the price per foot. (This is easily accomplished by moving the decimal point three places to the left.)

$$\$325 \text{ mbf} = \$0.325 \text{ per board foot}$$

Next, multiply this price by the number of board feet.

$$0.325 \times \frac{20 \times 2 \times 8 \times 12}{12} = \$104.00$$

EXAMPLE 11-3

If 1 × 8 pine boards are quoted at \$0.620 per linear foot, what is the price per board foot?

Solution: A price per board foot, as the unit term (price/board foot) suggests, may be found by dividing the price by the number of board feet. In this problem, the price for 1 linear foot is known. It is necessary to determine the number of board feet in 1 linear foot of 1 × 8. Using the board-foot formula,

$$\frac{1 \times 8 \times 1}{12} = \tfrac{2}{3} \text{ board foot}$$

Now divide the price of \$0.620 by the number of board feet—namely $\frac{2}{3}$.

$$\frac{0.620}{\frac{2}{3}} = \$0.93$$

Therefore, a price of \$0.620 per linear foot of 1 × 8 is equivalent to a price of \$0.93 per board foot.

It is sometimes confusing to the student to know which numbers are to be multiplied or divided when making conversions such as the one illustrated above. Let the units themselves describe how to handle the numbers. A "price per board foot" may be abbreviated as "\$/bf." This unit suggests that price is divided by board feet. A price per linear foot may be abbreviated as "\$/lf." This unit suggests that the price should be divided by the number of linear feet. Many terms like these appear in mathematics, physics, and everyday life. Some other examples are "miles per hour" or "m/h," "cost per gallon" or "\$/gal," "miles per gallon" or "mi/gal," "pounds per square inch" or "lb/in.^{2}." In each case, the units describe how the number is determined. For example, if you wish to calculate gasoline consumption in miles per gallon, "mi/gal" suggests that you divide miles by the number of gallons used.

EXAMPLE 11-4

A pine board 6-ft-long 1×8 is priced at \$3.84. Find the price per board foot.

***Solution*:** First, find the number of board feet.

$$\frac{1 \times 8 \times 6}{12} = 4 \text{ board feet}$$

Second, divide the price by the number of board feet.

$$\frac{\$3.84}{4} = \$0.96$$

Therefore, the price is 96 cents per board foot. Notice that this could also be quoted as \$960 mbf.

REVIEW EXERCISES

11-1. Eight feet long 2×4's are offered at \$1.96 each. Find the price per board foot.

11-2. 1×3 strapping is offered at $7\frac{1}{2}$ cents per linear foot. Find the price per board foot.

11-3. If 2×8's are quoted at \$345 mbf, what is the price per linear foot?

11-4. A piece of 1×10 pine 10 ft long is marked \$7.33. Find the price per board foot.

11-5. Knothole Lumber Company advertises the following prices for lumber. All quotes are prices per board foot.

2×4's: 26 cents up to and including 12 ft;
30 cents for over 12-ft lengths

2×6's: same as 2×4's

2×8's: 32.5 cents for 8-ft lengths; all others 36 cents

2×10's: 42 cents up to and including 12-ft lengths; all others 45 cents

2×12's: 49.5 cents for 8- and 10-ft lengths; all others 52 cents

Find the total price for the following list of materials.

(a) 2×4's: 325/8's; 44/12's
(b) 2×6's: 84/8's; 26/12's; 12/14's
(c) 2×8's: 6/14's
(d) 2×10's: 8/10's; 64/14's
(e) 2×12's: 3/8's; 9/12's; 6/16's
(f) Total price for materials (a) to (e)

(*Note:* The term "325/8's" as used in describing the 2×4's and similar terms describe the "number of pieces/lengths.")

11-6. A rectangular building measures 26 ft × 44 ft. Compare the materials cost of subflooring this area using $\frac{1}{2}$-in. CDX plywood at \$9.75 per sheet to the cost using No. 4 grade boards priced at \$265 mbf. (Whole sheets of 4 ft × 8 ft plywood will be purchased; add 10% to the board requirements for waste.)

11-7. 1×3 strapping is quoted at $3\frac{1}{2}$ cents per linear foot. Find the price of 256 linear feet.

11-8. Find the board foot price of the strapping in Problem 11-7.

11-9. 1 × 6 select pine is offered at $945 mbf by one supplier, while another quotes $0.65 per linear foot. Which is less expensive; how much less?

11-10. A newspaper advertisement offered 2 × 10's at the following prices per piece:

8 ft: $4.27
10 ft: $5.33
12 ft: $6.90
14 ft: $8.40
16 ft: $10.13

(a) Find the price per board foot in each case.
(b) Is the price per board foot the same for all lengths?
(c) Suggest a reason why the price of lumber increases with length.

chapter 12

Footings, Foundations, and Slabs

OBJECTIVES

Upon completing this chapter, the student will be able to:

1. Calculate the amount of concrete required for a foundation footing.
2. Calculate the amount of concrete required for a foundation wall.
3. Calculate the amount of concrete required for a slab.

The dimension in a building that is most important in calculating its concrete needs is the perimeter. The floor area also serves in an important way. The reason the perimeter is important is that the footing and the foundation may be considered as one long rectangle, rather than each wall being a separate rectangular section. Essentially, think of unwrapping a footing or foundation by straightening out the corners and reorienting the structure as a long rectangle. The length of this rectangle is the perimeter of the building. This procedure results in a certain amount of approximating the actual volume of concrete. However, an example will show that the error is negligible.

Consider the foundation shown in Figure 12-1. The outside dimensions of the

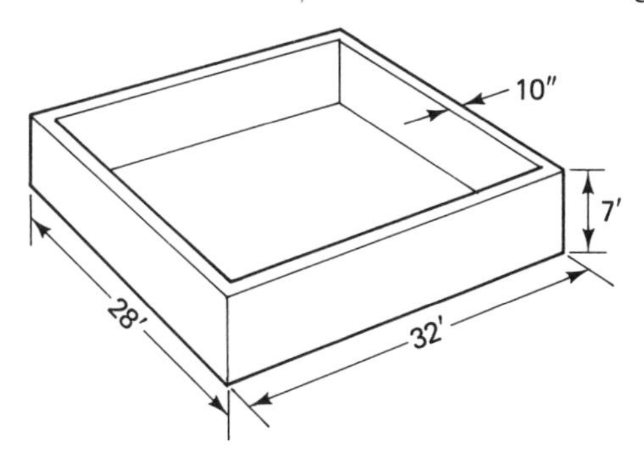

Figure 12-1

foundation are 28 ft × 32 ft. The walls are 7 ft high and 10 in. thick. The amount of concrete needed for this foundation will be calculated in two ways—an exact method, followed by an approximation method for comparison.

To get the exact solution, divide the foundation into four rectangular solids. The front and rear sections are identical. The dimensions of these sections are 32 ft long by 7 ft high and 10 in. thick (see Figure 12-2). The volume of the front and rear sections may be calculated as follows:

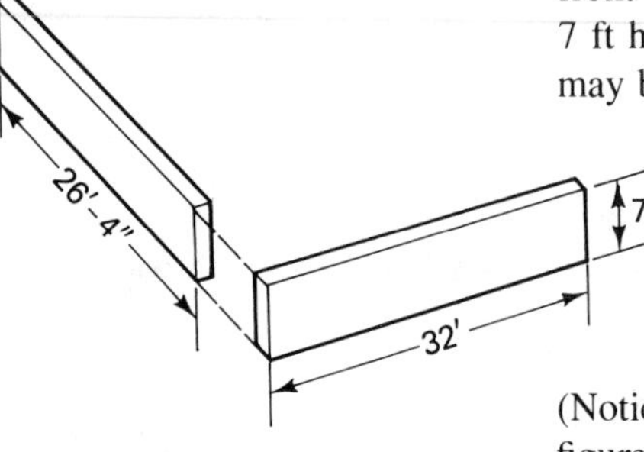

Figure 12-2

$$V = 32 \times 7 \times \frac{10}{12}$$

$$= 186.67 \text{ ft}^3$$

(Notice that the 10-in. dimension is divided by 12, converting it to feet.) Double this figure to take both sections into account.

$$2 \times 186.67 = 373.34 \text{ ft}^3$$

Now, for the end walls, notice that the length must be reduced by twice the wall thickness—a total of 20 in. The length then becomes

$$28 \text{ ft} - 20 \text{ in.} = 26 \text{ ft } 4 \text{ in. or } 26.33 \text{ ft}$$

The volume of an end section is

$$V = 26.33 \times 7 \times \frac{10}{12}$$

$$= 153.59 \text{ ft}^3$$

Doubling this to take into account both ends, we have a total volume of

$$2 \times 153.59 = 307.18 \text{ ft}^3$$

The total volume of the front, rear, and two ends is

$$373.33 + 307.18 = 680.51 \text{ ft}^3$$

Now, convert this to cubic yards, the standard unit of measure for concrete. Since there are 27 ft^3 in 1 yd^3 we divide the number of cubic feet by 27.

$$V = \frac{680.51}{27}$$

$$= 25.20 \text{ yd}^3$$

Now calculate the volume by an approximation method. This technique assumes the foundation to be "unwrapped" at the corners and considered as one long rectangular solid (see Figure 12-3). The length of this wall equals the perimeter of the building, that being

$$P = 2(32 + 28)$$

$$= 120 \text{ ft}$$

Figure 12-3

The total volume, V, of the foundation is

$$V = \text{perimeter} \times \text{height} \times \text{thickness}$$

$$= 120 \times 7 \times \frac{10}{12}$$

$$= 700 \text{ ft}^3$$

Convert this to cubic yards:

$$V = \frac{700}{27}$$

$$= 25.93 \text{ yd}^3$$

Comparing the results of the two methods, there is a difference of 0.73 yd^3—approximately $\frac{3}{4}$ yd^3. (A study of Figure 12-3 will show the source of this error to be the triangular prisms formed in each corner when the foundation is unwrapped. The approximation method assumes that these prisms are filled in.) When considered in light of the fact that it is better to have a slight excess of concrete than not to have enough, the approximation method of calculating concrete has to take preference. Bear in mind, too, that the measuring of concrete at a ready-mix batch plant involves some error due to the variable moisture content of sand and aggregate as it is weighed. The final argument for using the approximation method is its relative simplicity.

Should it be desirable to use a precise method, as an alternative to the technique used above, the inside and outside perimeters of the foundation could be averaged. Then this result could be used as the length of the "unfolded" rectangular solid and multiplied by the height and width.

It should be noted that the size requirements for foundation footings, wall widths, and floor slab thicknesses are dictated by the soil classification, use and size of the building to be placed on the footings and walls, and so on. It is not an objective of this book to cover this matter. Students interested in pursuing the matter of sizing should consult references that address specifications.

When calculating the amount of concrete necessary for footings, treat them in the same manner as foundations. The question of accuracy is left to the student as an exercise (see Problem 12-1). Suffice it to say that the outside dimensions of the foundation that will set on the footing may be used in calculating the volume of concrete needed.

EXAMPLE 12-1

How many cubic yards of concrete is needed for the foundation footing of a house with dimensions 28 ft × 32 ft? The footing is to be 20 in. wide and 8 in. thick.

***Solution*:** Calculate the perimeter of the building. Then "unwrap" the footing and consider it as a long, rectangular solid having a length equal to the perimeter of the building, a width of 20 in. ($\frac{20}{12}$ ft), and a height of 8 in. ($\frac{8}{12}$ ft). Remember that all dimensions used in the volume formula must be expressed in the same units.

The perimeter of the building is

$$P = 2(28 + 32)$$

$$= 120 \text{ ft}$$

Next, the volume is

$$V = 120 \times \frac{20}{12} \times \frac{8}{12}$$

$$= 133.33 \text{ ft}^3$$

Convert to cubic yards:

$$V = \frac{133.33}{27}$$

$$= 4.94 \text{ yd}^3$$

Next, consider the slab of concrete that will serve as the cellar floor in the building described in Example 12-1. Again, the question of accuracy is left to the student to be examined in an exercise (see Problem 12-2). The length and width of the cellar floor may be approximated using the outside dimensions of the building.

EXAMPLE 12-2

Find the number of cubic yards of concrete needed for the cellar floor in a house having outside dimensions of 28 ft × 32 ft if the floor is to be 3 in. thick.

***Solution*:** The floor is a rectangular solid. Find its volume in cubic feet. Then convert to cubic yards.

$$V = 28 \times 32 \times \frac{3}{12}$$

$$= 224 \text{ ft}^3$$

Convert to cubic yards:

$$V = \frac{224}{27}$$

$$= 8.30 \text{ yd}^3$$

In summary, footings, foundations, and slabs are, for the most part, rectangular solids. Calculating the quantity of necessary concrete amounts to using the formula for a rectangular solid, $V = l \times w \times h$. Remember to be consistent with the units. Dimensions in feet are frequently convenient. The problem can then be completed by converting to cubic yards, accomplished by dividing cubic feet by 27.

REVIEW EXERCISES

12-1. Calculate the number of cubic yards of concrete needed for the footing of a house measuring 28 ft by 32 ft. The foundation is to be 10 in. thick. The footing is to be 20 in. wide and 8 in. high. Adjust the dimensions of the footing to fit the foundation and calculate the answer accurately. Then compare the results to the result of Example 12-1.

12-2. Calculate the number of cubic yards of concrete needed for the cellar floor in the 28 ft × 32 ft house, using accurate methods. The slab is 3 in. thick. Compare the result with the approximate method used in Example 12-2.

12-3. Below are the outlines of several foundations. Dimensions are noted, including wall thickness and height. Assume the footings to be 8 in. high and twice the thickness of the foundation wall; the cellar floor is to be 3 in. thick. For each figure, calculate the amount of concrete needed for the footing, the walls, and the floor. Include a total figure.

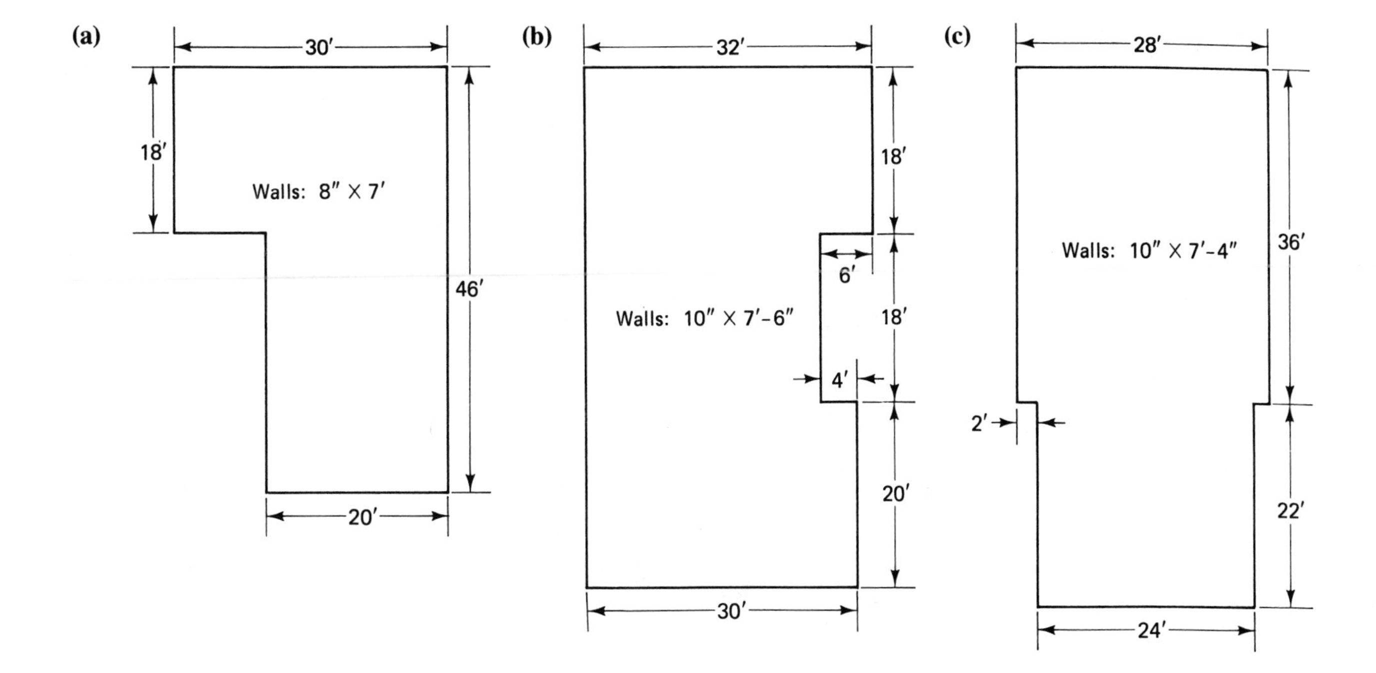
(a)
30′
18′
Walls: 8″ X 7′
46′
20′
(b)
32′
18′
6′
Walls: 10″ X 7′-6″
18′
4′
20′
30′
(c)
28′
Walls: 10″ X 7′-4″
36′
2′
22′
24′

chapter 13

Girders, Sill Plates, Bridging, Floor Joists, and Floor Covering

OBJECTIVES

Upon completing this chapter, the student will be able to:

1. Determine an economical materials list for a built-up girder.
2. Determine an economical materials list for sill plates.
3. Calculate the number and length of floor joists and floor joist headers for a particular framing plan.
4. Calculate the amount of bridging material needed for a particular framing plan.
5. Calculate the amount of floor covering material needed for a certain area.

A girder may be made of one of several types of materials. The material may be reinforced concrete, steel, solid wood, or built-up laminations of wood. The latter type is under consideration here.

The proper size for a girder is dependent on several factors. Included among these: the type of lumber making up the girder, the load it is to carry, the distance between supports, and the distance between other load-bearing girders or walls. Structural tables are available for sizing a girder. The objective here is to determine the materials required to construct a built-up girder once the specifications have been determined from a proper source.

Since a butt joint in a laminated girder is a point of weakness, column supports should be placed where the joints occur. In certain situations, space utilization may play a role in positioning column supports. Therefore, both the desired support positions and the butt joint positions are of consideration in constructing a built-up girder. Another factor that determines support position is the distance required between supports as per design specifications. Two guidelines that should be kept in mind when designing the materials layout for a girder are:

1. Butt joints should be supported.
2. Joints should be no closer to each other than 4 ft.

The example that follows assumes a particular distance between supports.

EXAMPLE 13-1

Determine the length of materials needed for a built-up girder made of three laminations of 2 × 10's. Support posts are to be placed every 8 ft and the overall length is 40 ft.

Solution: Begin by sketching a plan view of the girder. Mark the support positions at 8-ft intervals. Then sketch the possible butt-joint positions, making note of the support positions. To the extent possible, butt joints should occur at the support post. It is acceptable to have two butt joints opposite each other, as long as they are separated by a lamination and are supported by a post. In this case, lengths of 8 ft and 16 ft work well (see Figure 13-1). Thus materials for this girder could be

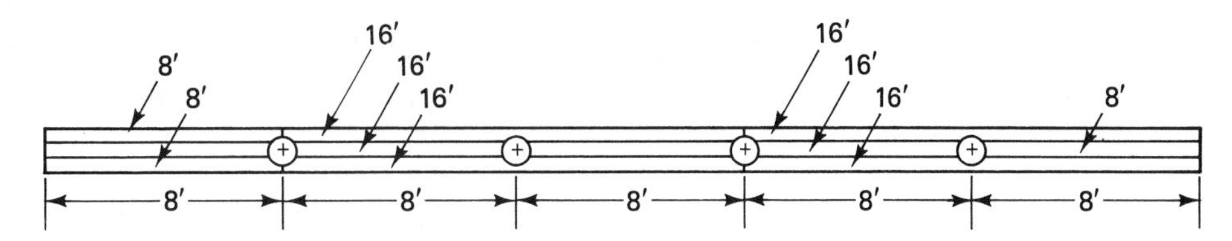

Figure 13-1

2 × 10's: 6 pcs. 16 ft; 3 pcs. 8 ft

As a check on the amount of material to be ordered, calculate the total length of material and compare it to three times the length of the girder. In this case,

$$(6 \times 16) + (3 \times 8) = 96 + 24 = 120 \text{ ft}$$

$$3 \times 40 = 120 \text{ ft}$$

It should be noted that other layouts exist which are equally correct. After adhering to butt-joint placement, economy is the next consideration. To the extent possible, materials should be ordered in lengths that minimize waste.

Sill plates are ordinarily made of a single layer of 2 × 6 or 2 × 8 lumber. It is advisable to place a layer of sill seal, an unfaced fiberglass insulation product, along the perimeter of the foundation. The sill plate is then laid on the sill seal and fastened to the top of the foundation by anchor bolts that have been embedded in the concrete at intervals of approximately 4 ft. The length of the wall segments will dictate the economical lengths of material to order for sill plates.

EXAMPLE 13-2

Figure 13-2 shows a plan view outline of a foundation. Determine an economic order for the 2 × 6 sill plates.

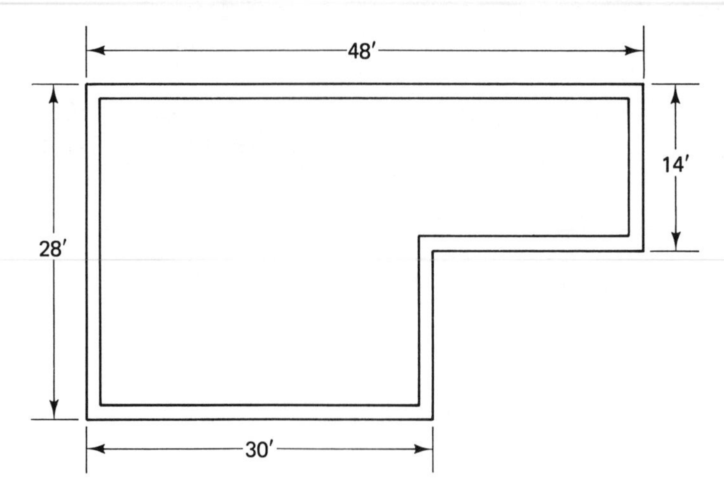

Figure 13-2

***Solution*:** The 48-ft wall along the rear may be covered with four pieces, 12 ft long. (Three 16's will work, also; however, a premium price may be charged for longer pieces.) The left and right ends may be covered with four pieces 14 ft long. The 30-ft and 18-ft segments along the front are the equivalent of the rear section and, therefore, may be covered with four 12-ft pieces. In summary,

2 × 6's: 8/12's; 4/14's

A check on the materials needed for sill plates is performed by comparing the total linear feet of calculated material to the perimeter of the building. In this case,

$$\text{perimeter} = 2(28 + 48) = 152 \text{ ft}$$

Calculated materials:

$$(8 \times 12) + (4 \times 14) = 96 + 56 = 152 \text{ ft of } 2 \times 6$$

Floor-joist sizing is determined by the type of lumber, the distance between joists, the span, and the load the joists are to carry. The joist size will be indicated on a set of plans that has been prepared by a designer. In the absence of a set of prepared set of plans, structural tables should be consulted for proper sizing.

The number of joists is determined by the spacing. A distance of 16 in., center to center, is common; however, spacing 24 in. o.c. (on center) or 12 in. o.c. may be dictated by design considerations. When the spacing is to be 16 in. o.c., the number of joists may be determined by the following formula:

$$\text{NJ} = \tfrac{3}{4}L + 1$$

where NJ is the number of joists and L is the length (in feet) to be covered. The reason this formula works is illustrated in the following example.

EXAMPLE 13-3

Determine the number of joists 16 in. o.c. needed to cover a distance of 40 ft.

***Solution*:** The problem could be solved by first converting 40 ft to inches; then the number of 16-in. spaces could be determined.

$$40 \times 12 = 480 \text{ in.}$$

$$\frac{480}{16} = 30 \text{ spaces}$$

There will be one more joist than the number of spaces. That is the reason for adding one in the formula.

In Example 13-3, the length, 40 ft, was multiplied by 12 and then divided by 16. The result would be the same if the length were multiplied by the fraction $\frac{12}{16}$, or $\frac{3}{4}$, in reduced form. This is the reason for the factor of $\frac{3}{4}$ in the formula.

If joists are to be spaced 12 in. o.c., the number of joists equals the length in feet, plus one. That is,

$$NJ = L + 1$$

If joists are to be spaced 24 in. o.c., the number of joists is one-half the length in feet, plus one. That is,

$$NJ = \frac{L}{2} + 1$$

When the calculation of the number of joists results in a fraction, the result should be rounded up. A fractional space will require a joist. Other adjustments occur for end trimmers, stairwells, or other alterations, such as extra joists needed for load bearing.

EXAMPLE 13-4

The floor framing for the building outlined in Figure 13-3 calls for 2 × 10's placed 16 in. o.c. How many are needed if a girder is centered as shown? Assume that the joists will overlap on the girder.

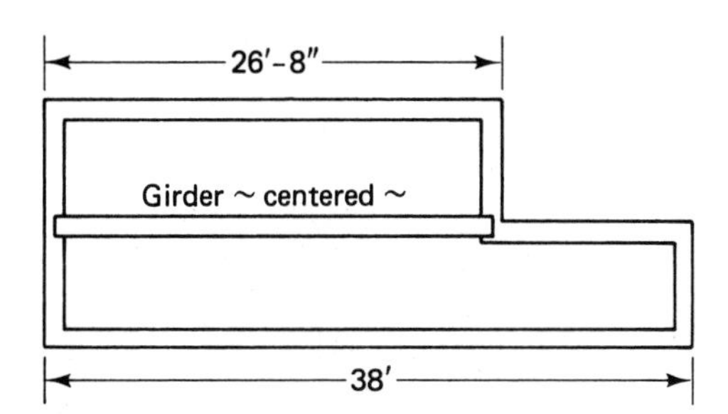

Figure 13-3

***Solution*:** Along the front, the length to be covered is 38 ft.

$$\begin{aligned} NJ &= \tfrac{3}{4} \times 38 + 1 \\ &= 28.5 + 1 \\ &= 29 + 1 \\ &= 30 \text{ joists} \end{aligned}$$

Along the rear, the length is 26 ft 8 in. or 26.67 ft.

$$\begin{aligned} NJ &= \tfrac{3}{4} \times 26.67 + 1 \\ &= 20 + 1 \\ &= 21 \text{ joists} \end{aligned}$$

Notice that in both of the calculations above, a fractional number of joists is rounded to the next-higher whole number. Therefore, a total of 51 floor joists are needed.

In addition to floor joists, floor-joist headers should be included at this point. Recalling the material lengths used for sill plates along the front and rear will be of assistance. This foundation plan would call for headers along the 38-ft length twice. Two 12-ft pieces and one 14-ft will cover 38-ft with no waste. Therefore, the floor-joist headers will be 4/12's and 2/14's.

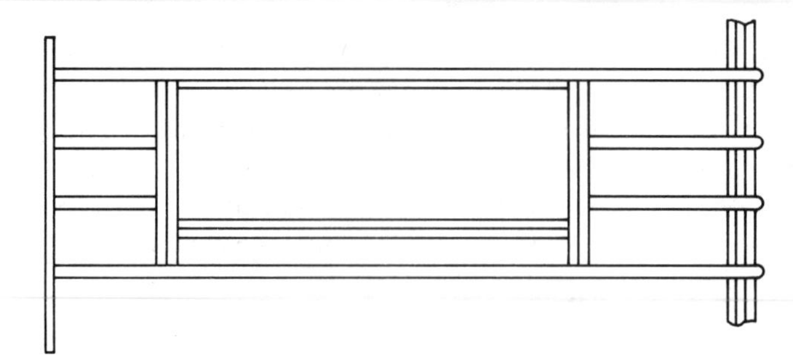

Figure 13-4

Figure 13-4 shows the framing for a stairwell. Adjustments to the framing materials needed are made according to the stairwell plan. If the stairwell is conventional and joists are spaced 16 in. o.c., parts of two joists will be eliminated for a 36-in.-wide well. But these should not be eliminated from the count, since the joists along the sides of the well will be doubled. Furthermore, both ends of the stairwell will be double headed. In general, one or two floor joists should be added for framing a standard stairwell.

Bridging material may be purchased in the form of steel strips designed for this purpose. If wooden bridging is to be used, 1 × 3 strapping may be purchased. This material comes in random lengths and is usually available up to 16 ft. A row of bridging should be placed midway along the span in cases where the span exceeds 8 ft. The following example determines the amount of material needed for a single piece of bridging between a pair of 2 × 10's placed 16 in. o.c.

EXAMPLE 13-5

Calculate the length of material needed to cover the inside diagonal between 2 × 10's (actual width $9\frac{1}{2}$ in) that have been placed 16 in. o.c. (see Figure 13-5).

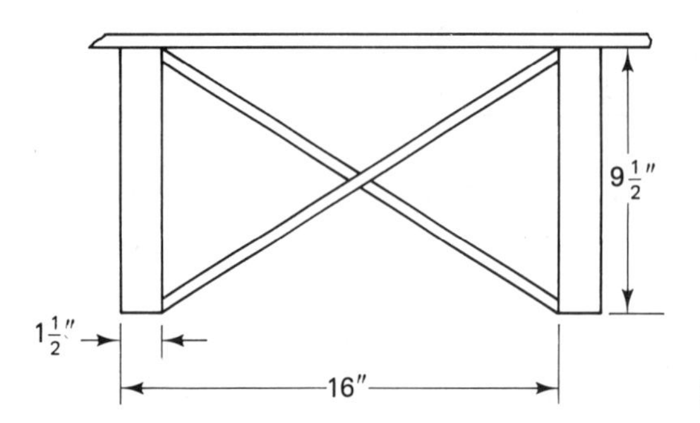

Figure 13-5

Solution: The joists are $1\frac{1}{2}$ thick. This makes the inside distance between joists $14\frac{1}{2}$ in. The depth of the 2 × 10 is $9\frac{1}{2}$ in. Use the Pythagorean theorem. The diagonal, representing the bridging, is the hypotenuse of a right triangle; the legs are the distance between joists and the joist width.

$$L^2 = 9.5^2 + 14.5^2$$
$$= 90.25 + 210.25$$
$$= 300.50$$
$$L = \sqrt{300.50}$$
$$= 17.3 \text{ in.}$$

This example could be repeated for floor joists of other sizes and spacing. However, the cost of 1 × 3 strapping does not warrant expending much effort toward precise calculations. The solution of 17.3 in. could be rounded to 18 in.; another 18 in. will make up the second member of the "X" that forms the bridging between joists. This amounts to a total of 36 in. or 3 ft of material for each space to be bridged.

Since the spacing between floor joists may be 12 in. o.c., 16 in. o.c., 24 in. o.c., and so on, it seems reasonable to establish an estimating factor that will accommodate all these variables. Multiplying the number of feet of length to be bridged by three produces a reasonable estimate of bridging material. That is,

$$B = 3L,$$

where B is the bridging material in feet and L is the length of a row of bridging.

EXAMPLE 13-6

Determine the amount of 1 × 3 strapping needed to bridge the floor joists from the plan shown in Figure 13-2.

***Solution*:** Assuming a girder is placed midway between the front wall (30 ft) and the rear wall (48 ft), two rows of bridging are required: one row parallel to the front wall, 7 ft back; another parallel to the rear wall, 7 ft in. The total length to be bridged is

$$30 + 48 = 78 \text{ ft}$$

The material needed is

$$B = 3 \times 78$$

$$= 234 \text{ ft}$$

The final problem considered in this chapter is the amount of floor covering material needed for a certain area. The typical subflooring material comes in sheets measuring 4 ft × 8 ft, whether the material to be used is plywood of various thicknesses, particle board, or waferboard. Since 4 ft × 8 ft covers an area of 32 ft^2, the number of sheets needed may be determined by dividing the floor area in square feet by 32 and rounding to the next higher whole number.

For example, a house measuring 34 ft × 26 ft has an area of

$$A = 34 \times 26$$

$$= 884 \text{ ft}^2$$

Dividing this area by 32 and rounding up, we get 28 sheets.

$$\frac{884}{32} = 27.625 \text{ or } 28$$

REVIEW EXERCISES

13-1. A built-up girder is to be constructed of three laminations of 2 × 12's. Steel column supports will be placed every 10 ft. The overall length is 52 ft. Determine the materials needed for an economical layout.

13-2. How many board feet of material will there be in the girder described in Problem 13-1?

13-3. On the following page are the outlines of several foundations. Girder positions are shown by dashed lines. Girder support positions are noted. For each foundation, determine the following materials requirements.
(1) The materials needed for an economical built-up girder design (assume three laminations for each girder)
(2) The number of pieces and their lengths for 2 × 6 sill plates
(3) The number of pieces and lengths for floor joists and floor joist headers
(4) The total number of feet of 1 × 3 bridging material

(5) The number of 4 ft × 8 ft sheathing required for subflooring

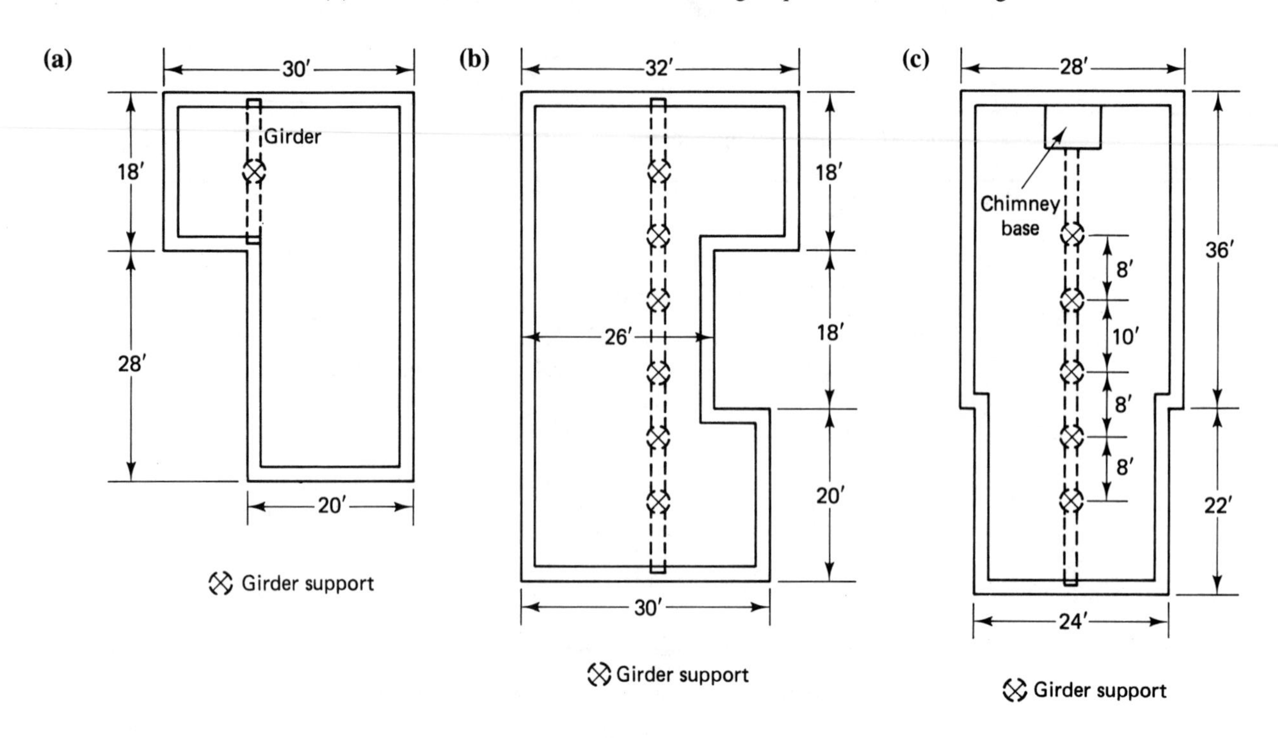

chapter 14

Wall Framing

OBJECTIVES

Upon completing this chapter, the student will be able to:

1. Calculate the number of studs needed to frame the exterior walls of a structure.
2. Calculate the number of studs needed to frame the interior walls of a structure.
3. Calculate the amount of material needed for shoes and plates.
4. Compare various methods of wall framing with respect to costs and energy efficiency.

A builder must develop accurate and efficient methods for estimating the costs of construction. The accuracy of his estimate of the quantity of materials affects profit not only in terms of a price quoted to a customer, but in terms of having sufficient materials on the job site to maintain the flow of construction.

The perimeter of a building is of primary importance in determining the quantity of materials needed to frame the exterior walls of a structure. Consider a simple rectangular structure with foundation size 24 ft × 32 ft (see Figure 14-1). The perimeter of this structure is

$$P = 2(24 + 32)$$
$$= 112 \text{ ft}$$

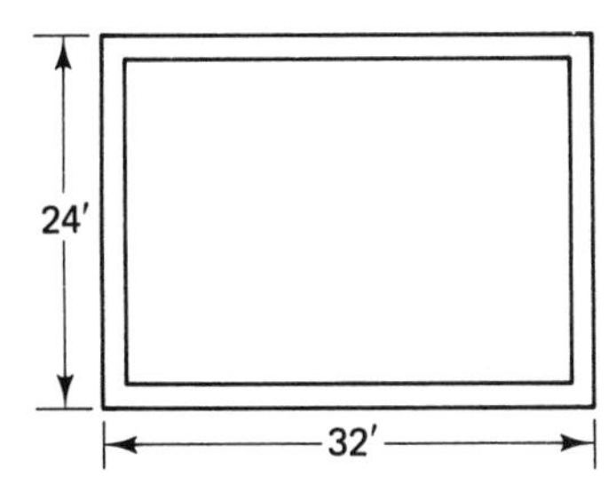

Figure 14-1

The materials needed for the single shoe (also called sole plate) and double top plate is three times the perimeter. In this case,

$$3 \times 112 = 336 \text{ linear feet}$$

In general, materials needed for the shoe and plates, S, is given by

$$S = 3 \times P$$

where P is the perimeter of the structure in feet.

When selecting the length of pieces making up this order, both economy and convenience should be considered. Material lengths should be matched to the length of the wall segments to be constructed with the objective of minimizing waste. However, the lengths should also be determined by what the carpenter finds convenient in terms of handling wall segments that will be prefabricated on the floor with the intent of standing them in place. The number of workers on hand is a factor.

The amount of material may be checked by comparing the value of three times the perimeter to the sum of the lengths of all pieces. For the building in Figure 14-1, assume that 12-ft lengths are chosen for the shoe and plate material, with 8-ft lengths to complete the 32-ft segments. This would require 24 pcs. 12 ft and 6 pcs. 8 ft.

Checking the total length against the perimeter, the total length is

$$(24 \times 12) + (6 \times 8) = 288 + 48 = 336 \text{ linear feet}$$

The formula established above yields

$$\begin{aligned} S &= 3 \times P \\ &= 3 \times 112 \\ &= 336 \text{ ft} \end{aligned}$$

The number of studs needed depends on the spacing and the number of openings for doors and windows. The common stud spacing in residential construction is 16 in. o.c.. This is the necessary spacing with 2 × 4 framing material. 2 × 6 framed walls are becoming common in areas of the country where the benefits of extra insulation are significant. When 2 × 6 framing material is used, sufficient structural strength is gained with 24-in. o.c. stud spacing. However, an argument in favor of 16-in. o.c. spacing can be made in terms of the application of drywall. It is possible to detect slight waves in drywall applied to studs spaced 24 in. o.c. (The author has had no significant problem with drywall using 24-in. o.c. spacing. However, the builder may wish to use 5/8-in. thick drywall over studs spaced 24 in. o.c.)

An efficient way of counting the number of studs needed is to apply a factor to the perimeter of the structure; then adjustments are made for openings, corners, odd lengths, and interior partition starters.

EXAMPLE 14-1

Determine the number of studs needed to frame the exterior walls of the first story of the structure shown in Figure 14-1. The spacing is to be 16 in. o.c.

***Solution*:** The method used is the same as that established for counting floor joists spaced 16 in. o.c. (see Chapter 3), with some modification. A basic count is first established by multiplying the perimeter by $\frac{3}{4}$. (Recall that $\frac{3}{4}$ is the reduced fraction formed when a number is multiplied by 12, to convert feet to inches, and then divided by 16, to determine the number of 16-in. spaces.) In Figure 14-1, the perimeter is 112 ft. Letting NS represent the number of studs,

$$\begin{aligned} \text{NS} &= \tfrac{3}{4} \times 112 \\ &= 84 \text{ studs} \end{aligned}$$

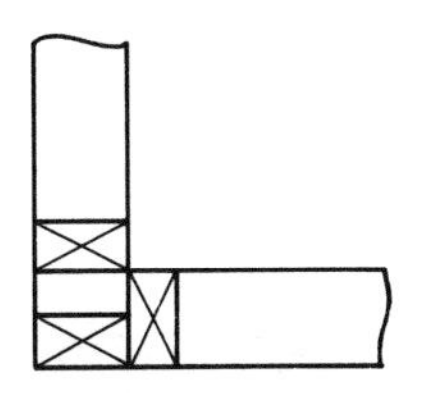

Figure 14-2

Recall that this method actually counts spaces, not studs. Therefore, one stud must be added for each wall segment to account for the "starter" stud. In addition, one stud must be added because of the manner in which corners are usually framed (see Figure 14-2). For each wall segment, then, add 2 studs. This increases the count for Figure 14-1 by 2 studs for each of the four wall segments, a total of

$$4 \times 2 = 8 \text{ studs}$$

and brings the total count to

$$84 + 8 = 92 \text{ studs}$$

(Checking this against the total found by counting the studs needed for each wall segment will prove the accuracy from using the perimeter.)

One other adjustment to the basic count may be necessary when the length of a wall is not evenly divisible by 4. This means that a fractional space exists. This space requires a stud and one should be added.

EXAMPLE 14-2

Determine the number of studs needed for a wall 22 ft 6 in. long. The spacing is 16 in. o.c.

***Solution*:** Use the formula $NS = \frac{3}{4} \times P$ with 22 ft 6 in. or 22.5 ft in place of P.

$$NS = \frac{3}{4} \times 22.5$$

$$= 16.875$$

This is increased to 17 studs; 2 additional studs are needed for the starter and the corner—a total of 19 studs.

The estimator should use the perimeter to establish a basic count; then examine the plan for wall segments that are not divisible by 4, adding one stud for each such segment.

When the spacing is 24 in. o.c., the factor applied to the perimeter becomes $\frac{1}{2}$ instead of $\frac{3}{4}$. This is because the number of 2-ft spaces dictates the number of studs. Essentially, the perimeter in feet is divided by 2, the divisor being 2 ft. Adjustments to the basic count are otherwise the same as with other spacing: add 1 for the starter; add 1 for the corner; add 1 for an odd length of wall—this being a wall segment not evenly divisible by 2.

EXAMPLE 14-3

Determine the number of studs needed 24 in. o.c. to frame the exterior walls of the building shown in Figure 14-3.

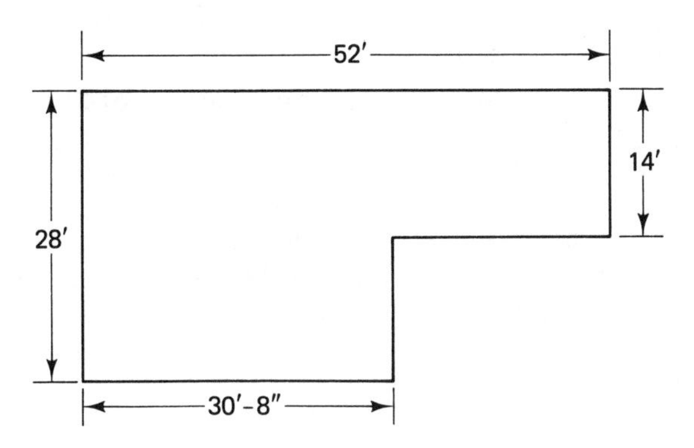

Figure 14-3

Solution: The perimeter is

$$P = 2(28 + 52) = 2(80) = 160 \text{ ft}$$

The basic count of the number of studs, NS, is

$$\text{NS} = \frac{P}{2} = \frac{160}{2} = 80 \text{ studs}$$

The adjustments are as follows:

6 wall segments	6 studs
6 corners	6 studs
2 odd lengths (30 ft 8 in. and 21 ft 4 in.)	2 studs
Total adjustments	14 studs
Total studs	94 studs

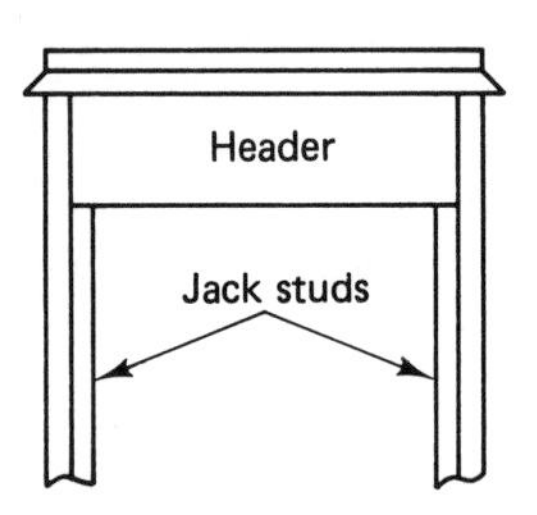

Figure 14-4

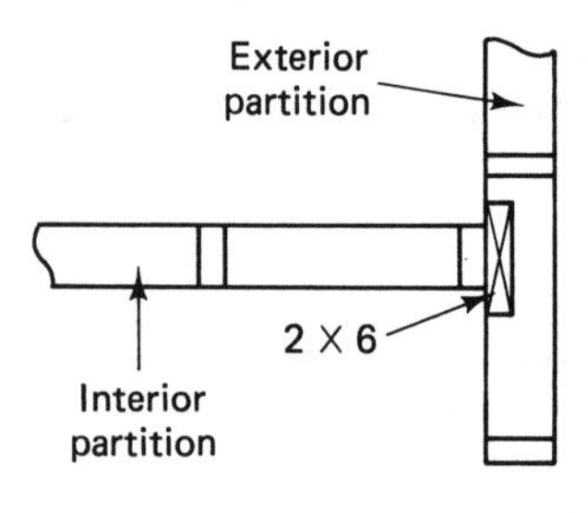

Figure 14-5

Wall openings for doors and windows should also be considered as adjustments to the number of studs. Standard door and window widths require an average opening of approximately 3 ft. Depending on where the openings occur with respect to stud location, two or three studs would normally be eliminated. Stud spacing also has some effect on the number to be eliminated. On the other hand, jack studs will be placed on each side of the opening to support the header (see Figure 14-4). The studs to be eliminated, then, become the jack studs, resulting in no adjustment to the count. (The estimator may find it more economical to replace these studs with 7-ft studs for use as jacks, resulting in less material waste.) For wider openings, such as those caused by sliding patio doors, the estimator may wish to eliminate from 2 to 4 studs, depending on spacing and door width.

The framing under a window does require additional material. For the average-sized window, the addition of 2 studs will ordinarily be sufficient to frame the sill and its supports.

The final adjustment to the number of studs in the exterior walls results from the intersection with interior partitions. Locating a 2 × 6 stud at the centerline of an interior partition serves as a corner for drywall nailing (see Figure 14-5). Therefore, add one 2 × 6 for each intersection with an interior wall.

Summarizing the stud count for exterior walls:

1. Use $S = \frac{3}{4} \times P$ for 16 in. o.c., $S = P/2$ for 24 in. o.c.
2. Add 2 studs per wall segment.
3. Add 1 stud for each odd length of wall.
4. Add 2 studs per average-sized window.
5. Add 1 2 × 6 for each interior partition intersection.

EXAMPLE 14-4

Determine the number of studs and the shoe and plate materials for the exterior walls of the house shown in Figure 14-6. The framing will be with 2 × 6-16 in. o.c.

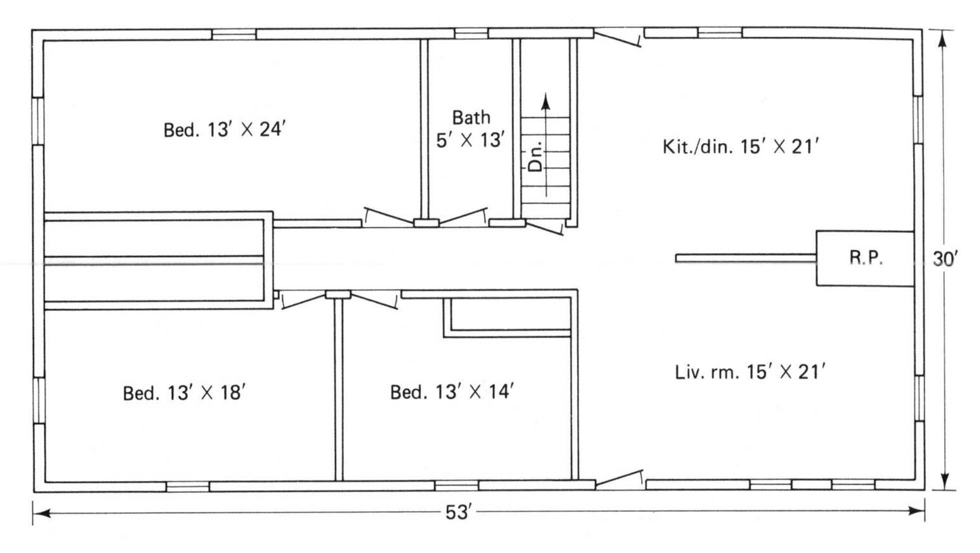

Figure 14-6

***Solution*:** First, determine the perimeter.

$$P = 2(53 + 30)$$
$$= 2(83)$$
$$= 166 \text{ ft}$$

The basic count is

$$NS = \tfrac{3}{4} \times 166$$
$$= 124.5 \text{ or } 125 \text{ studs}$$

The adjustments are as follows:

Starters and corners	8 studs
Odd lengths	4 studs
11 windows	22 studs
Interior partitions	8 studs
Total adjustments	42 studs
Total studs	167 studs

An accurate count of the number of studs needed to frame the interior partitions would consume enough time to question the value of such an exercise. Reasonable results may be obtained by totaling the linear footage of walls, closets included, and counting one stud for each linear foot. Openings are ignored, as are corners, starters, and partition intersections. Shoe and plate materials will be three times the total wall length.

A reasonable way to count the total linear footage of interior partitions is to survey the floor plan summing the lengths of all wall segments running parallel to the front wall. Then sum the lengths of those wall segments parallel to a side wall. Particular note might be taken of those wall segments which, while appearing intermittently throughout the length of the house, form the equivalent of the length of the house. It is simpler to take the length of the house as this total, rather than summing several lengths that produce the same result. The same analysis may be performed using wall segments that parallel the side walls.

Thus, to estimate the materials needed for interior partitions, determine the total linear feet of walls. Shoe and plate material will require three times this figure. The number of studs will be equal to the number of linear feet of walls.

EXAMPLE 14-5

Determine the materials needed to frame the interior partitions for the plan shown in Figure 14-6. The material used will be 2 × 4's placed 16 in. o.c.

***Solution*:** It is left to the reader to verify the following estimates of total wall length.

Parallel to the front (53 ft) wall: 93 ft

Parallel to the side wall: 75 ft

Total length of interior partitions: 168 ft

This would require 3 × 168 = 504 linear feet for shoes and plates and 168 studs.

The final consideration in this chapter is material needed for window and door headers. Sizing of headers depends on the span and load they are to bear. Refer to structural tables for proper sizing. As to length of materials needed, the total width of openings serves as a guide. If headers are to be constructed by doubling the spanning material (2 × 6's, 2 × 8's, etc.), the sum of the widths would, of course, be doubled. The length of materials ordered should be chosen so as to minimize waste. For average-sized doors and windows, material ordered in 12-ft lengths often serves economically.

A final comment deserves space here. As this is written, the building industry is experimenting with various insulating techniques in reaction to drastic price increases in heating costs during recent years. Insulation is given special consideration in Chapter 22. However, builders interested in energy conserving construction methods will certainly want to make cost comparisons of various framing methods. Among the techniques presently being tried are conventional 2 × 4-16 in. o.c. framing with both fiberglass and rigid insulation, 2 × 6 framing with fiberglass (and sometimes rigid, as well) insulation, and double 2 × 4 framing with two walls spaced to allow 12 in. of fiberglass insulation. The relative costs of various framing methods will be considered in the exercises that follow.

REVIEW EXERCISES

14-1. Determine the number of studs placed 16 in. o.c. needed to frame a wall segment 26 ft 8 in. long.

14-2. Answer Problem 14-1 if the studs are placed 24 in. o.c.

14-3. In Example 14-3, the number of studs placed 24 in. o.c., as well as the shoe and plate materials, were determined for the plan shown in Figure 14-3. Assume that the framing is to be done with 2 × 4's placed 16 in. o.c.

- **(a)** Determine the materials needed for the shoe and plates.
- **(b)** Determine the number of studs.
- **(c)** Find the number of board feet in the material totals for both the 2 × 4 and the 2 × 6 framing.
- **(d)** Assume a price per board foot of $0.30 for both 2 × 4's and 2 × 6's and compare the cost of constructing the exterior walls by both methods.

14-4. Refer to the plan in Figure 14-6 and determine the following materials requirements.

- **(a)** Materials for the shoe and plates of the exterior walls. Framing will be with 2 × 4's-16 in. o.c.
- **(b)** Studs for the exterior walls, including adjustments for openings, corners, starters, and interior partition intersections.

(c) Shoe and plates for the interior walls. Framing will be with 2×4's-16 in. o.c.

(d) Studs for the interior partitions.

(e) Assume that headers are to be constructed of doubled 2×6's. Estimate the necessary materials for door and window headers.

The remaining problems compare the costs of various framing methods and should all be completed.

14-5. In Problem 14-4, determine the materials needed to frame the exterior walls if the framing is done with 2×6's-24 in. o.c. How many board feet is this?

14-6. How many board feet of lumber are needed if the exterior walls in Figure 14-6 are framed with 2×6's-16 in. o.c.?

14-7. Using the results from framing the exterior walls of Figure 14-6 with 2×4's-16 in. o.c., 2×6's both 16 and 24 in. o.c., find the materials cost in each case. Use a price of $0.30 per board foot.

14-8. One method for framing exterior walls mentioned in this chapter called for two separate walls of 2×4's placed 16 in. o.c. With this technique, the outermost wall is framed conventionally, except that the 2×6 stud used for an interior partition intersection is omitted. The inner wall is placed 12 in., outside to outside. 6 in. of fiberglass insulation is placed in each wall, providing a total of 12 in. of insulation. Interior partition intersection studs are placed on the inner wall. Headers for openings in the inner wall may be of 2×4 material and serve primarily as fillers; the roof load is carried by the outer wall.

Determine the extra materials needed for this method of framing the exterior walls in Figure 14-6. Use the same price of $0.30 per board foot and compare the costs to the other methods in Problems 14-5 to 14-7.

chapter 15

Roofs I: Common Rafters

OBJECTIVES

Upon completing this chapter, the student will be able to:

1. Calculate the length of a common rafter.
2. Calculate the distance from the ridge to the bird's-mouth cut along a rafter.
3. Calculate the distance from the rafter tail to the bird's-mouth cut.
4. Determine the total rise.
5. Determine unit rise and unit line length.
6. Determine the pitch of an existing roof.

There are many types of rafters, as shown in Figure 15-1. The mathematical basis for most of the calculations necessary to rafters comes from proportions and the properties of similar triangles. Recall that similar triangles are those having equal angles. When two triangles are similar, knowing the lengths of sides in one triangle is sufficient to determine the unknown lengths of sides in the other. For example, Figure 15-2 shows two similar triangles with known lengths labeled. To determine the length of side X, a proportion is set up and solved.

$$\frac{X}{22} = \frac{12}{14}$$

$$14X = 22 \times 12$$

$$X = 18.86$$

The general principal to be recalled here is that when two triangles are similar, corresponding parts form equal ratios.

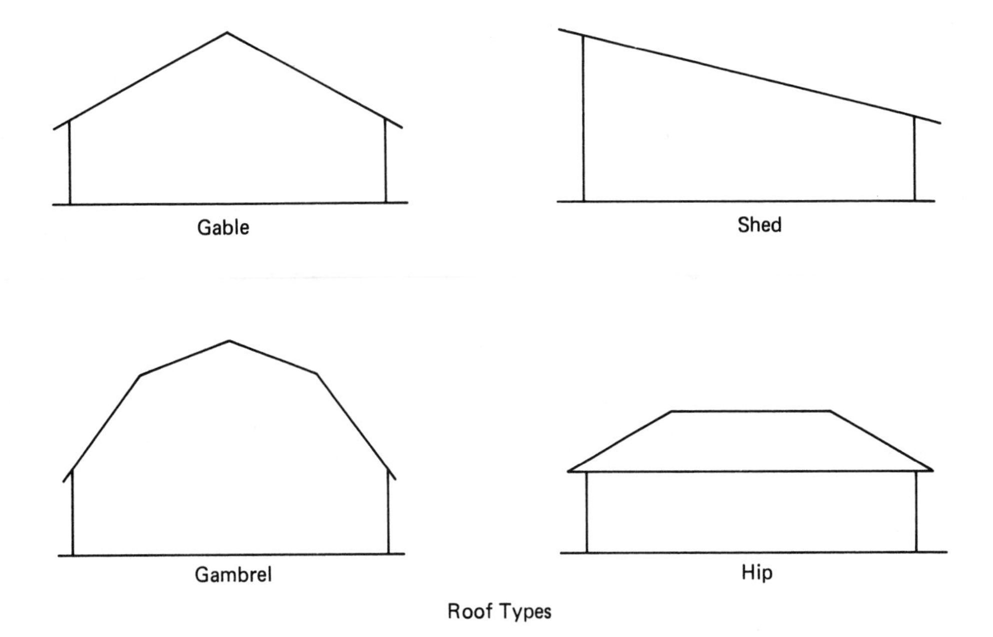

Roof Types

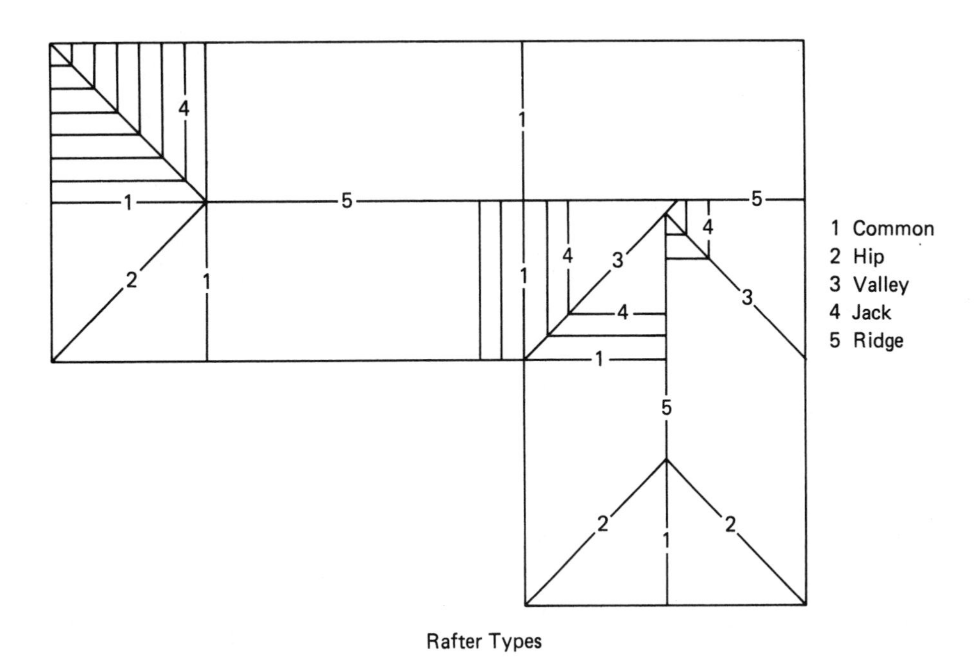

Rafter Types

Figure 15-1

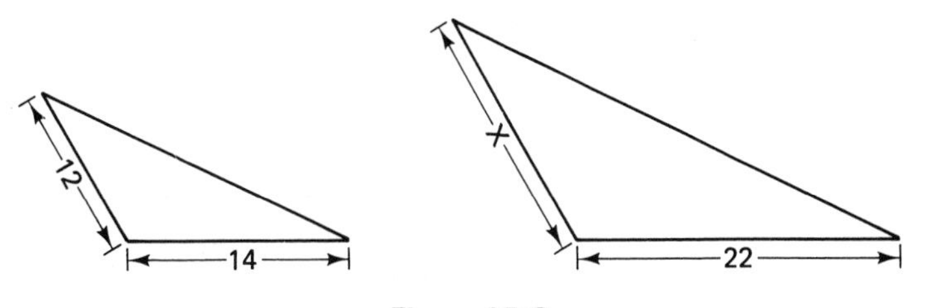

Figure 15-2

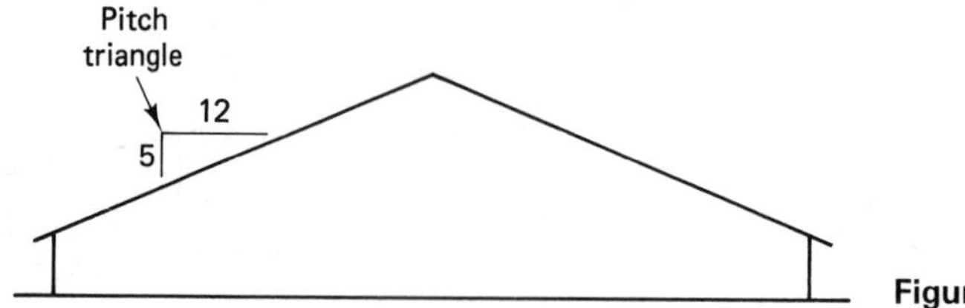

Figure 15-3

The slope of a roof is usually described on a blueprint by means of a **pitch triangle**, displayed along a roof line. Figure 15-3 shows a pitch triangle for a roof with a "5/12 pitch." This means that the roof rises 5 in. for every 12 in. or 1 ft of run. The run is always a horizontal distance.

The pitch triangle is a right triangle. The horizontal leg is always 12 and represents 12 in. 12 in. is called the **unit run**, abbreviated URu. The vertical leg, labeled "5" in Figure 15-2, describes the amount of rise in 12 in. of run. This is called the **unit rise**, abbreviated URi. Figure 15-4a, b, and c shows pitch triangles having unit rises of 4, 9, and 12, respectively. Notice that each has a run of 12.

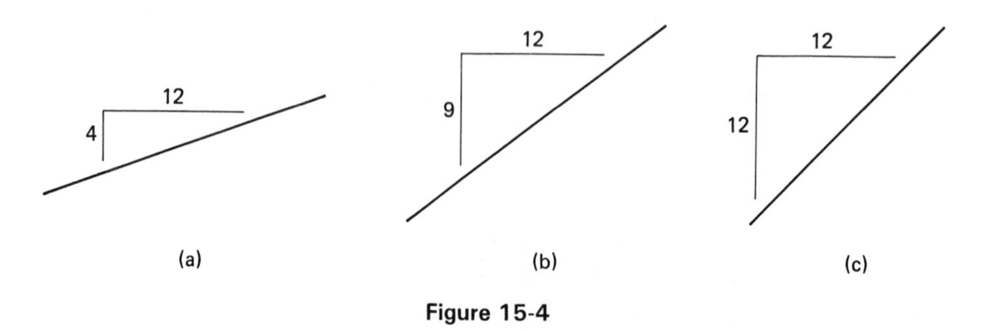

Figure 15-4

The hypotenuse of a pitch triangle serves an important role in the mathematics of roof rafters. This is called the **unit line length**, ULL. Its value can be determined using the Pythagorean theorem, since the unit line length is the hypotenuse of the right triangle. For a pitch triangle showing a 5/12 slope, the unit line length will be

$$\begin{aligned} \text{ULL} &= \sqrt{5^2 + 12^2} \\ &= \sqrt{25 + 144} \\ &= \sqrt{169} \\ &= 13 \end{aligned}$$

Verify that the unit line lengths for the pitch triangles shown in Figure 15.3a, b, and c are 12.65, 15, and 16.97, respectively.

In summary, the concepts used to describe the slope of a roof are:

Pitch triangle: right triangle found on a blueprint describing the amount of rise in a roof for 12 in. of run

Unit rise (URi): vertical leg of the pitch triangle, it describes the number of inches of rise for each 12 in. of run

Unit run (URu): horizontal leg of the pitch triangle, it always equals 12

Unit line length (ULL): The hypotenuse of the pitch triangle, describing the amount of travel along a rafter for each 12 in. of run

Comment: The terms "pitch" and "slope" when found in literature related to building construction are sometimes defined in a manner different from their use in the trade. "Pitch" is defined as the amount of rise divided by twice the span. However, for our purposes (and for the most part) these terms will be synonymous and will refer to the unit rise. Thus a "5-in. pitch" or a "5/12 pitch" or a "5-in. slope," and so on, will all mean a unit rise of 5 in.

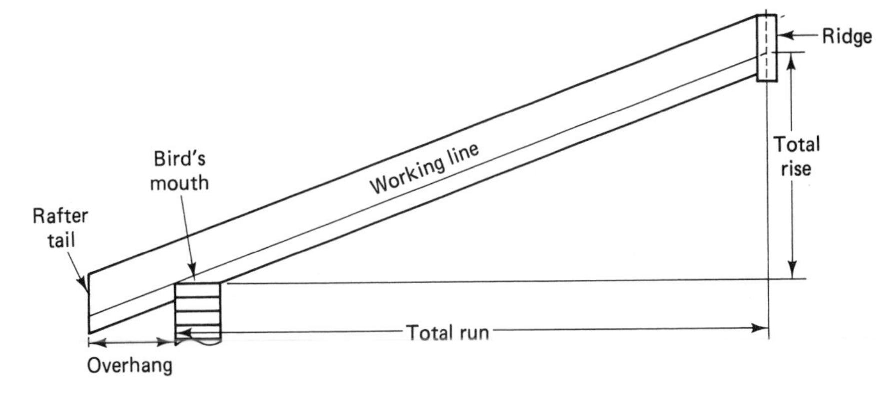

Figure 15-5

In Figure 15-5 are shown the working lines on which measurements are based when a common rafter is laid out. Notice that a right triangle is formed by the total run, the total rise and the line along the rafter itself extending from the bird's-mouth corner to the ridge. (The *bird's mouth* is the cut made to provide a notch where the rafter rests on the plate.) This triangle and the pitch triangle are similar triangles. (On a blueprint, in order to align the triangles so that corresponding parts appear in the same relative position, the pitch triangle has to be rotated top to bottom and left to right from its usual position along the slope of the roof.) It is important to note which sides are corresponding sides between the two triangles. From the pitch triangle to the actual roof, the unit line length corresponds to the rafter, the unit rise to the total rise, and the unit run to the total run.

Once the relation between the pitch triangle and the actual roof is understood, rafter length and total rise can be determined for any particular roof by using proportions involving a given unit rise and the run of the rafter.

EXAMPLE 15-1

Determine the common rafter length in a roof having a 5/12 pitch and a run of 14 ft.

***Solution*:** First, determine the unit line length. (Recall that this was found to be 13 for a 5/12 roof.) Second, set up a proportion involving the unit line length, the rafter length, and two other related, known parts. One possible proportion is

$$\frac{\text{RL}}{\text{ULL}} = \frac{\text{run}}{\text{URu}}$$

where RL is the rafter length and "run" is the run of the common rafter.

$$\frac{\text{RL}}{13} = \frac{14}{12}$$

$$\text{RL} = \frac{14 \times 13}{12}$$

$$= 15.167 \text{ ft or } 15 \text{ ft } 2 \text{ in.}$$

The procedure described in Example 15-1 may be shortened to a simpler formula. Notice in the next-to-last step, "12" appears as a divisor. Analysis of the units will show that the "12" also serves to convert the length to feet from inches. Elimination of this divisor will cause no harm and will simplify the formula to

$$\text{RL} = \text{ULL} \times \text{run}$$

Use of this formula will produce a rafter length in inches, convenient for layout on the rafter material. The rafter length may be easily converted to feet (by dividing by 12) for purposes of determining the length of material to be ordered.

EXAMPLE 15-2

Determine the rafter length and the total rise for a roof having a 6/12 pitch and a run of 15 ft 8 in.

***Solution*:** First, determine the unit line length.

$$\text{ULL} = \sqrt{6^2 + 12^2}$$
$$= 13.416$$

Second, determine the rafter length.

$$\text{RL} = \text{ULL} \times \text{run}$$
$$= 13.416 \times 15.667$$
$$= 210.188 \text{ in. or } 17.516 \text{ ft } (17 \text{ ft } 6\tfrac{3}{16} \text{ in.})$$

Total rise may be found either by a proportion or simply by multiplying the unit rise by the run. The roof rises 6 in. for each foot of run. This is the same principle on which the formula for rafter length was derived.

$$\text{Total rise} = \text{unit rise} \times \text{run}$$
$$= 6 \times 15.667$$
$$= 94 \text{ in.}$$

An examination of Figure 15-5 will reveal that total rise does not reflect actual headroom. Headroom is decreased both by the amount of material between the working line along the rafter and its lower edge and by the ceiling joists (and other materials making up possible flooring) which lie above the elevation of the top plate.

Overhang must be considered when calculating rafter length. We discuss overhangs more thoroughly in Chapter 16. The term ''overhang'' will mean the horizontal projection of the overhang. The horizontal projection will be measured from the vertical plane of the plumb cut on the bird's mouth to the vertical plane of the plumb cut on the tail of the rafter (see Figure 15-5.)

That length of rafter extending from the corner of the bird's-mouth cut to the tail of the rafter may be determined using the rafter-length formula established above; that is,

$$\text{RL} = \text{ULL} \times \text{run}$$

In this case, ''run'' will be the overhang, expressed in feet.

EXAMPLE 15-3

Determine the additional length of rafter needed for an 18-in. overhang on a 6/12 pitched roof.

***Solution*:** Notice that 18 in. equals 1.5 ft.

$$\text{RL} = \text{ULL} \times \text{run}$$
$$= 13.416 \times 1.5$$
$$= 20.125 \text{ or } 20\tfrac{1}{8} \text{ in.}$$

Another approach to this problem is to include the overhang as part of the run and calculate the total rafter length to include the overhang. The run of the rafter will be added to the overhang, both being expressed in feet. That is,

$$\text{RL} = \text{ULL} \times (\text{run} + \text{overhang})$$

To illustrate the use of this formula, refer to Examples 15-2 and 15-3, in which ULL = 13.416, run = 15.667, and overhang = 1.5. The formula above becomes

$$\text{RL} = 13.416 \times (15.667 + 1.5)$$
$$= 230.313 \text{ or } 230 \tfrac{5}{16} \text{ in.}$$

This is the same result when the answers to Examples 15-2 and 15-3 are added. That is,

$$210.188 + 20.125 = 230.313$$

When laying out a rafter pattern, three lengths should be calculated and checked against one another. These are:

1. Total rafter length (including overhang)
2. Distance from ridge to plumb cut of bird's mouth
3. Distance from plumb cut of bird's mouth to tail

After determining the lengths described above, a rafter pattern is laid out and cut for use in laying out the other common rafters. When the pattern is cut, certain adjustments are considered.

One of these adjustments is the position of the plumb cut at the ridge. If a ridge board is used, the plumb cut is moved to a position parallel to its original position, back toward the tail of the rafter a horizontal distance one-half the thickness of the ridge board. In other words, the plumb cut is moved back one-half the ridge-board thickness as measured along a line perpendicular to the plumb cut (see Figure 15-6).

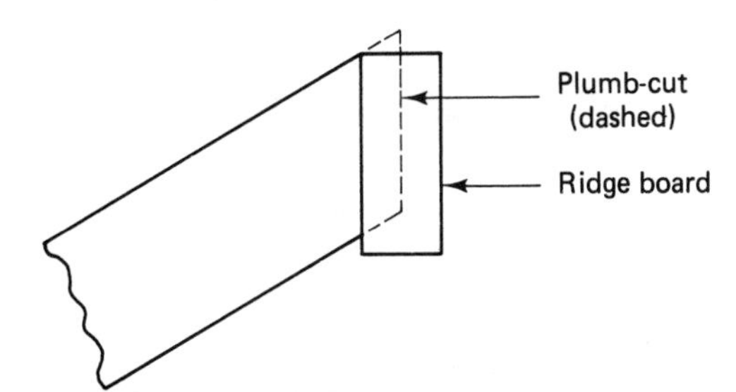

Figure 15-6

Another adjustment may take place at the tail of the rafter. A plumb cut is usually made here. Design of the overhang may dictate other types of cuts at the tail. Furthermore, some carpenters prefer to defer the tail cuts until the rafters are in place. Then a chalk line may be snapped to assure a straight overhang and compensate for any irregularities caused by cutting or placing the rafters.

Finally, consideration should be given to the depth of the bird's-mouth cut. Too much depth will weaken the overhang support. The seat cut should be of sufficient length to allow adequate nailing of the rafter to the plate. On the other hand, the corner of the bird's mouth should not project into the rafter more than one-third of its width.

To summarize rafter layout:

1. Determine the total rafter length, the distance from the ridge to the bird's mouth, and the distance from the bird's mouth to the tail.
2. Establish the depth of the bird's mouth and draw the working line along the rafter.
3. At the ridge end of the rafter, draw the plumb-cut line.
4. From the intersection of the working line with the ridge plumb cut, lay out the distance to the bird's mouth.
5. Draw the seat cut and plumb cut of the bird's mouth so that they intersect on the working line.

6. From the point of intersection of the bird's mouth with the working line, measure along the working line the distance to the rafter tail.
7. Draw the plumb-cut line for the tail to intersect the working line the proper distance from the bird's mouth. (This step may be postponed until the rafters are in place.)
8. Check the overall rafter length.
9. Make any compensating cuts at the ridge and/or tail.

Summarizing the mathematics covered to this point:

1. $\text{ULL} = \sqrt{\text{URi}^2 + 12^2}$, where

   ULL = unit line length

   URi = unit rise, the number of inches of rise per foot of run, and 12 is the unit run

2. $\text{RL} = \text{ULL} \times \text{run}$, where

   RL = rafter length, in inches

   ULL = unit line length

   run = run of the rafter, in feet, and may be chosen to include the overhang

3. $\text{TRi} = \text{URi} \times \text{run}$, where

   TRi = total rise, in inches

   URi = unit rise, in inches

   run = run of the rafter, in feet

The unit rise or pitch of a roof is not a known factor in every situation. In particular, suppose that an addition to an existing structure is planned and the roof lines are to be matched. The pitch of the existing roof must be determined. This may be done by either direct or indirect measurement.

One technique is to measure the total run and total rise of the existing roof. With this information, the total rise and total run can be proportioned to the unit rise and unit run, with the unit rise serving as the unknown.

EXAMPLE 15-4

The total rise and total run of a common rafter have been measured and found to be 5 ft 8 in. and 13 ft 9 in., respectively. Find the unit rise.

***Solution*:** Set up a proportion as follows:

$$\frac{\text{URi}}{12} = \frac{68}{165}$$

$$\text{URi} = 4.95 \text{ in.}$$

which converts to $4\frac{15}{16}$ in.

Notice that the total rise and total run were expressed in inches here. They could just as well have been expressed in decimal feet with no change in the outcome. In Example 15-4, if the measurements were actually taken, one would have to take notice of the lines along which the rafter was actually laid out. Two other methods for determining an existing pitch, perhaps more practical, are discussed.

Using a framing square and level, position one blade of the square so that the 12-in. mark coincides with the lower edge of the rafter. Holding the square in this position, place the level along this blade and move it to the level position. Read the unit rise on the corresponding edge of the other blade where it contacts the lower edge of the rafter (see Figure 15-7).

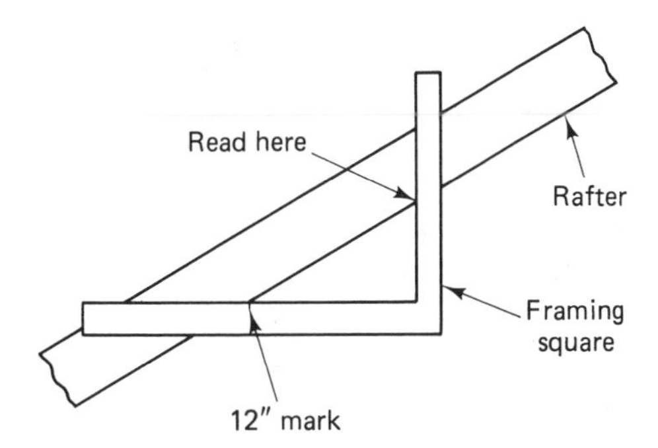

Figure 15-7

The last method for determining the pitch of an existing roof makes use of a transit (or any other type of angle finder) and trigonometry. (For a more thorough discussion of trigonometry, see Appendix B.)The **angle of elevation** is an angle made with the horizontal by rotating upward. The angle a roof makes with the horizontal is an angle of elevation. A transit is set up at a position where the line of sight is directly along the upward slope of the eave on the gable end of the building (see Figure 15-8). The angle of elevation may be read directly from the transit. A trigonometric function called the **tangent** of an angle, abbreviated "tan," measures the ratio of rise to run for an angle of elevation. If X represents the angle of elevation,

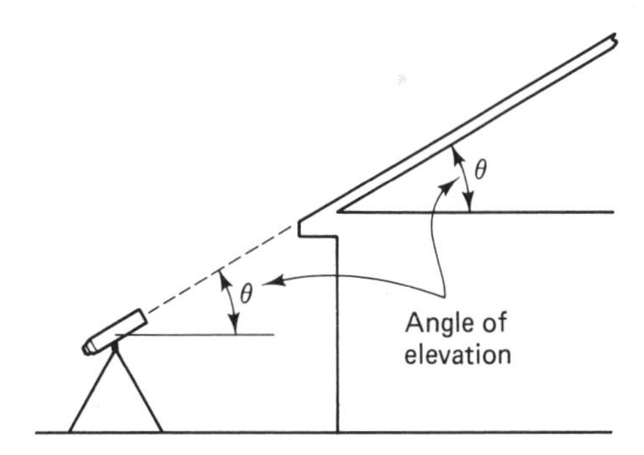

Figure 15-8

$$\tan X = \frac{\text{unit rise}}{\text{unit run}}$$

Values for tangents of angles may be found in a book of tables of trigonometric functions or from a scientific calculator.

EXAMPLE 15-5

Suppose that a transit has been set up and the angle of elevation of a roof measured. The angle was 22.6°. Find the unit rise for this roof.

Solution

$$\tan X = \frac{\text{URi}}{12}$$

$$\tan 22.6° = \frac{\text{URi}}{12}$$

$$0.4163 = \frac{URi}{12}$$

$$URi = 12 \times 0.4163$$

$$= 4.995 \quad \text{or} \quad 5, \text{rounded}$$

In Example 15-5, as with Examples 15-3 and 15-4, the unit rise can be used to determine the unit line length.

To conclude this chapter a final example is considered involving an unknown roof pitch. In this case, suppose that a roof is to be constructed with an off-center ridge—typical of a saltbox house. Assume that the pitch on the front side of the roof is known, as is the horizontal distance from the front of the building to the vertical plane of the ridge. The problem is to find the pitch of the back side of the roof so that rafter lengths may be calculated and proper cuts made.

EXAMPLE 15-6

A small building 12 ft wide is to be constructed having a roof with its ridge located 4 ft from the front of the building. The short side of the roof is to have a 10/12 pitch. Determine the unit rise on the rear section (see Figure 15-9).

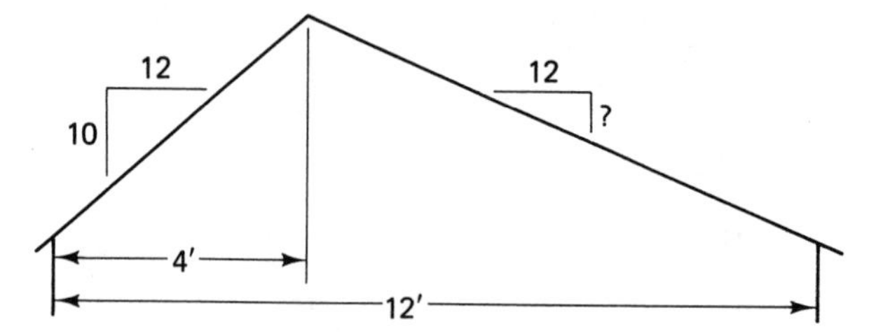

Figure 15-9

Solution: First, determine the total rise for the short roof segment.

$$TRi = URi \times \text{run}$$

$$= 10 \times 4$$

$$= 40 \text{ in.}$$

Second, set up and solve a proportion relating total rise and total run to unit rise and 12, the unit run.

$$\frac{URi}{12} = \frac{TRi}{TRu}$$

$$\frac{URi}{12} = \frac{40}{96}$$

$$URi = 5 \text{ in.}$$

Notice that the total rise and total run are expressed in the same units. Notice, too, that the total run used here is that of the rafter whose unit rise is being determined. Variations on this problem will be found among the exercises.

REVIEW EXERCISES

15-1. Listed below are various values for unit rise. For each, calculate the unit line length value correct to three decimal places.

Unit rise

(a) 3
(b) 7
(c) $4\frac{1}{2}$

(d) 10
(e) 9
(f) 15
(g) 12

15-2. How many decimal places should be used in the unit line length to obtain a tolerance of plus or minus $\frac{1}{16}$ in. in a rafter length?

15-3. Calculate the rafter length for each of the following unit rises and runs. Express answers in feet and inches, correct to the nearest $\frac{1}{16}$ in.

	Unit rise	Run
(a)	5	15 ft
(b)	$7\frac{1}{2}$	13 ft 6 in.
(c)	8	14 ft 8 in.
(d)	4	11 ft 5 in.
(e)	12	5 ft 7 in.
(f)	$9\frac{1}{4}$	10 ft $6\frac{1}{2}$ in.
(g)	$3\frac{1}{2}$	7 ft 4 in.

15-4. For each of the following situations, determine **(1)** the overall rafter length, **(2)** the length of rafter from the ridge to the bird's-mouth corner, and **(3)** the length of rafter from the bird's-mouth corner to the rafter tail. Express all answers in inches, correct to the nearest $\frac{1}{16}$ in.

	Unit rise	Run	Overhang
(a)	5	14 ft 4 in.	18 in.
(b)	7	8 ft 8 in.	6 in.
(c)	6	14 ft	36 in.
(d)	12	15 ft 9 in.	1 ft 4 in.
(e)	$8\frac{1}{2}$	7 ft 5 in.	8 in.

15-5. The total rise and total run (exclusive of overhang) are given for several roofs. In each case, calculate the unit rise, correct to the nearest $\frac{1}{8}$ in.

	Rise	Run
(a)	6 ft 3 in.	13 ft 5 in.
(b)	4 ft 9 in.	14 ft
(c)	14 ft	14 ft
(d)	9 ft 8 in.	16 ft 10 in.
(e)	5 ft 5 in.	11 ft 4 in.

15-6. The angle of elevation of various roofs has been determined and listed below. In each case, determine the unit rise, correct to the nearest $\frac{1}{16}$ in.

(a) 18.4° **(b)** 26.6° **(c)** 36.9°
(d) 30.3° **(e)** 38.4° **(f)** 45°

15-7. The shed roof shown here spans 20 ft. The front elevation is 3 ft 8 in. higher than the rear. What is the unit rise?

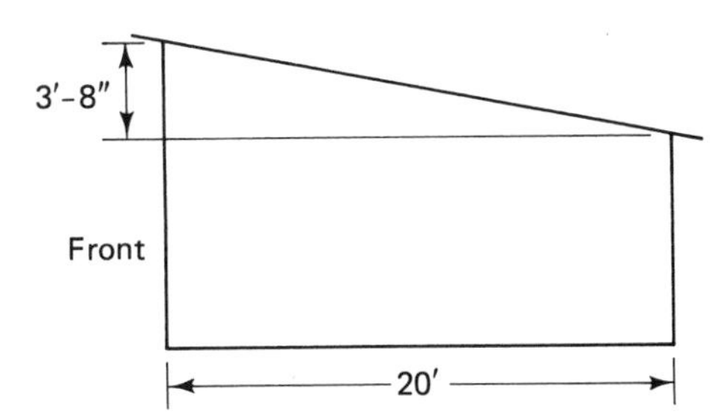

15-8. Refer to the building shown in the following figure. The building has an off-centered ridge located 8 ft from the front. The front roof section has a 9/12 pitch. The total building width is 22 ft.

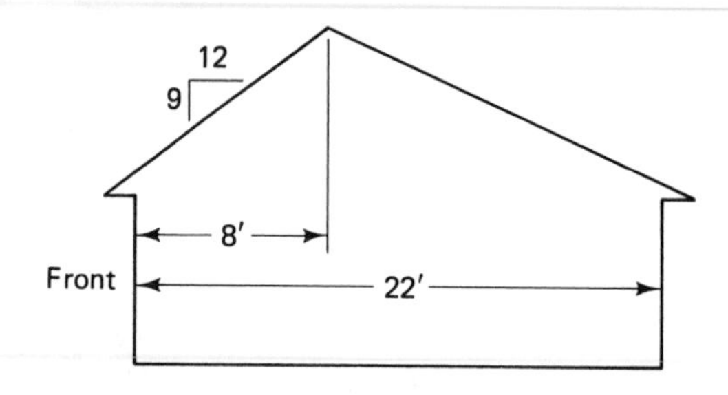

Determine the following dimensions.

(a) Unit rise for the rear roof section

(b) Unit line length for the rear

(c) Total rafter length for the rear section, including a 16-in. overhang

(d) Total rafter length for the front section, including a 16-in. overhang

(e) Distance along both rafters from the ridge to the bird's-mouth corner

(f) Distance along both rafters from the bird's-mouth corner to the plumb cut at the tail

15-9. Refer to the following figure. The front section of the roof has a 10-in. unit rise. Notice the different elevations of the front and rear walls.

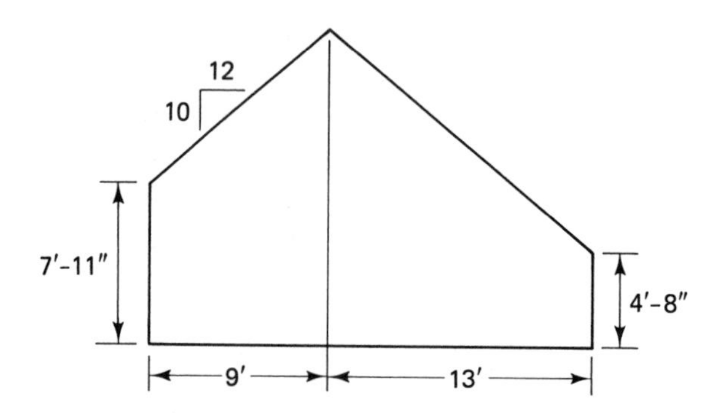

Determine the following dimensions.

(a) Total rise for both the front and rear rafters

(b) Unit rise for the rear rafter

(c) Unit line length for both the front and rear rafters

(d) Total rafter lengths for both front and rear (no overhang; answer in feet and inches, nearest $\frac{1}{16}$ in.)

15-10. The second story of a Cape Cod–style house is to be framed with knee walls 6 ft high. The width of the house is 28 ft and the roof has a 10-in. pitch. Assume that 1 ft of height is occupied by the second floor joists and that part of the rafter below the working line. How wide will the room be?

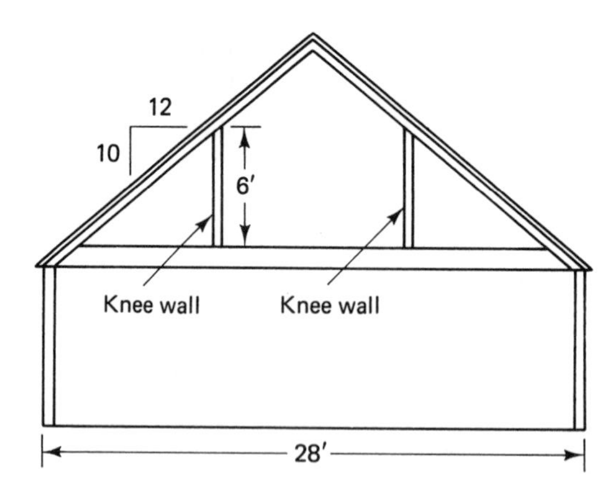

15-11. Refer to Problem 15-10. Suppose that it is desired to have a width of 20 ft for the second-story room. Assuming that the same 1 ft of height is lost from the theoretical rise, what will be the height of the knee wall?

chapter 16

Overhangs

OBJECTIVES

Upon completing this chapter, the student will be able to:

1. Calculate the vertical projection of an overhang for a given horizontal projection.
2. Calculate the horizontal projection of an overhang for a given vertical projection.
3. Determine the amount of overhang necessary for a given solar exposure.
4. Determine the amount of material needed to frame and enclose an overhang.

Overhang on a building serves various functions. Included among these are aesthetics, solar exposure, and rainwater control. The appearance of a building can be altered considerably by the style and amount of overhang. A little extra money spent on expanding the overhang of a small ranch house can greatly enhance its eye appeal. On the other hand, the classic Cape Cod–style house is quite attractive with very little overhang. The steepness of a roof affects the amount of overhang. In general, the steeper the roof, the shorter the overhang, avoiding interference with window and door tops.

Today's emphasis on solar considerations in building design has resulted in considering the sun's angle at various times of the year. The amount of overhang may be designed so as to minimize solar exposure during summer months and maximize it during winter months.

In Chapter 15 we established the mathematics related to problems considered here.

A proportion is a useful tool for answering questions related to overhangs. Since the overhang is usually an extension of the roof rafters, the amount of horizontal and vertical projection is dictated by the pitch of the roof. Recall that the unit rise, URi, and the unit run (which is always 12) control the roof angle. So, too, do these values control the overhang.

EXAMPLE 16-1

A roof has a 5/12 pitch. If a 16-in. horizontal projection of the overhang is desired, what will be the vertical projection?

Comment: The term *horizontal projection* will refer to the distance that the overhang projects outward perpendicular to the framed exterior wall. The term *vertical projection* will refer to the distance the overhang projects vertically, downward from the top plate (see Figure 16-1).

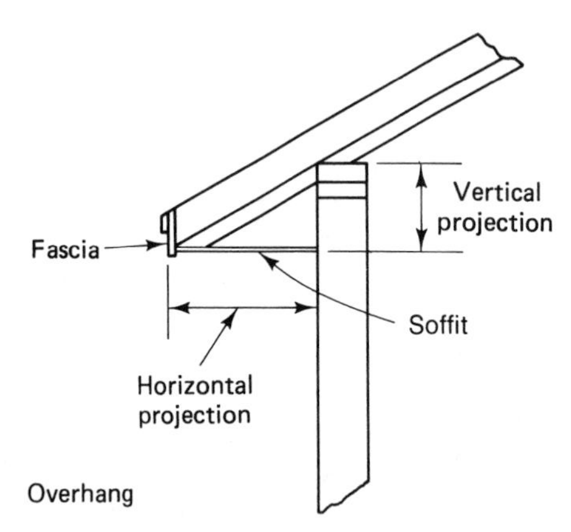

Overhang

Figure 16-1

Solution: Set up a proportion relating the unit rise and unit run to the vertical and horizontal projections, respectively.

$$\frac{\text{URi}}{12} = \frac{\text{vertical projection}}{\text{horizontal}}$$

$$\frac{5}{12} = \frac{\text{vertical projection}}{16}$$

$$\text{vertical projection} = 6.67 \quad \text{or} \quad 6\tfrac{11}{16} \text{ in.}$$

A careful examination of Figure 16-1 reveals that the vertical projection is actually measured with relation to the working line of the rafter. Different types of cuts made on the tail of the rafter may necessitate adjustments to calculated projection lengths.

The material used to cover the underside of an overhang is called the **soffit.** The material used to cover the vertical face of the overhang is called the **fascia** (see Figure 16-1). If the distance between door tops and/or window tops and the overhang is under consideration, the thickness of the soffit material must be included in the vertical projection of an overhang.

On certain types of houses, the exterior appearance is enhanced if the horizontal projection of the overhang returns to the front of the building at the door and window tops. If this is a design requirement, the elevation of the top of the door frame trim must be determined. The difference between this elevation and the elevation of the top plate will dictate the vertical projection, soffit thickness included (see Figure 16-2).

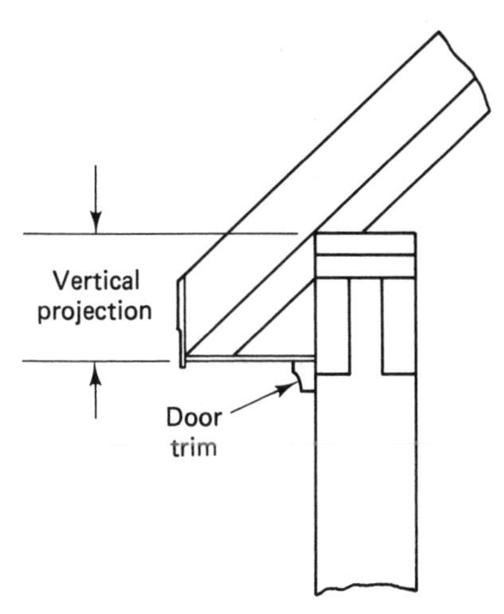

Figure 16-2

EXAMPLE 16-2

The top of the exterior door trim will be $12\frac{3}{4}$ in. below the top plate of the framed exterior wall. The soffit material will be $\frac{3}{8}$-in. A-C exterior-grade plywood. The roof carries a 6-in. pitch. Find the horizontal projection of the overhang (see Figure 16-3).

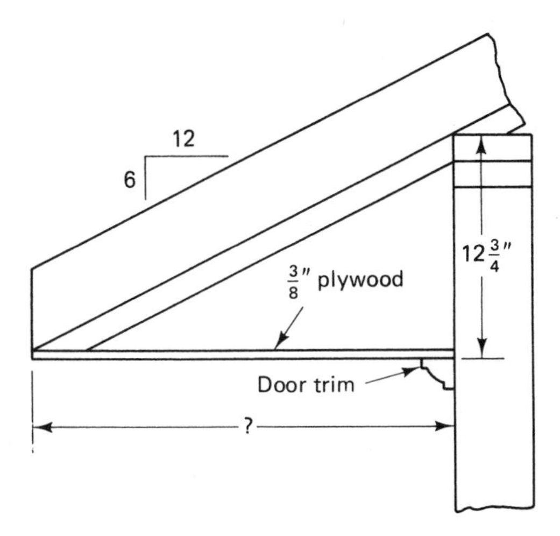

Figure 16-3

***Solution*:** The vertical projection as determined from the working line of the rafter will be $12\frac{3}{8}$ in. This is the result of subtracting the soffit thickness from the $12\frac{3}{4}$ in. available for the vertical projection. A proportion will determine the horizontal projection.

$$\frac{\text{URi}}{12} = \frac{\text{vertical projection}}{\text{horizontal projection}}$$

$$\frac{6}{12} = \frac{12.375}{\text{horizontal projection}}$$

$$\text{horizontal projection} = 24.75 \quad \text{or} \quad 24\tfrac{3}{4} \text{ in.}$$

Examples 16-1 and 16-2 assume that the roof pitch dictates the overhang design. In most cases this is true. However, under some circumstances design experimentation may be desirable, letting the amount of overhang dictate the pitch of the roof. This may be the case when solar exposure is a factor. The mathematics does not change in this situation—only the unknown changes.

For example, if it has been determined that the top of the door trim will be met by the horizontal projection return and that a certain amount of horizontal projection is optimal for solar exposure, these are the terms that will be proportioned to the unit rise and unit run. The unit rise will be the unknown and the pitch thus determined.

EXAMPLE 16-3

Solar analysis has determined that a horizontal projection of 18 in. will optimize exposure to the sun during the various seasons. Aesthetic quality requires the horizontal projection return 4 in. above the door trim elevation. Analysis of the framing plan reveals that the horizontal projection meets the wall $10\frac{1}{2}$ in. below the top plate. Find the unit rise.

Solution: A proportion is set up as shown below.

$$\frac{\text{URi}}{12} = \frac{10.5}{18}$$

$$\text{URi} = 7$$

Therefore, this calls for a 7/12 pitch.

It has been mentioned that overhangs can be designed to permit maximum solar gain during the winter months and afford shading during the summer months. The path of the sun as it travels across the sky reaches its highest altitude on June 21 and its lowest December 21. The altitude (height above the horizon) and azimuth (position from east to west) of the sun vary with the time of year at various latitudes. Specific information about these matters is available from various sources. Consult the Bibliography.

Since the path the sun travels across the sky changes each day, overhang design is usually based on the location of the sun on a particular day of the year. The day chosen might be when the sun reaches its highest or lowest elevation. A day partway between these dates may be preferred. The midsummer point, when temperatures are approaching their highest levels, may be the time when maximum shading is desired. A similar argument might be made for selecting a design date for solar gain between December 21 and June 21.

The collection of necessary data and calculations determining the sun's angles are matters not under consideration here. However, once these factors have been determined, how they relate to overhang design is a question to be pursued.

In Chapter 15, trigonometry was used to determine the pitch of a roof from its angle of elevation. The tangent function was the tool. Recall that the tangent of an angle expresses the ratio of rise to run. Thus an angle of 30° has a tangent of 0.5774. That is,

$$\tan 30° = 0.5774$$

Recall, also, that this value may be proportioned to the ratio of rise to run. If run is the unit run of 12, the unit rise can be determined.

$$\frac{\text{URi}}{12} = 0.5774$$

$$\text{URi} = 6.929 \quad \text{or} \quad 6\tfrac{15}{16} \text{ in.}$$

EXAMPLE 16-4

It has been determined that the elevation angle of the sun is 72° on the date that has been selected as the basis for solar shading. At this time of day, it is desired to keep the sun's rays below a horizontal line two-thirds of the distance from the top to bottom of the windows on the exposed side of the house. The window sizes

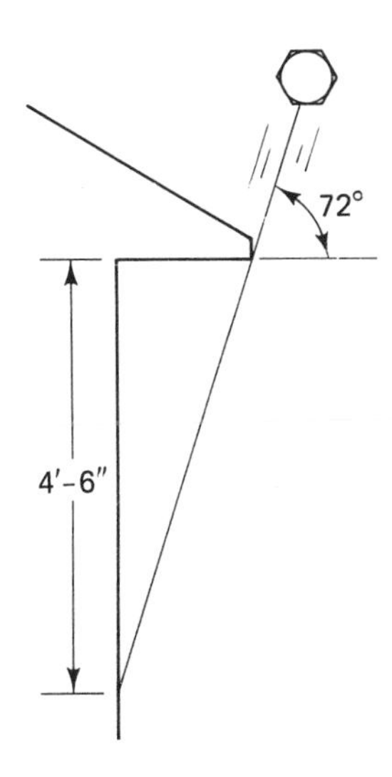

Figure 16-4

set this line 4 ft 6 in. below the soffit (see Figure 16-4). What should the horizontal projection be?

Solution: In the study of plane geometry it is established that when two lines intersect, vertical angles (opposing angles) are equal. In Figure 16-4 the 72° angle of elevation of the sun is transferred to the shade angle with reference to the soffit. The distance from the base of the soffit to the horizontal shade line on the windows becomes the amount of rise and the horizontal projection of the overhang becomes the run. Using the tangent function gives us

$$\tan 72^\circ = \frac{\text{rise}}{\text{run}}$$

$$\tan 72^\circ = \frac{4.5}{\text{run}}$$

$$3.0777 = \frac{4.5}{\text{run}}$$

$$\text{run} = \frac{4.5}{3.0777}$$

$$= 1.462 \text{ ft} \quad \text{or} \quad 1 \text{ ft } 5\tfrac{9}{16} \text{ in.}$$

In summary, problems dealing with overhangs are readily solved using a proper proportion or the tangent function. Care should be taken in setting up the proportion so that the terms are consistent; that is, terms describing rise should correspond with vertical distances and terms describing run should correspond with horizontal distances. In general, the following relations are used:

$$\frac{\text{unit rise}}{12} = \frac{\text{vertical projection}}{\text{horizontal projection}}$$

$$\tan X = \frac{\text{rise}}{\text{run}}$$

Materials needed to frame and enclose overhangs vary with type and style of overhang. This makes the subject difficult to generalize. Some house styles, such as a steep-roofed cape, require only that the rafter ends be covered with a fascia board. A ranch-style house might require more elaborate construction of a horizontal return that is to be covered with both soffit and fascia material. There may or may not be an overhang along the sloping eave of a roof edge.

Despite these and other variables, once the style of overhang has been determined, the efficient estimator will make use of many predetermined measurements and calculations to facilitate determining the materials needed. Two of the more frequently used measurements used in the estimation of overhang materials are the length of the building and the rafter length. Overhang width is used in conjunction with these dimensions to determine a total of square feet of soffit material needed. The same measurements are used to determine amounts of fascia material. The width of fascia material is dictated by the length of the plumb cut on the end of a roof rafter.

The estimator will find it useful to calculate the entire distance around the roof edge. This figure can be used to determine total lengths of fascia and/or trim boards needed to cover those areas. If the horizontal projection is the same on all roof edges, this figure can be multiplied by the width of the horizontal projection to obtain a total area to be covered with soffit material. That total can be translated into a number of pieces of sheet goods if appropriate. Some of these same dimensions will be useful in estimating materials needed to frame the overhang.

REVIEW EXERCISES

16-1. Listed below are various roof pitches and a given horizontal projection of overhang. In each case, determine the vertical projection.

	Pitch	Horizontal projection
(a)	5/12	16 in.
(b)	7/12	9 in.
(c)	12/12	24 in.
(d)	12/12	5 in.
(e)	10/12	8 in.
(f)	6/12	2 ft 4 in.

16-2. Listed below are various roof pitches and a given vertical projection of overhang. In each case, determine the horizontal projection.

	Pitch	Vertical projection
(a)	8/12	12 in.
(b)	9/12	16 in.
(c)	4/12	2 ft 6 in.
(d)	15/12	5 in.
(e)	10/12	8 in.
(f)	5/12	1 ft 9 in.

16-3. Plans call for the return of the horizontal projection of an overhang to meet the wall at a point $11\frac{3}{8}$ in. below the top plate. The roof has an 8-in. pitch. Find the length of the horizontal projection of the overhang.

16-4. The top of the exterior trim around a door is $7\frac{5}{8}$ in. below the top plate. The soffit is $\frac{3}{8}$-in. A-C plywood. The overhang is to return to the top of the door trim. The roof carries a 6/12 pitch. What is the length of the horizontal projection of the overhang?

16-5. Plans call for the horizontal projection of the overhang to meet the exterior wall at a point $8\frac{1}{2}$ in. below the top plate. Determine the roof pitch for the following horizontal projection overhang measures.

(a) 6 in. **(b)** 10 in. **(c)** 18 in. **(d)** 24 in.

16-6. Plans specify a 10/12 pitch and a 6-in. horizontal projection of the overhang. The rafter tails are to be left uncut, except for proper length. The rafter material is 2 × 10, with an actual measurement of $9\frac{1}{2}$ in. The working line is located one-third of the distance from the rafter bottom. If the lowest point on the rafter tail is projected horizontally back to the wall, how far below the top plate will it fall? See Figure.

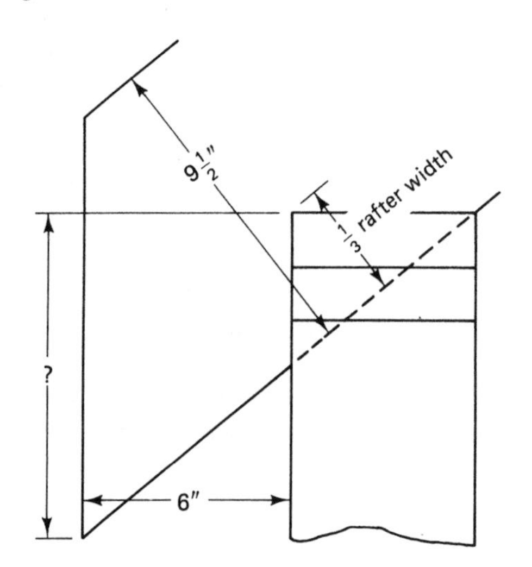

16-7. Repeat Problem 16-6 if the roof has a 7/12 pitch, the horizontal projection measures 16 in., and the rafters are 2 × 8's, measuring $7\frac{1}{2}$ in.

16-8. A roof has a 9/12 pitch and a horizontal projection of the overhang measuring 8 in. When the angle of elevation of the sun is 65°, how long will the shadow of the overhang be as measured from the return of the overhang downward?

16-9. Solar analysis has determined that when the angle of elevation of the sun is 70°, the overhang should cast a shadow $3\frac{1}{2}$ ft below the line where the horizontal projection meets the wall. How long (to the nearest $\frac{1}{8}$ in.) should the horizontal projection be?

16-10. In Problem 16-9, if the roof has a 5/12 pitch, what will be the distance from the top plate to the line where the horizontal projection returns?

16-11. In Problem 16-9, if the horizontal projection returns to a line 12 in. below the top plate, what is the roof pitch?

chapter 17

Roofs II: Hip, Valley, and Jack Rafters

OBJECTIVES

Upon completing this chapter, the student will be able to:

1. Calculate the length of a conventional hip rafter.
2. Calculate the length of an unconventional hip rafter.
3. Determine the angles for cuts at the ridge, bird's mouth, and tail of both conventional and unconventional hip rafters.
4. Calculate the length of a valley rafter.
5. Determine the angles for cuts at the ridge, bird's mouth, and tail of a valley rafter.
6. Calculate the incremental change in the length of jack rafters.

Hip rafters may be classified as **conventional** or **unconventional.** A conventional hip rafter is preferred in construction—both for simplicity and aesthetic appeal. Some situations, however, may dictate the use of an unconventional hip rafter. For example, in attaching a porch to the face of a house, the roof design may require an unconventional hip to avoid interference with a second-story window.

The difference between a conventional and an unconventional hip rafter is the angle the run of the hip makes with the two adjacent walls forming the corner of the building. The angle referred to lies in the plane across the top plates. With a conventional hip rafter, this angle is 45°. An unconventional hip rafter may describe any other angle between its run and one of the walls forming the corner of the building (see Figure 17-1).

With a conventional hip roof, the unit rise and unit run for the hip rafter are determined from the common rafter. Beginning at a corner of the building, the first full-length common rafter will be placed a distance along the top plate equal to one-half the building

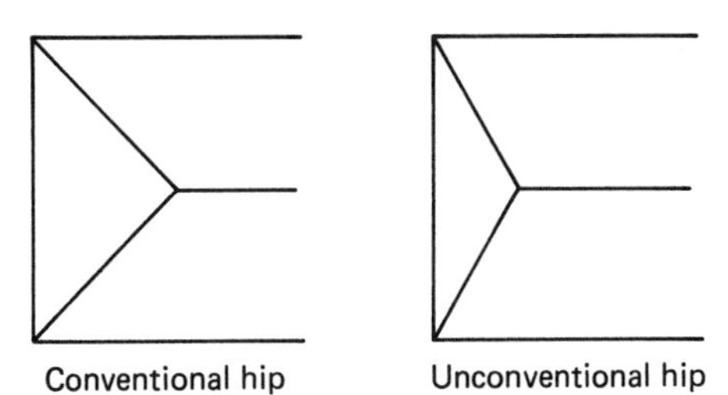

Figure 17-1

width. This distance is also the run of the common rafter (see Figure 17-2). The hip rafter will extend from the corner of the building to the point on the ridge where the common rafter is attached. (Bear in mind that all references to distances are with respect to the working lines on the rafters. In a plan view of a roof, the working lines appear along the center of the rafter's top edge.)

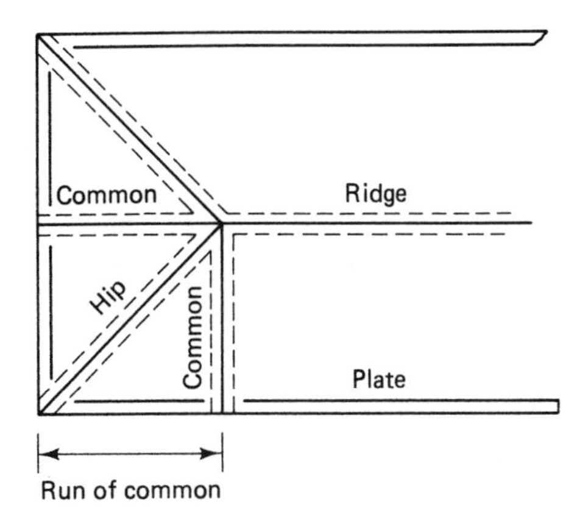

Figure 17-2

An examination of Figure 17-2 will reveal that a right triangle is formed by the hip rafter, the common rafter, and the distance along the plate from the corner to the common rafter (that is, the run of the common). It will be convenient later when unconventional hips are considered to think of this distance along the plate from the corner to the first full-length common as the **setback.** For a conventional hip roof, then, the setback equals the run of the common.

This right triangle may be used as the basis for calculating the length of the hip rafter. The hip rafter is the hypotenuse while the common rafter and its run are the legs. Using the Pythagorean rule,

$$\text{HRL} = \sqrt{\text{CRL}^2 + \text{SB}^2}$$

where HRL is the hip rafter length, CRL the common rafter length, and SB the setback (run of common in this case). The units should all be feet or all be inches.

EXAMPLE 17-1

Find the length of a hip rafter if the common rafter measures 15 ft 2 in. and the run of the common measures 14 ft.

***Solution*:** Use the right-triangle approach:

$$\begin{aligned}\text{HRL} &= \sqrt{\text{CRL}^2 + \text{SB}^2}\\ &= \sqrt{15.167^2 + 14^2}\\ &= \sqrt{230.038 + 196}\\ &= \sqrt{426.038}\\ &= 20.641 \text{ ft} \quad \text{or} \quad 20 \text{ ft } 7\tfrac{11}{16} \text{ in.}\end{aligned}$$

In addition to determining the length of the hip rafter, the ratio of rise to run (slope) must be found in order to make the proper plumb cuts at the tail and ridge, as well as the bird's-mouth cut. It should be noted that the slope of the hip rafter is not the same as that of the common. This becomes obvious when it is noted that the hip must travel a greater distance than the common in order to rise to the same elevation.

Since both the common and the hip have the same total rise, the unit rise is taken as the same for both and the run of the common used to calculate the run of the hip. The run of the hip becomes its unit run.

For a conventional hip roof, the relation between the run of the hip and the run of the common is described by the 45° right triangle. Indeed, the setback (which equals the run of the common), the run of the common, and the run of the hip form a 45° right triangle, the run of the hip being the hypotenuse. Recall that the ratio of the sides in a 45° right triangle is $1:1:\sqrt{2}$. Therefore, multiplying the unit run for the common by 1.414 (which is $\sqrt{2}$) will result in the unit run for the hip. Thus

$$1.414 \times 12 = 16.968$$

Many carpenters round 16.968 to 16.97 or even 17. For purposes of using a framing square to draw the lines for cuts on a hip rafter, the ratio 5/17 is sufficiently accurate. However, when calculating a unit line length to be used in determining rafter length, 16.97 should be used. A significant error is incurred if 17 is used instead. The end result is that when, for example, the common rafter has a pitch of 5/12, the hip rafter will have a pitch of 5/17 (or 5/16.97). A common rafter with a 7/12 pitch would result in a hip with a 7/17 pitch.

In summary, to calculate a conventional hip rafter length, use

$$\text{HRL} = \sqrt{\text{CRL}^2 + \text{SB}^2}$$

where HRL is the hip rafter length in feet, CRL the common rafter length in feet, and SB the setback, the run of the common in this case, in feet. To make the proper cuts on the hip rafter, use

$$\frac{\text{URi}}{17}$$

where URi is the unit rise of the common rafter.

EXAMPLE 17-2

A conventional hip roof is to have a 6/12 slope. The run of the common rafter is 12 ft 6 in. Determine (a) the length of the common rafter, (b) the length of the hip rafter, and (c) the slope of the hip rafter.

***Solution*:** (a) For a 6/12 slope, the unit line length is

$$\begin{aligned} \text{ULL} &= \sqrt{6^2 + 12^2} \\ &= 13.416 \end{aligned}$$

The common rafter length, CRL, is

$$\begin{aligned} \text{CRL} &= \text{ULL} \times \text{run} \\ &= 13.416 \times 12.5 \\ &= 167.7 \text{ in. or } 13 \text{ ft } 11\tfrac{11}{16} \text{ in.} \end{aligned}$$

(b) The length of the hip rafter (HRL) is found by solving the right triangle. The hip rafter is the hypotenuse, the common rafter and its run (setback) are the legs.

$$\begin{aligned} HRL &= \sqrt{CRL^2 + SB^2} \\ &= \sqrt{13.975^2 + 12.5^2} \\ &= \sqrt{351.551} \\ &= 18.750 \text{ ft} \quad \text{or} \quad 18 \text{ ft } 9 \text{ in.} \end{aligned}$$

(c) The pitch of the hip rafter is 6/17. Remember that with a conventional hip rafter the unit rise is the same as that for the common and the unit run becomes a constant of 17 (that is, $\sqrt{2} \times 12$).

A second method for calculating the length of a conventional hip rafter is sometimes preferred. It is consistent with the method developed in Chapter 15. In this method the unit line length (ULL) for the hip rafter is calculated from the pitch triangle. Then this is multiplied by the run of the common rafter.

EXAMPLE 17-3

Referring to Example 17-2, the unit line length for the hip rafter is

$$\begin{aligned} ULL &= \sqrt{URi^2 + URu^2} \\ &= \sqrt{6^2 + 16.97^2} \\ &= \sqrt{323.98} \\ &= 18 \end{aligned}$$

The unit line length is now multiplied by the run of the common, CRu.

$$\begin{aligned} HRL &= ULL \times CRu \\ &= 18 \times 12.5 \\ &= 225 \text{ in.} \quad \text{or} \quad 18 \text{ ft } 9 \text{ in.} \end{aligned}$$

Which of these two methods is used to calculate the length of a hip rafter is, of course, the choice of the carpenter. However, the first method, in which the hip rafter is treated as the hypotenuse of a right triangle with the setback and the run of the common as legs, is more general. Because of that it lends itself to determining the length of an unconventional hip rafter. The second method cannot be used on an unconventional hip since the 45° right triangle does not exist.

EXAMPLE 17-4

Find the length of a hip rafter if the common rafter is to have a 9/12 pitch, a run of 15 ft, and the setback is 12 ft (see Figure 17-3).

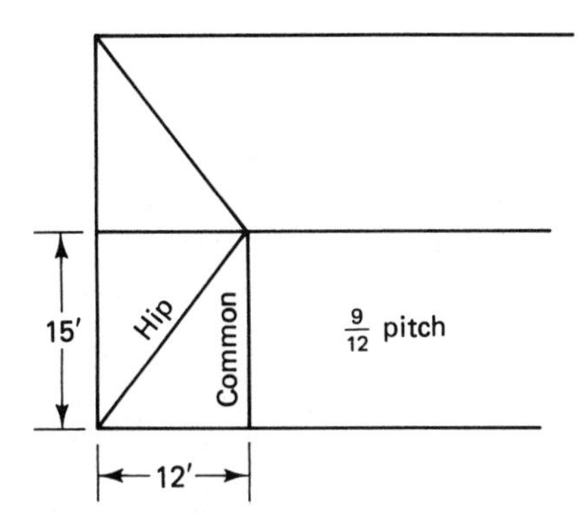

Figure 17-3

***Solution*:** First, the length of the common rafter is found. For a 9/12 pitch, the unit line length is 15. With a run of 15 ft the common rafter length is

$$\begin{aligned} \text{CRL} &= \text{ULL} \times \text{run} \\ &= 15 \times 15 \\ &= 225 \text{ in. or } 18 \text{ ft } 9 \text{ in.} \end{aligned}$$

Second, the hip rafter length is calculated using the right-triangle approach. The hip is the hypotenuse and the legs are the setback and the common rafter.

$$\begin{aligned} \text{HRL} &= \sqrt{\text{CRL}^2 + \text{SB}^2} \\ &= \sqrt{18.75^2 + 12^2} \\ &= \sqrt{495.56} \\ &= 22.261 \text{ ft or } 22 \text{ ft } 3\tfrac{1}{8} \text{ in.} \end{aligned}$$

Valley rafters are placed where two intersecting roofs meet. There are ways of constructing this intersection so that no valley rafter is needed. Essentially, this is accomplished by completing one roof plane, properly supported, to the point where it has been sheathed. The intersecting roof is then constructed and extended to meet the face of the sheathed section.

Assuming that valley rafters are to be used, the method for determining their length depends on whether the adjacent roofs have the same pitch. If this is the case, the valley rafter will have a run that makes a 45° angle with the plate, as did the hip rafter. This is called a **conventional valley rafter.** Nevertheless, some differences from the hip rafter calculations occur. In cases where the adjacent roofs have different pitches, a right-triangle approach is used. This is called an **unconventional valley rafter.**

The conventional valley rafter is most common and also has a more pleasing appearance than the unconventional valley rafter. If the adjacent building sections have the same width, their ridges will have the same elevation and will meet at a point. On the other hand, different widths will cause one ridge to run into the face of the other roof or rise above and terminate at a point above the other ridge (see Figure 17-4).

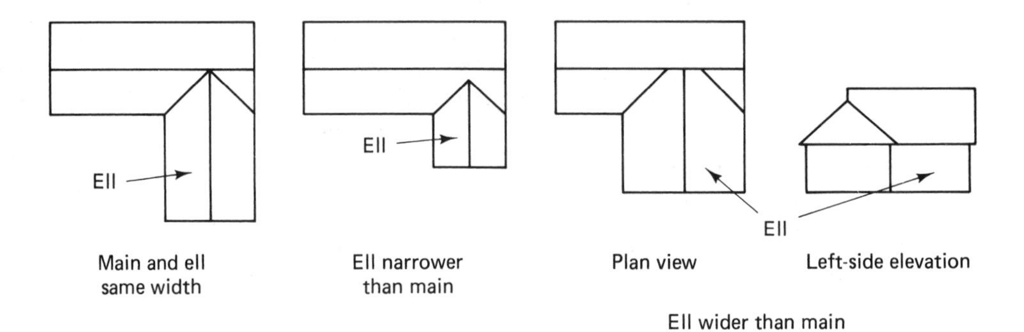

Figure 17-4

A standard way of placing the conventional valley rafters when the ell is narrower than the main building is shown in Figure 17-5. Notice that a long valley rafter extends to the main ridge, while a valley rafter of normal length meets the long valley at the ridge of the ell. The longer valley rafter must extend to meet the main ridge. This will happen if the same approach is used as for a conventional hip rafter. The valley rafter length VRL, then, is

$$\text{VRL} = \text{ULL} \times \text{CRu}$$

where $\text{ULL} = \sqrt{\text{URi}^2 + 16.97^2}$ and CRu is the run of the common on the main section.

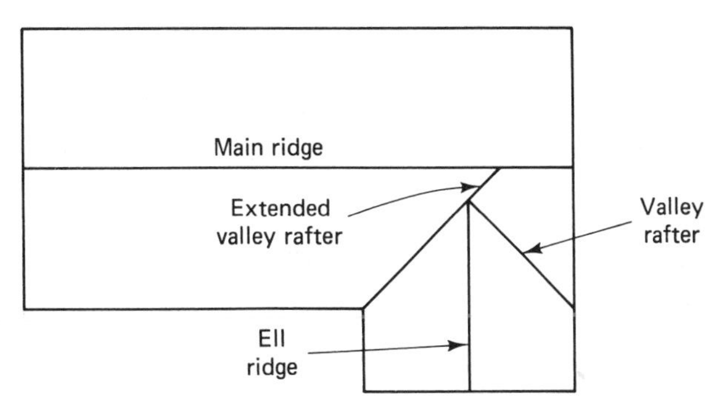

Figure 17-5

For the shorter valley rafter, the same formula is used with the exception that CRu represents the run of the common on the ell.

EXAMPLE 17-5

Referring to Figure 17-5, suppose that the main building is 26 ft wide and the ell is 20 ft wide. Both roofs have a 5/12 slope. Find the lengths of both the long and the short valley rafters.

***Solution*:** Determine the unit line length for the valley rafters:

$$\text{ULL} = \sqrt{5^2 + 16.97^2}$$
$$= \sqrt{312.98}$$
$$= 17.69$$

For the long valley rafter,

$$\text{VRL} = 17.69 \times 13$$
$$= 229.97 \text{ in.} \quad \text{or} \quad 19 \text{ ft } 2 \text{ in.}$$

For the short valley rafter,

$$\text{VRL} = 17.69 \times 10$$
$$= 176.9 \text{ in.} \quad \text{or} \quad 14 \text{ ft } 8\tfrac{7}{8} \text{ in.}$$

At the ridge, the long valley rafter will have two types of angles making up the plumb cut. The plumb cut is made using the 5/17 line. But the rafter meets the main ridge at a 45° angle. Therefore, the plumb cut travels at a 45° angle through the lumber. Of course, the length of this rafter must be reduced by one-half the thickness of the ridge board. This reduction is measured parallel to the 45° line. The short rafter meets the long rafter at a 90° angle. Only the 5/17 angle is scribed; however, one-half the thickness of the long rafter is cut from the length of the short valley rafter. Valley rafter lengths are extended for overhangs by applying the formulas described above, using the horizontal projection of the overhang in place of the run of the common.

Unconventional valley rafters are created by several situations. The common ingredient that is present in all unconventional valleys is a different pitch on the adjacent roof sections. One situation in which this can occur is an ell of different width than the main building whose ridge intersects the main ridge. For example, suppose that a 22-ft-wide ell is attached to a 28-ft-wide building having a 5/12 slope, so that the roof ridges meet at the same elevation (see Figure 17-6). The total rise of the main roof is

$$\text{TRi} = \text{URi} \times \text{run}$$
$$= 5 \times 14$$
$$= 70 \text{ in.} \quad \text{or} \quad 5.83 \text{ ft}$$

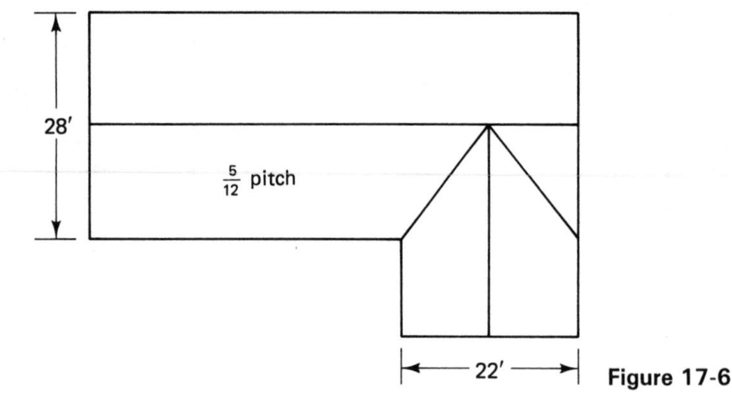

Figure 17-6

The ell must have the same total rise but with a total run of 11 ft. Proportioning this to determine the unit rise for the ell,

$$\frac{\text{URi}}{\text{URu}} = \frac{\text{TRi}}{\text{TRu}}$$

$$\frac{\text{URi}}{12} = \frac{5.83}{11}$$

$$\text{URi} = 6.36 \quad \text{or} \quad 6\tfrac{3}{8} \text{ in.}$$

Thus, while the main section has a 5/12 pitch, the ell has a 6.36/12 pitch.

The simplest way to calculate the valley rafter length, VRL, in the unconventional case is to identify a right triangle in which the valley rafter serves as the hypotenuse and use the Pythagorean rule. For the building described in Figure 17-6, the right triangle is formed by the valley rafter, with the legs being the main common rafter length, MCRL, and the run of the ell common rafter, ECRu. For convenience the latter distance is transferred along the main ridge from the point where the last full-length common rafter meets the ridge to where the valley rafter meets the ridge (see Figure 17-7).

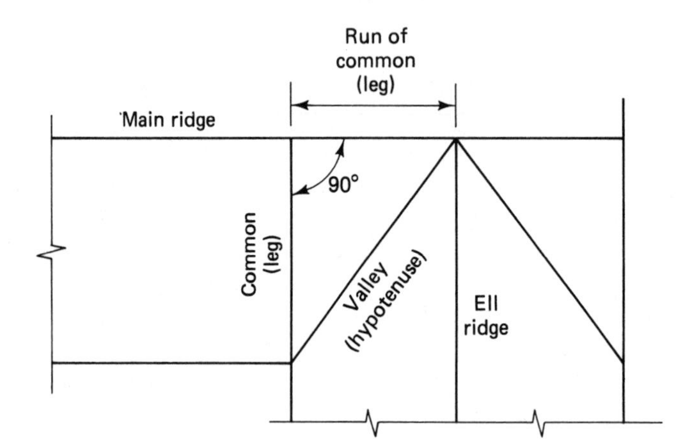

Figure 17-7

The importance of correctly identifying the parts of a right triangle in calculating any rafter length is worthy of mention. In general, more than one right triangle can be identified and used. In the preceding example, the legs could have been the common rafter for the ell and the run of the common for the main. (Still another right triangle exists. Can you find it?) In all cases, it is the rafter itself (valley, hip, or common) which serves as the hypotenuse of the right triangle.

Jack rafters are segments of common rafters whose length has been reduced by a hip or valley rafter cutting diagonally through their path. Figure 17-8 shows hip and valley jack rafters.

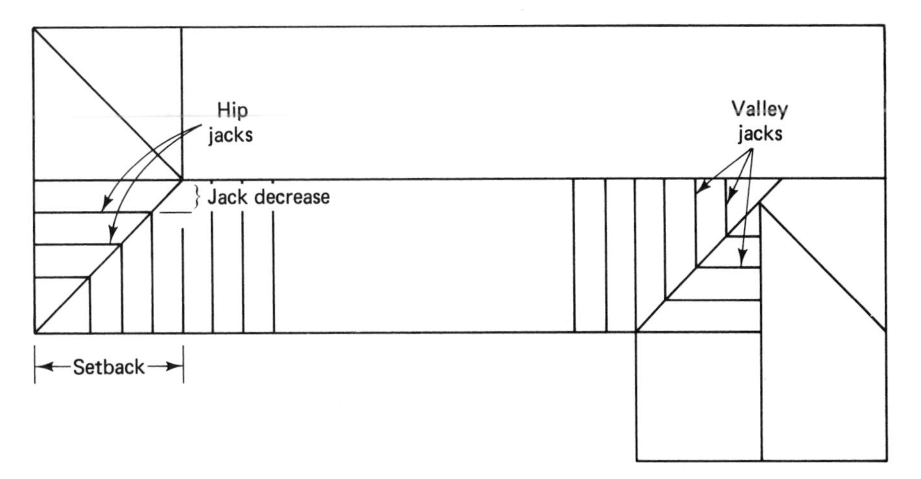

Figure 17-8

Since the spacing of rafters is uniform, the decrease in length of successive jack rafters is constant. Should it be necessary to place a jack rafter in a position that changes the spacing, the amount of decrease will change. However, the method for calculating the amount of decrease does not change.

Consider the jack rafters for a hip roof. The last full-length common rafter and the setback form the basis for a proportion from which the decrease in length of the first jack rafter can be determined. By the ''first jack rafter'' is meant that jack closest to the last full-length common rafter. The common rafter and the setback are proportioned to the decrease and the spacing, respectively.

$$\frac{\text{CRL}}{\text{SB}} = \frac{\text{decrease}}{\text{spacing}}$$

where CRL is the common rafter length, SB the setback, decrease the decrease in length, and spacing the spacing between adjacent rafters.

As is the case with any proportion, attention to the units is important. The common rafter length and the setback must be expressed in the same units—both in feet or both in inches. The spacing will likely be in inches (e.g., 16 in. o.c. or 24 in. o.c.). With the spacing in inches, the decrease will be in inches.

EXAMPLE 17-6

The full-length common rafter on a hip roof measures 15 ft 2 in. (15.17 ft) and the setback is 14 ft. The rafters are spaced 24 in. o.c. Determine the decrease in length for the jack rafters.

***Solution*:** Referring to the formula established above, CRL = 15.17, SB = 14, and spacing = 24. Solve for the decrease:

$$\frac{\text{CRL}}{\text{SB}} = \frac{\text{decrease}}{\text{spacing}}$$

$$\frac{15.17}{14} = \frac{\text{decrease}}{24}$$

$$\text{decrease} = 26 \text{ in.}$$

This means that with the last full-length common rafter measuring 15 ft 2 in., the first (and longest) jack rafter will measure 26 in. less or 13 ft. The next jack rafter will be 26 in. less than the first jack—or 10 ft 10 in., and so on.

Valley jack lengths may be calculated in a similar manner. The problem is solved using a proportion involving a common rafter, the distance along the ridge from the common rafter to the valley rafter, the decrease, and the spacing.

EXAMPLE 17-7

Determine the decrease in length of the valley jacks on the main roof described in Figure 17-6. Assume that the rafters are spaced 16 in. o.c.

***Solution*:** The common rafter length, CRL, is 15.17 ft; the distance along the ridge from the last common rafter to the valley is 11 ft (this replaces SB in the case of the hip roof); the spacing is 16 in. o.c.

$$\frac{15.17}{11} = \frac{\text{decrease}}{16}$$

$$\text{decrease} = 22.065 \quad \text{or} \quad 22\tfrac{1}{16} \text{ in.}$$

Here again, the length of the common rafter is used for reference; the first valley jack rafter will be $22\frac{1}{16}$ in. shorter than the common rafter—namely, $159\frac{15}{16}$ in., and each other jack rafter will be $22\frac{1}{16}$ in. shorter than the previous.

It should be noted that the valley jacks on the ell will have a different decrease than those on the main (see Problem 17-8).

In summary, the Pythagorean rule may be used to solve most roof problems. This, along with a properly set up proportion, is the most useful mathematical tool the carpenter will use in solving roof rafter problems.

REVIEW EXERCISES

17-1. Find the length of the conventional hip rafter for each of the following cases.
- **(a)** Common rafter length: 13 ft; run of common: 12 ft
- **(b)** Common rafter length: 17 ft 6 in.; run of common: 14 ft
- **(c)** Pitch: 6/12; run of common: 13 ft 6 in.
- **(d)** Pitch: 8/12; run of common: 15 ft 4 in.
- **(e)** Pitch: 10/12; run of common: 9 ft 8 in.

17-2. For each hip rafter in Problem 17-1, give the unit rise and unit run.

17-3. Find the length of the unconventional hip rafter for each of the following cases.

	Pitch	Run of common	Setback
(a)	7/12	13 ft	10 ft
(b)	9/12	14 ft 8 in.	12 ft
(c)	12/12	10 ft 9 in.	8 ft
(d)	6/12	4 ft	3 ft 6 in.

17-4. For each hip rafter in Problem 17-3, find the unit rise and unit run.

17-5. For each of the roofs described in Problem 17-1, determine the jack rafter decrease. Assume that the rafter spacing is 16 in. o.c.

17-6. For each of the roofs described in Problem 17-3, determine the jack rafter decrease. Assume that the rafter spacing is 24 in. o.c.

17-7. The following figure shows a front elevation and plan view of a porch that is to be attached to the front of a house. A 5/12 pitch is planned. The hip will be conventional.

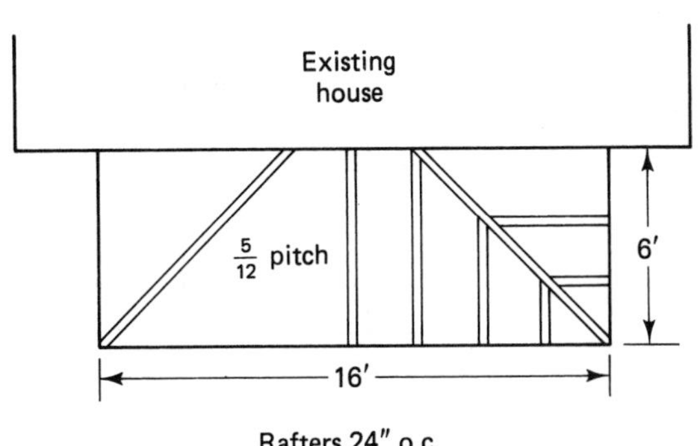

Plan view

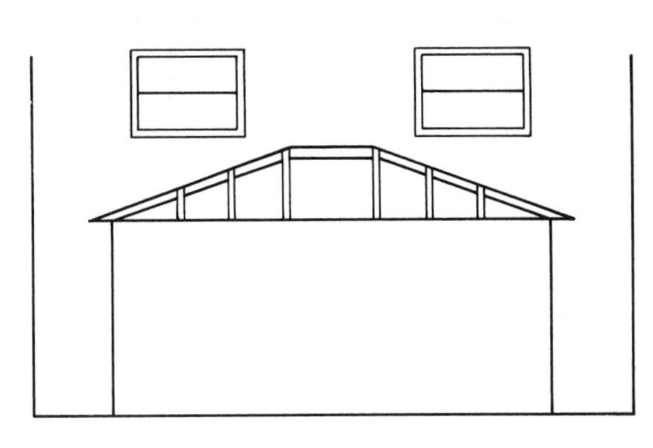

Front elevation

(a) If the bottom of the second-story windows is 10 in. above the top plate of the porch, will the roof remain below the bottom of the windows?

(b) Suppose a pitch is desired that will ensure that the porch roof meets the face of the house 6 in. below the window bottom. If the window bottom is 22 in. above the top plate of the porch, what pitch should be used?

(c) Determine the lengths of the common and hip rafters using the description of the roof in part (b).

(d) Determine the jack rafter decrease for both the front and side jacks in the roof as described in part (b).

17-8. Refer to Figure 17-6. Determine the decrease in length for the valley jack rafters on the ell. Assume 16 in. o.c. spacing.

17-9. An ell is attached to the main house in the manner shown in Figure 17-5. Below are given descriptions of several variations on this design. In each case determine **(1)** the common rafter length, **(2)** the valley rafter length(s), **(3)** the unit rise and unit run for the valley rafter, and **(4)** the jack rafter decrease for both the main and the ell roof sections.

(a) Main section width: 22 ft
Ell width: 22 ft
Pitch of both sections: 6/12

(b) Main section width: 30 ft
Ell width: 24 ft
Pitch of both sections: 5/12

(c) Main section width: 28 ft
Ell width: 16 ft
Pitch of main section: 8/12
Ridge of ell to intersect ridge of main section

The following is a summary problem that covers common, hip, valley, and jack rafter calculations.

17-10. The following figure shows the plan view and right-side elevation of a roof. Dimensions, roof pitches, and overhangs are noted on the drawings.

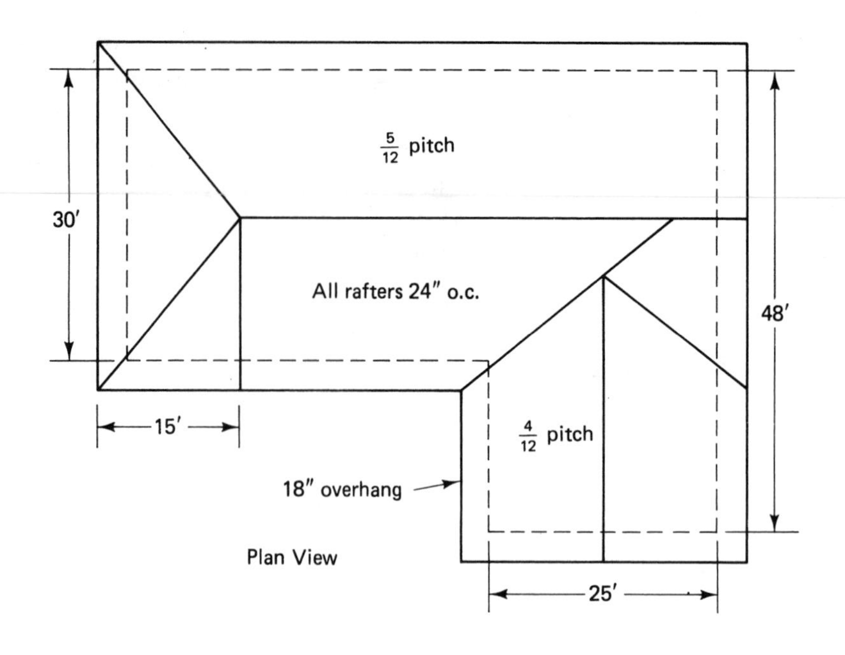

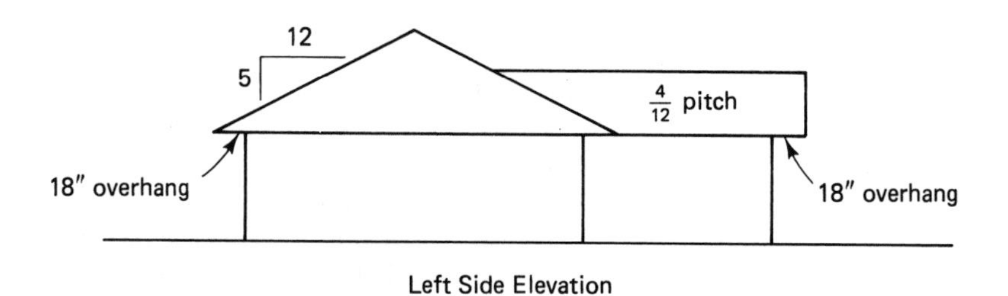

Determine the following dimensions.

(a) Common rafter length: overall length; distance from ridge to bird's-mouth corner; distance from bird's-mouth corner to tail

(b) Hip rafter length: overall length; distance from ridge to bird's-mouth corner; distance from bird's-mouth corner to tail

(c) Unit rise for the hip rafter

(d) Total rise for each roof section

(e) Valley rafter lengths (both valley rafters): overall length; distance from ridge to bird's-mouth corner; distance from bird's-mouth corner to tail

(f) Jack rafter decrease for hip and valley jacks

chapter 18

Trusses

OBJECTIVES

Upon completing this chapter, the student will be able to:

1. Determine the materials needed to construct a W-truss.
2. Lay out a W-truss.

Truss designs are many and varied—too much so to be given adequate treatment here. Good references are available on truss designs; several are included in the Bibliography.

In this chapter one of the truss designs more commonly used in residential construction is described. This is the **W-truss,** so named because of the "W" formed by the internal structural members. Figure 18-1 shows a typical W-truss. This type of

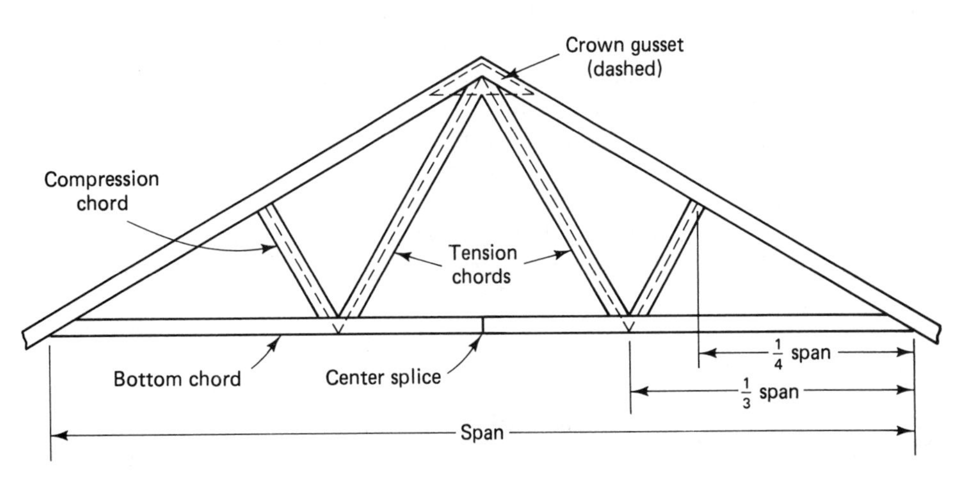

Figure 18-1

truss is suitable for spans up to 36 ft and lower pitches. Design books should be consulted for questions related to lumber sizing and spacing of trusses.

For the on-site carpenter, laying out a pattern for use in constructing trusses is a relatively simple matter. The open deck provided by the first or second story of a house provides an excellent work area. An examination of Figure 18-1 will reveal that the working lines of the roof rafter trace the lower edge of the bottom chord and the lower edge of the rafters.

To lay out the pattern for a truss, a chalk line is snapped marking the lower edge of the bottom chord. The end points of the span are marked on this line, as is the midpoint. At the midpoint of the bottom chord, a perpendicular is erected. The total rise is measured from the bottom chord line and marked on the perpendicular. This is the point at which the bottom edge of the two rafters will meet. (An alternative to erecting a perpendicular is to use two tapes. The end of one is held at the midpoint of the bottom chord, while the end of another is held at the end point of the span. The length of the rise on the tape from the midpoint is maneuvered to meet the length of the rafter on the tape from the end point of the span. Where these two tape lengths intersect marks the intersection of the lower edge of the rafters.)

Chalk lines representing the bottom edge of the rafters are then snapped from the total rise point on this perpendicular to both end points of the span on the bottom chord line. At this time the rafter lengths should be measured along the snapped lines and checked against the calculated rafter lengths. Any necessary corrections are now made.

Next, the contact points for the compression and tension chords are marked. The compression and tension chords meet on the bottom chord one-third of the distance from each end. The tension chord contacts the rafter at its midpoint. (This coincides with the vertical projection of the one-quarter point of the bottom chord—a fact established in studying the similar right triangles formed by the rafter, its run, and the total rise, in one case, and one-half the rafter, one-half the run, and the rise at that point, in the other case.)

The members of the first truss are now ready to be cut and laid in place. Care should be taken that each piece is marked in place, cut, and then replaced. Each piece is temporarily nailed in place. Any bows in the material should be removed and the pieces made to conform to the lines as the temporary nailing takes place.

After all members of the truss are fixed in place, pattern guides are nailed on each side of each piece. These should be nailed firmly as they will serve as the constraints to which all the trusses will be made to conform. Now the individual pieces of the first truss can be removed and used as patterns from which the other truss members will be cut.

Once all the truss members and gusset plates have been cut, assembly can begin. If assembly is taking place on site, it may be expedient to set each truss in place on the top plate as it is assembled. This will prevent crowding the layout area. On the other hand, if there is room in which to stack the trusses and a crane is available, the trusses could all be hoisted onto the plate in a cluster or clusters once their assembly has been completed.

It is during the estimating process that the mathematics of trusses is first encountered. Rafter lengths and various chord lengths are calculated at this time in order to determine the most efficient lumber lengths to order.

Rafter lengths are determined as described in Chapter 15. To review briefly, the rafter length, RL, is found by multiplying the unit rise, URi, by the run, Ru. If the truss is for a conventional ranch house, the run equals one-half the span. Thus

$$\text{RL} = \text{URi} \times \text{Ru}$$

The extra amount of material needed for the overhang will be included in the rafter length calculation if the horizontal projection is included in the run.

Materials for the bottom chord must, of course, have a total length equal to the span. The position of center splice is somewhat arbitrary. It is not necessary that it be located at the midpoint of the span. This reduces the potential waste in buildings whose span might call for odd lengths of lumber. For example, a 30-ft span might suggest using 15-ft lengths if the center splice were located at the midpoint. 16-ft lumber would be ordered with 1 ft wasted from each of the two pieces making up this chord. However, the chord could be made up of a 14-ft and a 16-ft piece with no significant waste.

The tension chord is twice the length of the compression chord in this type of W-truss. This fact is established by studying the two similar right triangles formed having these members as their hypotenuses. The tension chord is the hypotenuse of the right triangle whose legs are the total rise and one-sixth the span. The compression chord is the hypotenuse of the right triangle whose legs are one-half the total rise and one-twelfth the span. These are similar triangles because the legs are each twice the length of their corresponding parts in the other triangle. It follows, then, that the tension chord is twice the length of the compression chord. Therefore, if the length of either the compression or tension chord can be determined, so can the other.

To determine the length of the tension chord, examine the right triangle having it as its hypotenuse, the total rise, TRi, as one leg, and one-sixth the span, $S/6$, as the other leg. Because the terms used in the calculations need to be expressed in the same units, a conversion is in order. The term $S/6$ would ordinarily be in feet since S, the span, is usually expressed in feet. Converting $S/6$ to inches,

$$12 \times \frac{S}{6} = 2S$$

The other leg, that is, the total rise, may be expressed in terms of the unit rise, URi, and the run, which is one-half the span, as follows:

$$\text{URi} \times \frac{S}{2}$$

Using the Pythagorean rule, the length of the tension chord, T, becomes

$$T = \sqrt{(2S)^2 + \left(\text{URi} \times \frac{S}{2}\right)^2}$$

$$= \sqrt{4S^2 + \text{UR}_i^2 \times \frac{S^2}{4}}$$

$$= \sqrt{\frac{16S^2 + \text{URi}^2 S^2}{4}}$$

$$= \sqrt{\frac{S^2}{4}(16 + \text{URi}^2)}$$

$$= \frac{S}{2}\sqrt{16 + \text{URi}^2}$$

Notice that the units in the above terms were inches; therefore the value T will be in inches.

Once the length of the tension chord, T, has been determined, one-half of that value will be the length of the compression chord, C. That is,

$$C = \frac{T}{2}$$

EXAMPLE 18-1

Determine the lengths of the tension and compression chords for a W-truss to span 30 ft with a 5/12 pitch.

***Solution*:** The tension chord length is found using the formula

$$T = \frac{S}{2}\sqrt{16 + \text{URi}^2}$$

With $S = 30$ and $\text{URi} = 5$ the formula becomes

$$\begin{aligned} T &= \frac{30}{2}\sqrt{16 + 5^2} \\ &= 15\sqrt{16 + 25} \\ &= 15\sqrt{41} \\ &= 15 \times 6.403 \\ &= 96.047 \quad \text{or} \quad 96\tfrac{1}{16}\text{ in.} \end{aligned}$$

The length of the compression chord, C, is

$$\begin{aligned} C &= \frac{T}{2} \\ &= \frac{96.047}{2} \\ &= 48.024 \quad \text{or} \quad 48 \text{ in.} \end{aligned}$$

Theoretically, both of these lengths extend through the width of the bottom chord. If that member is made of 2 × 4 material, lengths of 8 and 4 ft for the tension and compression chords, respectively, will be sufficient. In fact, one 12-ft piece might be ordered from which one tension and one compression chord could be cut.

A complete materials list for one truss as described in Example 18-1 follows. The reader may wish to verify that the rafter materials will accommodate an 18-in. overhang.

Materials List

Specifications: One W-truss with 5/12 pitch, 30-ft span, 18-in. overhang
Rafters: 2 × 4's: 2/18′
Bottom chords: 2 × 4's: 1/16′, 1/14′
Tension chords: 2 × 4's: 2/8′
Compression chords: 2 × 4's: 1/8′

It is generally recommended that each joint be fastened with plywood gusset plates on both sides, glued and nailed. On the other hand, other types of fasteners are in use. These are in the form of cluster nails or gang nails.

If the exterior walls have been sheathed with 1/2-in. CDX plywood, the sections cut out for window and door openings will provide most of the materials needed for gusset plates. It has been the experience of this writer that those pieces along with two or three extra sheets is usually sufficient. Furthermore, it is not an efficient use of the estimator's time to determine accurately the total area of gusset plates toward the end of ordering an exact amount of plywood.

REVIEW EXERCISE

18-1. Listed below are the spans, pitches, and overhangs for several conventional ranch-type houses. In each case, determine a materials list for a W-truss. Include the economic lengths to order for the rafters, the bottom chords, the tension chords, and the compression chords.

	Span	Pitch	Overhang
(a)	28 ft	5/12	16 in.
(b)	15 ft	6/12	12 in.
(c)	24 ft	6/12	18 in.
(d)	22 ft	5/12	14 in.
(e)	26 ft	4/12	22 in.
(f)	25 ft 8 in.	$5\frac{1}{2}/12$	20 in.

chapter 19

Stairs

OBJECTIVES

Upon completing this chapter, the student will be able to calculate for a conventional set of stairs, as well as for a set of stairs having restricted landing areas and for stairs with intermediate landings:

1. The number of risers and runs.
2. The unit rise and the unit run.
3. The stairwell opening.
4. The total run.

Several factors must be considered when laying out a set of stairs. Included among these are:

Total distance between floors or landings
Head clearance
Number of steps
Height of each step
Width of each stair tread
Total horizontal distance covered by the set of stairs

In some situations there are restrictions imposed on the stair layout over which the carpenter has no control. The length of the stairwell opening and space available for a landing area are two possible restrictions. Space available for the run and landing area may necessitate an intermediate landing with a 90° or 180° turn. Spiral or other configurations may be dictated by other constraints. A thorough understanding of the mathematics related to stairs will enable the carpenter to work within any reasonable restrictions and construct a set of stairs that satisfies most design standards.

Listed below are some of the design standards for stairs:

Minimum finished width: 36 in.

Minimum landing area: 3 ft × 3 ft

Minimum head clearance: 6 ft 8 in.

Maximum unit rise: $8\frac{1}{4}$ in.

Minimum tread width: 9 in.

(For more detail regarding specifications, consult the standards in effect in your locale.) *A cardinal rule to be observed in designing a set of stairs is that all steps have the same height and tread width.*

Of convenience in certain calculations to follow is the concept of the **greatest integer.** The greatest integer associated with a decimal value, N, is the next-larger whole number. The notation used to denote the "greatest integer of N" is $\langle N \rangle$.

Examples

$$\langle 13.26 \rangle = 14 \quad \text{and} \quad \langle 11.78 \rangle = 12$$

The symbols defined below will be used in explaining the mathematics related to a set of stairs. All dimensions are in inches (see Figure 19-1).

Total rise, TRi: distance between finished floors

Total run, TRu: horizontal distance spanned by the set of stairs

Unit rise, URi: height of each step

Unit run, URu: width of each step (this is the framed width and does not refer to the finished tread width)

Number of risers, NRi: number of steps

Number of runs, NRu: usually one less than the number of risers (i.e., NRu = NRi − 1)

Head clearance, HC: vertical distance between the bottom of the floor joist (or stair well header) and the stair tread below

Stairwell opening, SO: length of the stairwell opening

Floor thickness, FT: vertical distance taken up by the floor joists and all layers of flooring

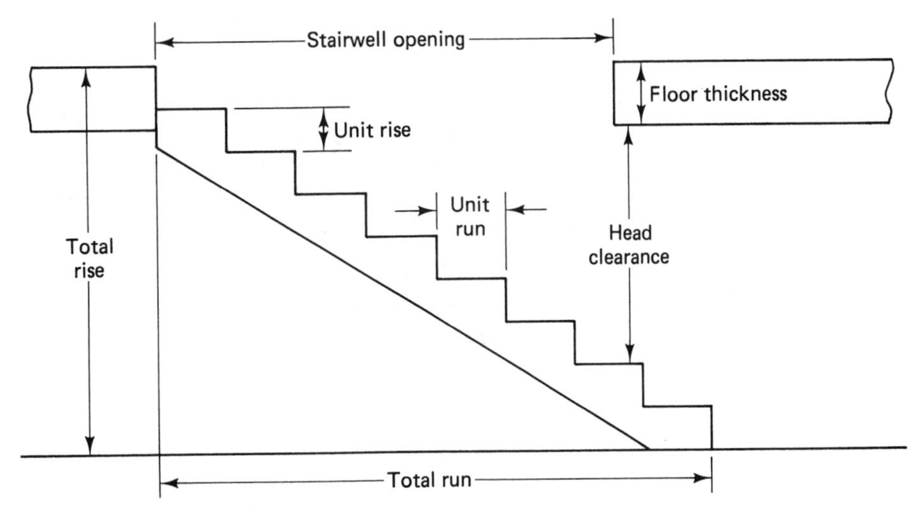

Figure 19-1

The calculations involved in a set of stairs may be approached in three stages:

1. Determining the number of risers, NRi, and unit rise, URi
2. Determining the stairwell opening, SO
3. Determining the total run, TRu

Determining Number of Risers, NRi, and Unit Rise, URi

The number of risers, NRi, is the greatest integer value when the total rise, TRi, is divided by 8. That is,

$$\text{NRi} = \left\langle \frac{\text{TRi}}{8} \right\rangle$$

The value 8 is used here since it is close to the desired unit rise. The value could be changed if conditions impose a greater or smaller unit rise on the set of stairs. The greatest integer value is taken since a whole number of stairs is desired. The next-lower integer value might be used should the quotient be very close to that lower value. Of course, the greater the divisor, the smaller the quotient. In this case, that means a greater number of risers will result in a smaller (and possibly more desirable) unit rise.

The unit rise, URi, is found by dividing the total rise, TRi, by the number of risers, NRi. That is,

$$\text{URi} = \frac{\text{TRi}}{\text{NRi}}$$

EXAMPLE 19-1

Find the number of risers and unit rise for a set of stairs having a total rise of 98 in.

Solution

$$\text{NRi} = \left\langle \frac{98}{8} \right\rangle$$

$$= \langle 12.25 \rangle$$

$$= 13$$

$$\text{URi} = \frac{98}{13}$$

$$= 7.538 \text{ in.}$$

Therefore, there will be 13 risers of 7.538 or $7\frac{9}{16}$ in. each.

EXAMPLE 19-2

The vertical distance from a cellar floor to the bottom of the 2 × 10 floor joist is 7 ft 8 in. The first floor is covered with $\frac{1}{2}$-in. plywood and $\frac{5}{8}$-in. particleboard. Determine the number of risers and the unit rise for a set of stairs from the first floor to the cellar (see Figure 19-2).

***Solution*:** The total rise consists of the floor thickness and the distance from the cellar floor to the bottom of the floor joist.

$$\text{TRi} = 7 \text{ ft } 8 \text{ in.} + 9\tfrac{1}{2} \text{ in.} + \tfrac{1}{2} \text{ in.} + \tfrac{5}{8} \text{ in.}$$

$$= 92 + 9.5 + 0.5 + 0.625, \text{ converting to inches}$$

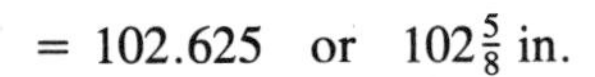

$$= 102.625 \quad \text{or} \quad 102\tfrac{5}{8} \text{ in.}$$

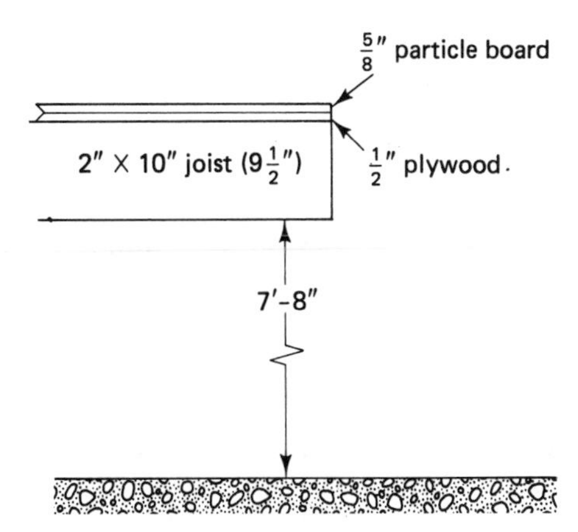

Figure 19-2

The number of risers becomes

$$\text{NRi} = \left\langle \frac{102.625}{8} \right\rangle$$
$$= \langle 12.83 \rangle$$
$$= 13$$

Now, the unit rise becomes

$$\text{URi} = \frac{102.625}{13}$$
$$= 7.894 \quad \text{or} \quad 7\tfrac{7}{8} \text{ in.}$$

Thus, there will be 13 risers of $7\frac{7}{8}$ in. each.

Determining the Stairwell Opening, SO

When the run of the stairs is not restricted by space limitations, the stairwell opening, SO, is determined by the horizontal distance needed to descend far enough to give head clearance, HC. The floor thickness, FT, must be considered, as well as the desired head clearance. The general procedure is to determine the number of risers needed to descend a known distance, translate this into a number of runs, then multiply by the unit run.

EXAMPLE 19-3

Determine the stairwell opening necessary to provide a head clearance of 6 ft 8 in. The floor thickness is $10\frac{5}{8}$ in. and the unit rise has been determined to be $7\frac{5}{8}$ in.

***Solution*:** The vertical distance that must be descended is 6 ft 8 in. plus the floor thickness, $10\frac{5}{8}$ in.—a total of 90.625 in. Divide this by the unit rise, $7\frac{5}{8}$, to get the number of risers needed to descend 90.625 in.

$$\frac{90.625}{7.625} = 11.89$$

Using the greatest integer value, 11.89 becomes 12. This means that on the twelfth riser from the top, one will have descended slightly more than $90\frac{5}{8}$ in., and considering the floor thickness, will have 6 ft 8 in. head clearance.

The steps taken here may be summarized as

$$\frac{\text{HC} + \text{FT}}{\text{URi}}$$

In a conventional set of stairs, the number of runs is one less than the number of risers (see Figure 19-3). That is,

$$\text{NRu} = \text{NRi} - 1$$

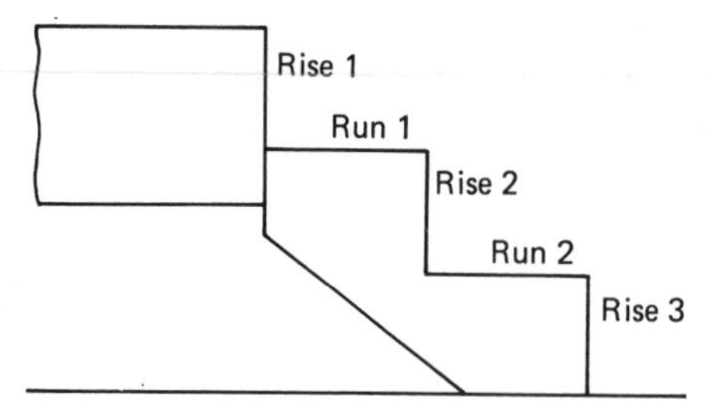

Figure 19-3

Since it takes 12 risers to descend the desired amount, the number of runs will be 11.

$$\text{NRu} = 12 - 1$$
$$= 11$$

Multiplying the number of runs by the unit run will result in the stairwell opening. That is,

$$\text{SO} = \text{NRu} \times \text{URu}$$

Assuming a unit run of 9 in., the stairwell opening in this example becomes

$$\text{SO} = 11 \times 9 = 99 \text{ in.}$$

The steps needed to calculate the stairwell opening are summarized in the following formula:

$$\text{SO} = \text{URu} \times \left(\frac{\text{HC} + \text{FT}}{\text{URi}} - 1\right)$$

Determining the Total Run, TRu

The total run of a set of stairs is found by multiplying the unit run, URu, by the number of runs, NRu. That is,

$$\text{TRu} = \text{URu} \times \text{NRu}$$

The total run can also be expressed in terms of the number of risers.

$$\text{TRu} = \text{URu} \times (\text{NRi} - 1)$$

EXAMPLE 19-4

Find the total run of a set of stairs having a 9-in. unit run and 13 risers.

***Solution*:** Using the latter formula gives us

$$\text{TRu} = \text{URu} \times (\text{NRi} - 1)$$
$$= 9 \times (13 - 1)$$
$$= 9 \times 12$$
$$= 108 \text{ in.}$$

Summarizing the procedure and formulas related to laying out a set of stairs:

1. Determine the number of risers and the unit rise.

$$\text{NRi} = \left\langle \frac{\text{TRi}}{8} \right\rangle \qquad \text{URi} = \frac{\text{TRi}}{\text{NRi}}$$

2. Determine the stairwell opening.

$$\text{SO} = 9 \times \left(\frac{\text{HC} + \text{FT}}{\text{URi}} - 1 \right)$$

3. Determine the total run.

$$\text{TRu} = \text{URu} \times (\text{NRi} - 1)$$

or

$$\text{TRu} = \text{URu} \times \text{NRu}$$

EXAMPLE 19-5

The distance from a cellar floor to the bottom of the floor joists above is 6 ft 11 in. The floor joists are 2 × 8's (actual width is $7\frac{3}{8}$ in.). The subflooring is $\frac{3}{4}$-in. boards and a layer of $\frac{5}{8}$-in. hardwood makes up the finish floor. Find the number of risers, the unit rise, the stairwell opening, and the total run for a set of stairs to run from the cellar to the first floor. Assume a head clearance of 6 ft 6 in. and a unit run of $8\frac{1}{2}$ in.

***Solution*:** Find the total rise.

$$\begin{aligned} \text{TRi} &= 6 \text{ ft } 11 \text{ in.} + 7\tfrac{3}{8} \text{ in.} + \tfrac{3}{4} \text{ in.} + \tfrac{5}{8} \text{ in.} \\ &= 83 \text{ in.} + 7.375 \text{ in.} + 0.75 \text{ in.} + 0.625 \text{ in.} \\ &= 91.75 \text{ in.} \end{aligned}$$

Find the number of risers.

$$\begin{aligned} \text{NRi} &= \left\langle \frac{91.75}{8} \right\rangle \\ &= 12 \end{aligned}$$

Find the unit rise.

$$\begin{aligned} \text{URi} &= \frac{91.75}{12} \\ &= 7.646 \quad \text{or} \quad 7\tfrac{5}{8} \text{ in.} \end{aligned}$$

Find the stairwell opening.

$$\begin{aligned} \text{SO} &= 8.5 \times \left(\frac{78 + 8.75}{7.646} - 1 \right) \\ &= 8.5 \times (11.3 - 1) \\ &= 8.5 \times (12 - 1) \\ &= 8.5 \times 11 \\ &= 93.5 \text{ in.} \end{aligned}$$

Find the total run.

$$\text{TRu} = 8.5 \times (12 - 1)$$
$$= 93.5 \text{ in.}$$

Consider next a situation in which the distance available for the total run is limited. This can occur due to the location of a wall that crowds the lower landing.

EXAMPLE 19-6

The head of a set of stairs is to be located a horizontal distance 10 ft 6 in. from the vertical plane of a lower level wall, toward which the stairs will run (see Figure 19-4). The total rise is $92\frac{1}{4}$ in., $10\frac{1}{4}$ in. of which is floor thickness. Design a set of stairs to fit in this space, determining the number of risers, the unit rise, and the unit run.

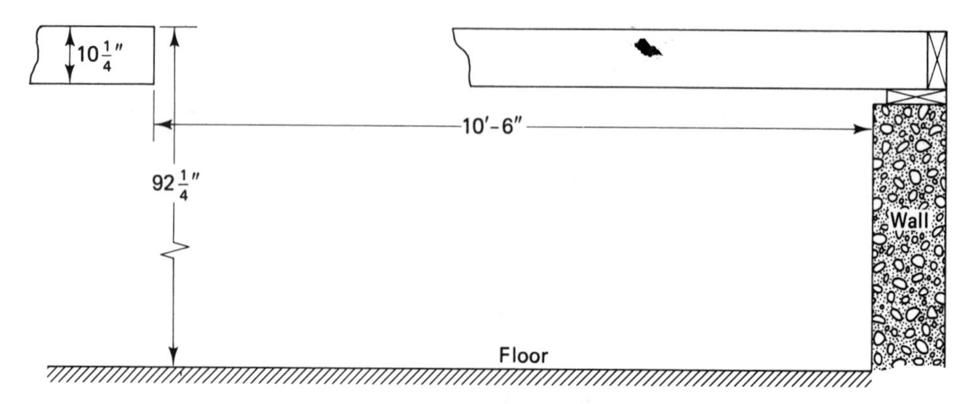

Figure 19-4

***Solution*:** The number of risers is

$$\text{NRi} = \left\langle \frac{92.25}{8} \right\rangle$$
$$= 12 \text{ risers}$$

Determine the unit rise:

$$\text{URi} = \frac{92.25}{12}$$
$$= 7.688 \text{ or } 7\tfrac{11}{16} \text{ in.}$$

Since there are 12 risers, there will be 11 runs. If a distance of 36 in. is reserved for the lower landing, 90 in. remain for the total run. This results in a unit run of

$$\text{URu} = \frac{90}{11}$$
$$= 8.182 \quad \text{or} \quad 8\tfrac{3}{16} \text{ in.}$$

Since $8\frac{3}{16}$ in. is less than the desired 9-in. minimum unit run, adjustments should be considered. A decrease in the length of the landing area should result in a corresponding increase in the total run and therefore a greater unit run. To accommodate a 9-in. unit run, 99 in. of total run would be needed ($9 \times 11 = 99$). This would leave only 27 in. for the length of the landing area. Some trial-and-error calculations might be made in this case with the objective of establishing a reasonable compromise between the landing area and the unit run.

In any situation involving limited space, the amount of head clearance should be explored, as well as the topics discussed previously. Limited space for the stairwell opening may result in restricted head clearance. The stair design is incomplete without considering the head clearance.

In cases where a trade-off between landing area and unit run is not feasible, an intermediate landing with a 90° or 180° turn could be considered. Of course, space must be available for the turn under consideration. In general, a landing replaces a step. Therefore, the number of risers does not change. On the other hand, the total run is extended by the landing. These layouts do not conserve space; they merely reorient the stairs to take advantage of available space in another direction (see Figure 19-5).

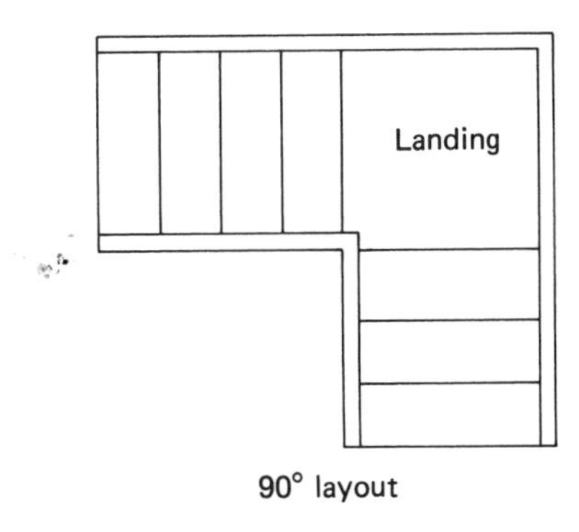

90° layout

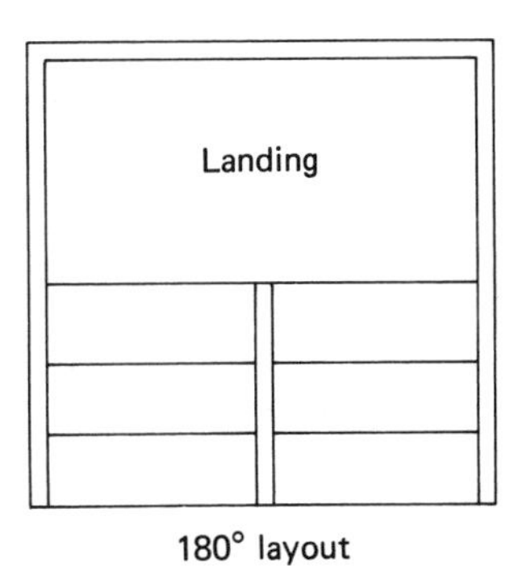

180° layout

Figure 19-5

REVIEW EXERCISES

19-1. A set of stairs is to be constructed from the first to the second story. The vertical distance from the first floor to the bottom of the second floor joist is 7 ft $6\frac{1}{2}$ in. The floor thickness is $10\frac{3}{8}$ in. Assume a unit run of $9\frac{1}{2}$ in. Determine the following dimensions.

(a) Total rise
(b) Number of risers
(c) Unit rise
(d) Number of runs
(e) Total run
(f) Stairwell opening
(g) Head clearance

19-2. Design a set of stairs for each of the following situations. Dimensions given are actual size. Assume a unit run of 9 in. Determine total run, number of risers, unit rise, number of runs, stairwell opening, and head clearance.

	Total rise	Floor joist	Sub-flooring	Finish flooring
(a)	7 ft 2 in.	$7\frac{3}{8}$ in.	$\frac{3}{4}$ in.	$\frac{5}{8}$ in.
(b)	6 ft 11 in.	$9\frac{1}{2}$ in.	$\frac{1}{2}$ in.	$\frac{5}{8}$ in.
(c)	7 ft 5 in.	$5\frac{1}{2}$ in.	$\frac{7}{16}$ in.	$\frac{3}{4}$ in.
(d)	6 ft 8 in.	$11\frac{1}{2}$ in.	$\frac{1}{2}$ in.	$\frac{3}{4}$ in.
(e)	8 ft 1 in.	$9\frac{1}{4}$ in.	$\frac{1}{2}$ in.	$\frac{3}{8}$ in.

19-3. The horizontal distance available for a set of stairs is 11 ft 8 in. The landing area at the bottom of the stairs is to be 36 in. The total rise is 8 ft $5\frac{1}{4}$ in. and the floor thickness is $9\frac{7}{8}$ in. The head clearance required is 6 ft 8 in. Determine the number of risers, the unit rise, the number of runs, the unit run, and the total run for a set of stairs to fit these conditions.

19-4. Design an L-shaped set of stairs to fit the following conditions: total rise, 8 ft 2 in., floor thickness, $10\frac{1}{4}$ in., head clearance, 6 ft 6 in. Determine the number of risers, unit rise, number of runs, unit run, and total run. The intermediate landing is to be 36 in. × 36 in. and is located at the midpoint of the total rise.

19-5. Design a set of stairs with a 180° turn. The stairs will be 36 in. wide and the intermediate landing area will be 36 in. × 72 in. The total rise is 7 ft $11\frac{1}{2}$ in. The floor thickness is $6\frac{1}{4}$ in. Head clearance is 6 ft 8 in.

19-6. Design an L-shaped set of stairs to fit the following specifications: total rise, 8 ft 6 in.; floor thickness, 10 in.; head clearance, 7 ft. Determine the number of risers, unit rise, number of runs, unit run, and total run. The intermediate landing is to be 36 in. × 36 in. and serves as the third run.

chapter 20

Framing and Covering Gable Ends; Exterior Trim

OBJECTIVES

Upon completing this chapter, the student will be able to:

1. Determine the materials needed to frame gable ends.
2. Determine the materials needed to sheath gable ends.
3. Determine the materials needed for siding gable ends.
4. Determine the materials needed for exterior trim.

The term **gable ends** as used in this chapter will mean that portion of a house between the slope of the roof and the line extending across the top plates. It is not meant to be restricted to the triangular ends of a gable-framed roof. The discussion here will apply to all types of roofs—gable, gambrel, and so on.

The length of jack studs needed to frame a gable end depends on the total rise of the roof. The total rise may be determined by multiplying the unit rise by the total run. That is,

$$\text{TRi} = \text{URi} \times \text{run}$$

where TRi is the total rise, in inches, URi is the unit rise, in inches per foot, and run is measured in feet.

EXAMPLE 20-1

A roof has a 6/12 pitch. The common rafter has a run of 15 ft. Find the total rise.

Solution: Using the formula established above,

$$\begin{aligned}\text{TRi} &= \text{URi} \times \text{run}\\ &= 6 \times 15\\ &= 90 \text{ in. or } 7 \text{ ft } 6 \text{ in.}\end{aligned}$$

Assume that the roof described in Example 20-1 is a gable roof on a ranch-type house (see Figure 20-1). The gable ends form isosceles triangles (two equal sides). An effective way to determine the number and length of studs needed to frame the gable ends is to picture both triangular ends together. These two triangles taken together form the equivalent of a rectangle having a width equal to the width of the house and a height equal to the total rise.

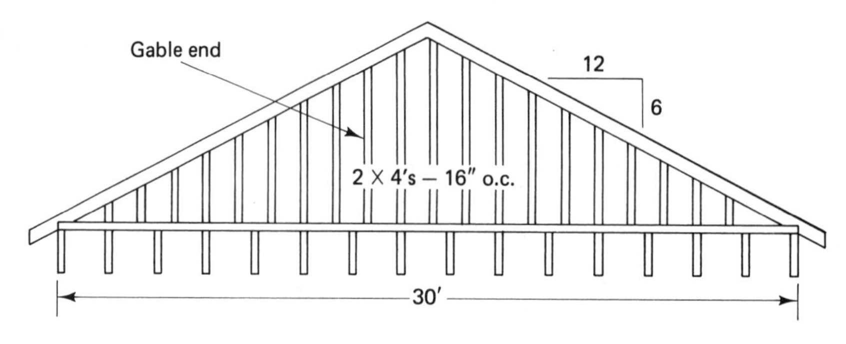

Figure 20-1

The length of materials for framing the gable ends is determined by the total rise. In this case 8-ft studs will suffice. The maximum length is 7 ft 6 in. Consider the succession of studs starting at the center of the gable end and moving outward. The amount cut from each stud will be usable at the shorter side of the triangle.

As to the number of studs, the spacing has to be considered. Assuming 16 in. o.c. in the case described in Example 20-1, the number of studs needed, NS, to cover a distance, L, is

$$\begin{aligned}\text{NS} &= \tfrac{3}{4} \times L\\ &= \tfrac{3}{4} \times 30\\ &= 22\tfrac{1}{2}\end{aligned}$$

(Recall from earlier discussions that the factor "$\frac{3}{4}$" stems from the reduced fraction $\frac{12}{16}$; multiplication by 12 converts the length from feet to inches, while dividing by 16 counts the number of 16-in. spaces.) Round this to 23 studs. There is no need for a "starter" as was the case in wall framing. 23 studs 8 ft long will be sufficient to frame both gable ends.

Adjustments to the basic count should be considered as necessary. Adjustments would be necessary in cases where openings are planned for windows, louvered vents, and so on.

The amount of sheathing for a gable end is determined by the area to be covered. If plywood or a similar product is to be used, the area, in square feet, of the gable end is divided by 32, the number of square feet in a sheet of plywood. That is,

$$\text{number of sheets} = \frac{\text{area}}{32}$$

In the case of a ranch house, the gable ends should be considered together, as was suggested for the stud count. Instead of two separate triangles, the two ends form the equivalent of one rectangle.

EXAMPLE 20-2

Determine the number of sheets of $\frac{1}{2}$-in. CDX plywood needed to cover the gable ends of the ranch house described in Example 20-1.

***Solution*:** The building width is 30 ft and the total rise is 7 ft 6 in. The gable ends considered together form the equivalent of a rectangle measuring 30 ft × 7.5 ft. This is a total area of

$$\text{area} = 30 \times 7.5$$
$$= 225 \text{ ft}^2$$

Dividing by 32, the number of square feet in a sheet of plywood,

$$\text{number of sheets} = \frac{225}{32}$$
$$= 7.03125 \quad \text{or} \quad 7 \text{ sheets}$$

If only one gable end is being considered, the procedures described above can be used with slight modification. Instead of having two triangles that form the equivalent of a rectangle, a single triangle exists. The number of studs is simply one-half the number previously determined. The area to be covered is also one-half that of the rectangle. Indeed, the formula for finding the area of a triangle is

$$\text{area} = \tfrac{1}{2}BH$$

where B is the base of the triangle and H is its height (altitude).

If boards are to be used for closing-in the gable ends, some adjustment to the quantity will be needed to account for the waste factor. Generally, the addition of 10 to 15% to the area will provide sufficient boarding material.

The procedures for determining materials needed to frame and cover the ends created by other types of roofs are similar to those described above. The estimator should take care to note sections that appear elsewhere in the structure and can be combined in some geometric form toward the end of simplifying calculations. Efforts to do this are rewarded in faster, more efficient, and more accurate estimating.

EXAMPLE 20-3

The building in Figure 20-2 shows a gambrel roof. The 15/12 pitch has a run of 6 ft; the 4/12 pitch has a run of 12 ft. The building width is 36 ft. Determine the following materials for both ends of the building: (a) number of studs and their lengths to frame the ends; (b) number of sheets of $\frac{1}{2}$-in. CDX sheathing to close in the ends; (c) number of squares of cedar shingles to side the ends.

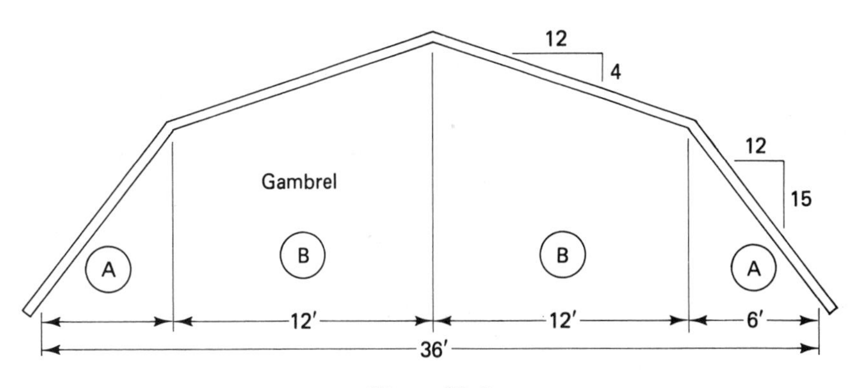

Figure 20-2

***Solution*:** (a) Assume that the stud spacing is 16 in. o.c. There are four triangular sections like that labeled ''A''—two on each end. Two of these sections may be considered together to form a rectangle. The width of this rectangle is 6 ft. Its height is the total rise of that section of roof, namely,

$$\text{TRi} = \text{URi} \times \text{run}$$
$$= 15 \times 6$$
$$= 90 \text{ in.} \quad \text{or} \quad 7.5 \text{ ft}$$

Material length of 8 ft is suggested.

As to the number of studs,

$$\text{number of studs} = \tfrac{3}{4} \times 6$$
$$= 4.5, \text{ rounded to } 5$$

Five 8-ft studs will frame two of these triangular sections. A total of 10 studs will be needed for both ends.

Section B appears four times. The maximum length needed here is found by determining the additional rise for this section and adding it to the 7.5-ft rise of section A.

$$\text{TRi} = \text{URi} \times \text{run}$$
$$= 4 \times 12$$
$$= 48 \text{ in.} \quad \text{or} \quad 4 \text{ ft}$$

Added to the 7.5-ft rise of section A, the maximum length is

$$7.5 + 4 = 11.5 \text{ ft}$$

The question as to what material lengths to order should not consume much time. It is possible to determine accurately a number of 8-, 10-, and 12-ft pieces needed for this job. It is probably just as efficient simply to order based on the maximum length needed. In this example, no more than 4 ft of material would be cut from any particular piece. This would likely find a use as blocking or spacing material.

The number of 12-ft studs needed to cover the 24-ft width of section B is

$$\text{number of studs} = \tfrac{3}{4} \times 24$$
$$= 18$$

Including the other end, a total of 36 studs each 12 ft long are needed.

(b) The total area of the ends is needed to determine the number of sheets of sheathing for closing in. For section A, two sections together form a rectangle and there are two such rectangles. Find the area of one rectangle and double it.

$$\text{Area} = 2 \times 6 \times 7.5$$
$$= 90 \text{ ft}^2$$

Section B is a trapezoid having parallel sides of 7.5 and 11.5 ft and a height of 12 ft. (Remember that the term ''height'' as used in the formula for area of a trapezoid refers to the perpendicular distance between the parallel sides.) For one section B, the area is

$$A = \frac{B + b}{2} \times H$$

where B and b represent the parallel sides and H the height.

$$A = \frac{7.5 + 11.5}{2} \times 12$$

$$= \frac{19}{2} \times 12$$

$$= 9.5 \times 12$$

$$= 114 \text{ ft}^2$$

Since there are four of these sections, the total area is

$$4 \times 114 = 456 \text{ ft}^2$$

Adding this to the total for section A, the combined area of the ends is

$$90 + 456 = 546 \text{ ft}^2$$

The number of sheets of sheathing required is

$$\text{number of sheets} = \frac{546}{32}$$

$$= 17.0625 \quad \text{or} \quad 17 \text{ sheets}$$

(c) The number of squares of cedar shingles is directly determined from the total area to be covered. A "square" of shingles is 100 ft^2. Dividing the total area by 100 results in

$$\frac{546}{100} = 5.46 \text{ squares}$$

Since most cedar shingles are bundled with three bundles to the square (that is, each bundle covers $33\frac{1}{3}$ ft^2), $5\frac{2}{3}$ squares or 17 bundles will suffice.

Quantities of material needed for exterior trim are difficult to generalize due to the various types of siding and methods of application. Aluminum and vinyl siding are specialty items. Jobbers specializing in the sale and application of these types of siding have knowledge of techniques for estimating materials needed for corners, door and window starter strips, soffit and fascia, and so on. Some other types of siding will require the use of corner boards and/or trim boards immediately beneath the soffit or bordering the roof line on a gable end.

Estimation of materials for trim boards is a simple task involving the summation of wall heights at corners, possibly building perimeters and distances along roof edges. The efficient estimator will make use of previously calculated rafter lengths where appropriate.

REVIEW EXERCISES

20-1. Listed below are the building width and roof pitch for several typical ranch-type houses. For both gable ends of each house, determine **(1)** the number of studs and their lengths, and **(2)** the number of sheets of sheathing needed to close in the ends.

	Building width	Pitch
(a)	24 ft	5/12
(b)	26 ft 8 in.	6/12
(c)	25 ft 2 in.	8/12
(d)	32 ft 6 in.	7/12
(e)	38 ft	5/12

20-2. The ends of several buildings are shown below with dimensions and roof pitches noted. In each case, for both ends, determine **(1)** the number of studs and their lengths, **(2)** the number of sheets of sheathing needed for closing in, **(3)** as an alternative, the number of board feet of boards needed to close-in, and **(4)** the number of bundles of cedar shingles needed to cover the ends (3 bundles = 1 square).

(a)

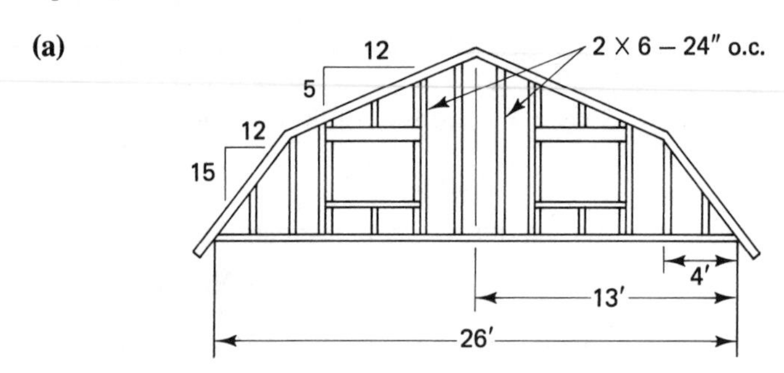

(b)

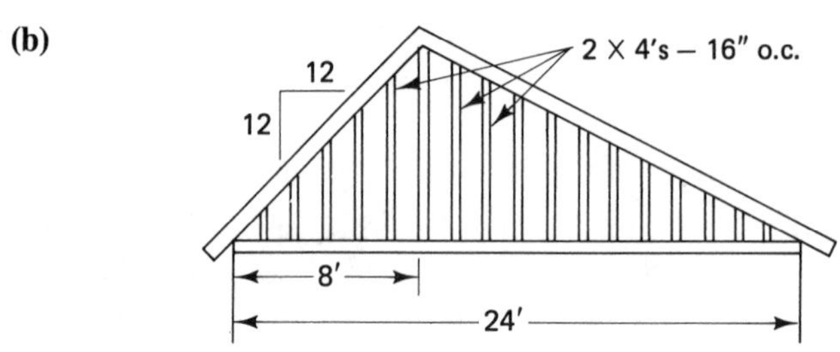

(c)

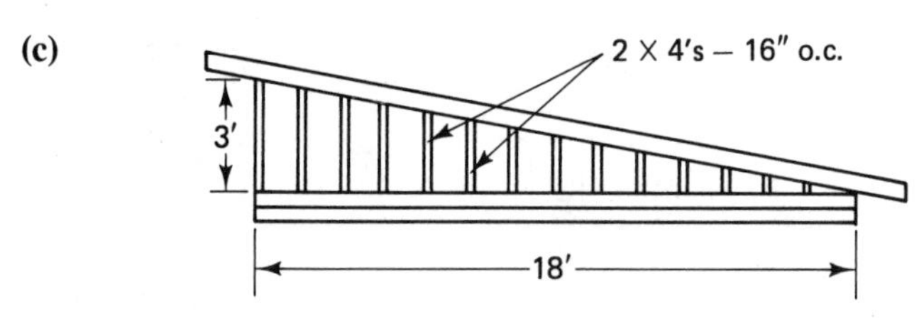

(d)

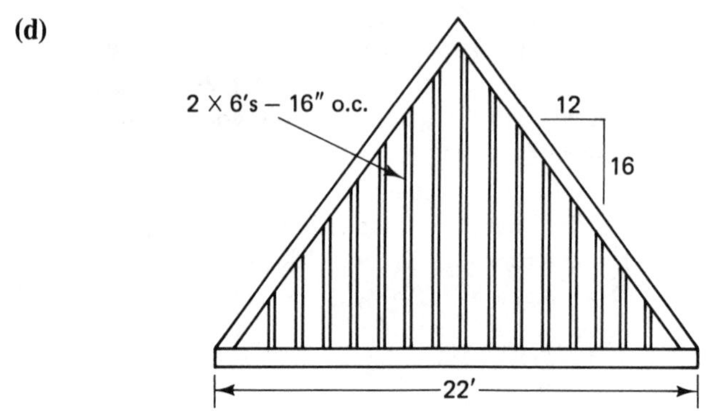

chapter 21

Wall and Roof Covering

OBJECTIVES

Upon completing this chapter, the student will be able to:

1. Determine the amount of materials needed to close-in the exterior walls.
2. Determine the amount of various types of siding needed to cover the exterior of a house.
3. Determine the amount of drywall needed for the interior of a house.
4. Determine the amount of materials needed to sheath a roof.
5. Determine the amount of roofing material (felt paper, drip edge, shingles) needed for any roofing job.

When an estimator begins the process of determining materials needed to construct a house, one of the measurements of the structure that will serve repeatedly is its perimeter. The perimeter is easily translated into exterior wall area by multiplying by the wall height. The estimator uses the perimeter and exterior wall area in determining amounts of various materials. Some of these are: wall framing, exterior sheathing, siding, insulation, and drywall. It is easy to understand why these measurements are so important to the estimator and should be at his fingertips.

Sheathing for the exterior walls may be determined using either a perimeter measurement or an area measurement. In fact, it is expedient in most cases to use both. To determine the number of sheets of sheathing needed to cover the exterior walls of a single story house, the perimeter may be divided by 4. This assumes that the 4 × 8 sheets will be applied with the 8-ft length in the vertical position. Dividing by 4 determines the number of 4-ft spaces there are around the exterior. However, it usually requires more than 8 ft to cover from approximately 1 in. below the foundation top to the top plate. Even when the interior ceiling height is to be 7 ft 6 in. (a common practice today to reduce the heating volume requirements), the added height caused by the floor framing and sill plate will generally exceed 8 ft. Therefore, a few sheets of sheathing must be added to cover the lower strip.

EXAMPLE 21-1

A single-story conventional ranch house, 28 ft × 44 ft, is to have a ceiling height of 7 ft 6 in. How many sheets of $\frac{1}{2}$-in. CDX plywood are needed to cover the exterior walls to at least 1 in. below the foundation top? Assume that the floor joists are 2 × 10's.

***Solution*:** The perimeter of the house is

$$P = 2(28 + 44)$$
$$= 2(72)$$
$$= 144 \text{ ft}$$

Dividing by 4 gives a number of sheets, NS, equal to

$$NS = \frac{144}{4}$$
$$= 36$$

The distance from 1 in. below the foundation top to the top plate is found by totaling the following: 7 ft 6 in. wall height, $\frac{1}{2}$-in. subflooring, $9\frac{1}{2}$-in. floor joists, $1\frac{1}{2}$-in. sill plate, and adding 1 in. below the foundation top. This gives a total of

$$7 \text{ ft } 6 \text{ in.} + \tfrac{1}{2} \text{ in.} + 9\tfrac{1}{2} \text{ in.} + 1\tfrac{1}{2} \text{ in.} + 1 \text{ in.} = 8 \text{ ft } 6\tfrac{1}{2} \text{ in.}$$

Assuming that the sheathing is applied to be level with the top plate, an additional $6\frac{1}{2}$ in. would remain to be covered around the lower perimeter. This is an area of

$$\frac{6.5}{12} \times 144 = 78 \text{ ft}^2$$

and would require extra sheathing amounting to

$$\frac{78}{32} = 2.4 \quad \text{or} \quad 3 \text{ sheets}$$

Thus a total of 39 sheets will be needed to cover the exterior walls from 1 in. below the foundation top to the top plate.

In Example 21-1, if the interior wall height was to be 8 ft instead of 7 ft 6 in., the border perimeter would have required in excess of 4 sheets; a total of 41 sheets would be necessary. In general, the estimator is on safe grounds by adding from 3 to 6 sheets of sheathing to the basic count.

Materials needed to cover the gable ends (and other types of roof ends) can be determined by using the area of the ends to be covered (see Chapter 20). The area is determined and divided by 32, the number of square feet in a sheet of 4 × 8 sheathing.

EXAMPLE 21-2

The 28 ft × 44 ft ranch house described in Example 21-1 has a 5/12 pitch with 16-in. overhangs front and rear. Determine the number of sheets of sheathing needed to cover the gable ends.

***Solution*:** The two ends together form the equivalent of a rectangle. The height of the rectangle is the total rise of the roof, as measured from the rafter tail. Including the overhang, the run is

$$16 \text{ in.} + 14 \text{ ft} = 15.33 \text{ ft}$$

The total rise, TRi, is

$$\begin{aligned} \text{TRi} &= \text{URi} \times \text{run} \\ &= 5 \times 15.33 \\ &= 76.65 \text{ in. or } 6.39 \text{ ft} \end{aligned}$$

Applying the concept that the two triangular ends form an equivalent rectangle, the total area, A, for both gable ends is

$$\begin{aligned} A &= 6.39 \times 2(14 \text{ ft} + 16 \text{ in.}) \\ &= 6.39 \times 30.67 \\ &= 195.98 \text{ ft}^2 \end{aligned}$$

The number of sheets of sheathing needed to cover this area is

$$\begin{aligned} \text{NS} &= \frac{195.98}{32} \\ &= 6.12 \text{ sheets} \end{aligned}$$

It is likely that 6 sheets will be sufficient.

Comment: This solution assumes a trussed roof in which the entire truss makes up the gable end to be covered. With conventional roof framing, slightly less material would be needed. The area to be covered in this case would have a run of only 14 ft and a rise calculated over this 14-ft run. The reader may wish to verify, however, that 6 sheets would still be needed.

In summary, determining amounts of material needed for exterior sheathing requires the use of perimeter and area. If boarding boards are to be used, the perimeter may still be used as a means of determining the area. That is,

$$\text{exterior wall area} = \text{perimeter} \times \text{wall height}$$

In those situations where boards are to be used, a waste factor should be applied—usually, 10 to 15% is sufficient. Areas of gable and other roof ends are the basis for determining the amount of sheathing needed to cover those sections.

Window and door areas are not usually subtracted when determining exterior sheathing. In the construction process, as wall segments are framed and covered on the deck for erection in place, it is more efficient simply to cover the entire wall with sheathing and cut the openings later. The material covering the openings is not usually wasted. It can be used for gusset plates if trusses are constructed on site. It can be used as shelving in the house.

In determining the amount of siding required to cover the exterior of a house, different methods are used for different types of siding. Applicators of aluminum and vinyl siding use special techniques to determine amounts of siding, corners, J-channel, and so on. Wood clapboards and wood shingles or shakes may be estimated using exterior area as a basis for calculation. The many varieties of plywood siding may be estimated using the same techniques as described above for sheathing. Vertical board siding and board-and-batten siding may be estimated using exterior area.

In general, all types of siding will be estimated based on exterior area. The estimator should make judgments as to waste factors, but there is ordinarily little waste in most types of siding. The estimator should also make judgments as to when areas for windows and doors are subtracted and when they are ignored. When estimating aluminum or vinyl siding, the openings will be considered. Special materials are used to line these openings. In cases where plywood-type siding or other solid sheets of siding are to be applied, openings should be ignored in order to provide continuity to the visual effects of the building. On the other hand, clapboard and wood shingle siding requirements may be reduced by the total area of openings.

When the total area of openings is to be subtracted from the total exterior area, it is not necessary to be precise. Exterior doors usually occupy an area measuring approximately 3 ft by $6\frac{2}{3}$ ft, for a total of $3 \times 6\frac{2}{3}$ or 20 ft^2. Sliding doors measure approximately $6\frac{2}{3}$ ft high and commonly are 6 or 8 ft wide—areas of 40 or 53 ft^2. A picture window occupies 36 to 40 ft^2. Most other windows can be estimated based on their glass size or rough-opening size available from order books. In older, existing houses a typical window might be measured and its area used to represent the others.

In summary, areas to be subtracted for typical openings may be estimated based on typical openings in the house being estimated. Some average opening areas are:

Door	20 ft^2
Sliding door	40 to 53 ft^2
Picture window	40 ft^2
Large window	18 ft^2
Average window	12 ft^2

EXAMPLE 21-3

The 28 ft × 44 ft ranch house described in Examples 21-1 and 21-2 is to be covered with cedar shingles. The house has two entry doors, a picture window, and nine average-size windows. Determine the quantity of cedar shingles needed for the siding.

Solution: In Examples 21-1 and 21-2 it had been determined that the sheathing required 36 sheets for the walls, 3 sheets for the lower trimming, and 6 sheets for the gable ends—a total of 45 sheets. The efficient estimator will use either this or a previously calculated exterior area total as the basis for determining siding needs. 45 sheets of sheathing is equivalent to a total area, A, of

$$A = 45 \times 32$$
$$= 1440 \text{ ft}^2$$

The total area of all openings is:

Doors:	$2 \times 20 =$	40
Picture window:		40
Windows:	$9 \times 12 =$	108
Total opening area:		188 ft^2

Determining the net area:

Gross area:	1440
Less area of openings:	188
Siding area:	1252 ft^2

Cedar shingles ordinarily come 4 bundles per square. The estimator might round this to 1300 ft^2 and order 13 squares, or 52 bundles for the job.

Drywall applicators generally prefer to get an accurate count of the number of sheets needed for a job after the interior framing has been completed. However, reliable estimates can be made based on area. Materials needed for the ceilings can be estimated based on the total ceiling area, which, of course, equals the floor area, except where sloped ceilings exist. Drywall for exterior walls can be estimated based on the interior wall area. The building perimeter multiplied by the interior wall height will give the interior wall area. An opening as large as a sliding door might be considered, but most openings should be ignored in drywall estimation. Proper application of drywall takes into consideration seam location and frequency. A better-appearing job results when a small quantity of drywall is traded for fewer seams.

As for the interior walls, recall that a total linear footage of interior partitions was determined for purposes of estimating framing lumber (see Chapter 14). The efficient estimator will make use of this figure in estimating drywall needs. The total linear footage of interior partitions multiplied by the wall height results in an amount of area. This is doubled, since both sides of each interior partition are covered.

Thus a total area to be drywall covered is determined by totaling the ceiling area, the exterior wall area, and the interior wall area. Drywall applicators will sometimes estimate a cost based on square footage and sometimes based on a number of sheets. Cost figures are available from any applicator. (For example, one drywall contractor might charge \$11 per 4 × 12 sheet to install, tape, mud, sand walls and ceilings, and spray the ceilings. This includes the cost of nails, screws, tape, joint compound, and labor. The sheets of drywall are not included. Thus a cost factor based on area could be determined by adding to \$11 the cost of a 4 × 12 sheet of $\frac{1}{2}$-in. sheetrock, currently approximately \$5.50, for a total of \$16.50 per sheet, then dividing by 48 ft^2, resulting in a cost of \$0.34375 per square foot.) If only the quantity of drywall is desired, the total area is divided by whatever area is contained in the size sheet to be used—32 for a 4 × 8 sheet, 48 for a 4 × 12 sheet, and so on.

EXAMPLE 21-4

Determine the number of 4 × 12 sheets of drywall required to cover the ceilings and walls of the 28 ft × 44 ft house described in Examples 21-1 to 21-3. Assume that the interior wall height is 8 ft. Assume that the interior partitions total 154 linear feet. Ignore all openings.

Solution: Determine the total area to be covered; then divide by 48 square feet per sheet.

Ceilings:	28×44	$= 1232$ ft^2
Exterior walls:	144×8	$= 1152$ ft^2
Interior walls:	$2 \times 154 \times 8$	$= 2464$ ft^2
Total area:		4848 ft^2

The number of 4 × 12 sheets needed to cover 4848 ft^2 is

$$\frac{4848}{48} = 101 \text{ sheets}$$

If a total cost is desired, assuming a rate of \$0.34375 per square foot as determined above, this job would cost

$$4848 \times 0.34375 = \$1666.50$$

The last two objectives of this section, dealing with sheathing and shingling a roof, may be considered together. Both rely on roof area.

A study of the various roof planes in Chapters 15 and 17 will show that most of the common roof sections are made up of one or more plane figures. Included among these are rectangles, parallelograms, triangles, and trapezoids. Other roofs may be divided into sections that form one or more of these figures. In certain unusual configurations, roof areas may be approximated by one of the common plane figures.

Some examples of plane figures that appear in roof sections include the rectangular plane of a ranch house roof or the face of any gambrel roof section, the parallelogram plane formed by a hip roof with attached ell, the trapezoidal plane on a conventional hip roof, and the triangular plane of the roof section above the narrower end of a hip roof or on a square hip roof.

Figure 21-1 shows each of the plane figures discussed above, as well as the area formula for that figure. An examination of Figure 21-1 will reveal that in each formula the value h is the length of a common rafter. This should suggest to the efficient estimator another value of which to make special note as calculations are being made.

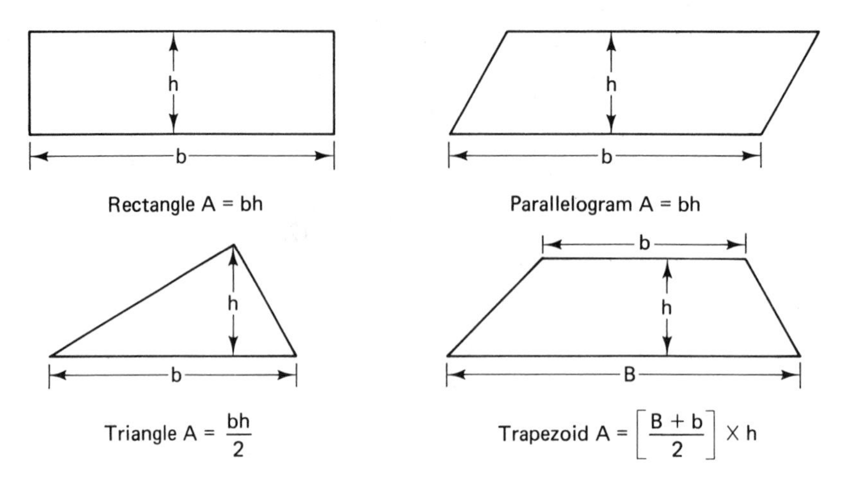

Figure 21-1

EXAMPLE 21-5

The 28 ft × 44 ft ranch house in earlier examples has a 5/12 pitch, a 16-in. overhang front and rear, and a 12-in. overhang on the ends. Determine (a) the number of sheets of $\frac{5}{8}$-in. CDX sheathing, and (b) the number of squares of shingles needed to cover the roof.

***Solution*:** First, determine the overall length of the common rafter, including overhang.

$$\begin{aligned} \text{RL} &= \text{ULL} \times \text{run} \\ &= 13 \times 15.33 \\ &= 199.29 \text{ in.} \quad \text{or} \quad 16.61 \text{ ft} \end{aligned}$$

Recall that a 5/12 pitch has a unit line length (ULL) of 13. The run is 14 ft for the building plus 16 in. or 1.33 ft for the overhang—a total of 15.33 ft.

The plane of the roof is a rectangle with a length, b, equal to the building length plus 1 ft overhang on each end—a total of 46 ft. Using the area formula gives us

$$\begin{aligned} A &= bh \\ &= 46 \times 16.61 \\ &= 764.06 \text{ ft}^2 \end{aligned}$$

This is doubled to account for both sides of the roof.

$$\text{Total area} = 1528.12 \text{ ft}^2$$

(a) Determining the amount of $\frac{5}{8}$-in. CDX sheathing, with each sheet covering 32 ft^2.

$$\frac{1528.12}{32} = 47.75 \text{ sheets}$$

(b) A "square" of shingles covers 100 ft^2. The number needed is found by dividing the roof area by 100 or, more simply, moving the decimal point in the area two places left. Thus

$$\frac{1528.12}{100} = 15.2812 \text{ squares}$$

Most shingles are packaged 3 bundles to the square. Certain extra-thick shingles are packaged 4 bundles to the square. This means that shingle orders may be rounded to the next higher $\frac{1}{3}$ square (0.33) or $\frac{1}{4}$ square (0.25).

The question arises as to how many extra bundles of shingles should be added for starters and ridge capping. (Aluminum vented ridge capping is frequently used today, providing much of the necessary attic venting area.) Shingles are laid with approximately 5 in. to the weather. The area one starter row covers along the lower edge determines the amount needed. Along the ridge, shingles are cut into thirds (12 in. each) and placed perpendicularly to the others. Here, again, 5 in. to the weather is the norm.

EXAMPLE 21-6

How many extra bundles of shingles are needed for (a) starters and (b) ridge capping in the 28 ft × 44 ft ranch house described in Example 21-5?

***Solution*:** (a) The area along one lower edge measures 5 in. × 46 ft. There are two edges for a total of

$$A = \frac{5}{12} \times 46 \times 2$$

$$= 38.33 \text{ ft}^2$$

This represents slightly more than 1 bundle.

(b) The length of the ridge will be covered 5 in. per piece, with each piece 12 in. or 1 ft wide. Essentially, this amounts to covering an area equal to the length of the ridge and 1 ft wide. Therefore, the length of the ridge also equals the number of square feet to be covered. In this case 46 ft of ridge length means 46 ft^2 or 0.46 square of shingles. All three quantities of shingles are added to the order. This roof called for 15.28 squares for the main roof, 0.38 square for the starters, and 0.46 square for capping—a total of

$$15.28 + 0.38 + 0.46 = 16.12 \text{ squares}$$

Assuming that the shingles are packaged 3 bundles per square, 49 bundles would be ordered.

In summary, the concept of area plays a most important role in estimating quantities of materials needed for covering walls, ceilings, and roofs of buildings. Perimeter multiplied by a wall height or divided by a width of material to be applied to walls provides an easy access to material quantities. The efficient estimator will maintain ready access to those building dimensions that determine the necessary perimeters and areas.

REVIEW EXERCISE

21-1. Sufficiently detailed plans for several houses are shown below. For each house, determine the following materials requirements.

(1) Number of sheets of sheathing needed to close in the exterior walls

(2) Number of square feet of siding **(i)** disregarding openings and **(ii)** deducting for openings (Assume windows to be average size. Sliding doors are 8 ft wide.)

(3) Number of sheets of 4 × 8 sheetrock for walls and ceilings

(4) Number of sheets of roof sheathing

(5) Number of squares of shingles (Assume that aluminum ridge cap is to be used on all except the ranch house; include a starter strip for all.)

(6) Metal drip edge comes in lengths of 10 ft. Assume that drip edge is to be applied to all roof edges along the rake as well as the eaves. Determine the number of pieces of drip edge needed for each roof.

(a)

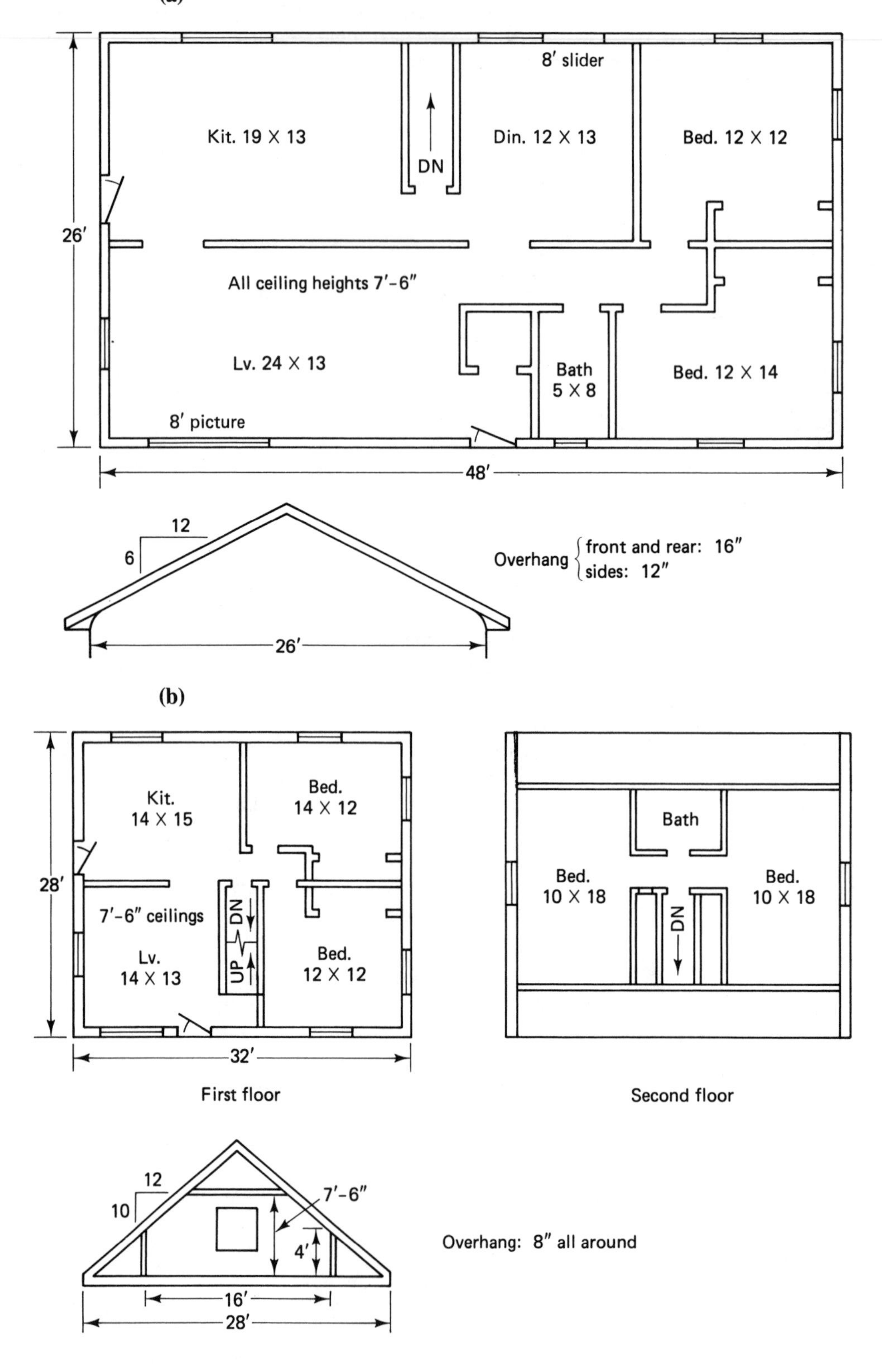

(c)

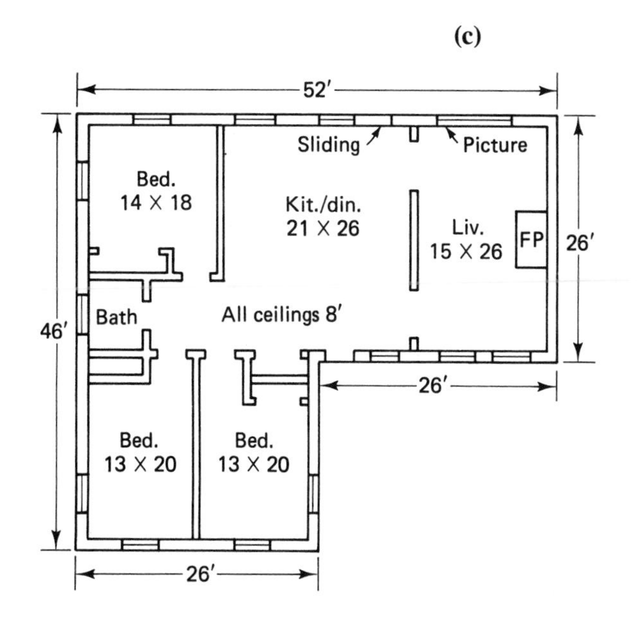
52′
Sliding
Picture
Bed.
14 X 18
Kit./din.
21 X 26
Liv.
15 X 26
FP
26′
46′
Bath
All ceilings 8′
26′
Bed.
13 X 20
Bed.
13 X 20
26′

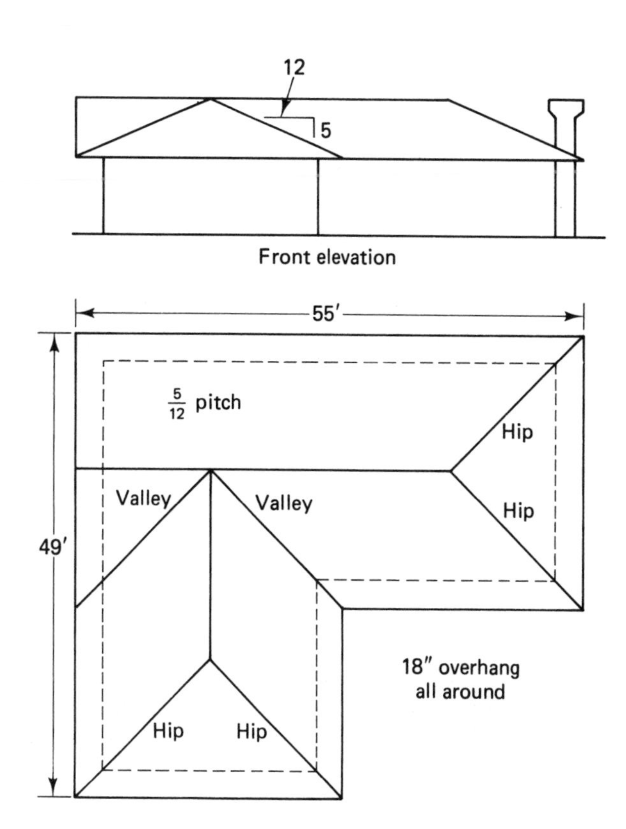
12
5
Front elevation
55′
5/12 pitch
Hip
Hip
Valley
Valley
49′
18″ overhang
all around
Hip
Hip

chapter 22

Insulation and Heat Loss

OBJECTIVES

Upon completing this chapter, the student will be able to:

1. Determine the quantities of various types of insulation needed for a given structure.
2. Determine R-factors provided by various combinations of insulating materials.
3. Make comparisons of various construction and insulating methods on the basis of cost-effectiveness.
4. Perform a heat-loss calculation for a structure.

Americans seem to need to be jolted into action when it comes to adjustments to their life-style. It sometimes takes events of near-crisis proportions to stimulate activities resulting in belt-tightening measures. Fortunately, it is at these times that American ingenuity enjoys a re-birthing and new ways of doing old things are discovered or invented.

Such was the case when the oil embargo took place and we suddenly found it necessary to spring into action to conserve oil, reducing our dependency on those countries. In a very brief period we learned to reduce drastically our gluttonous use of gasoline, electricity, and home heating oil. Car manufacturers sprang into action, building smaller, gas-conserving automobiles. Contractors experimented with ways of insulating older energy-hungry buildings. Thermostats were turned down and sweaters donned. Wood stoves and wood-burning furnaces were installed. Coal experienced rejuvenation as a heat source. Groundwater heat pumps and geothermal heat sources were given serious attention. The government got into the act with tax credits offered in exchange for money invested in certain energy-saving measures. Architects and home builders are still experimenting with various methods of insulating, with the objective of maximizing heat retention from any source. The search goes on.

At this writing, some facts related to energy-conservation methods have asserted themselves. Many more theories and techniques remain to be proven. Some techniques have been proven unworthy.

One fact remains obvious: Regardless of which source of heat is used to warm a house, every reasonable step should be taken to conserve that heat and retain it within the house. Heat retention should head the list of priorities for any building design. Closely following heat retention is the priority of heat source. (The authors recognize that heat loss is only a part of the overall problem of climate control. Certainly, elimination of excess, unwanted heat during the summer months is a part of the overall problem. However, proper insulation includes proper ventilation, with both concepts coordinated for maximum creature comfort.)

Toward the goal of heat retention, builders in the northeast have experimented with sundry techniques, finding flaws in some, utility in others. There follows a brief description of some of the framing and insulating methods for exterior walls recently tried and appropriate comments. The *R*-values mentioned ignore the insulating qualities of sheathing, siding, and so on.

1. *2 × 6 studs placed 16 in. o.c.* Fiberglass insulation provides *R*-19.
2. *2 × 6 studs placed 24 in. o.c.* Fiberglass insulation provides *R*-19.

Comments: The difference between the methods above is the number of studs used. Some builders prefer placing the studs 16 in. o.c. to provide greater support for drywall installation. Structurally, 24 in. o.c. provides no problem to the single-family residence designs. This writer has experienced no problems with drywall installation over 24 in. o.c. stud spacing. In any case, the extra costs of lumber, insulation, and labor should not be considered without also analyzing their benefit in reduced heating costs.

With either of the framing methods above, extra insulation may be provided by application of rigid insulation to the walls. One inch of rigid Styrofoam can provide an extra *R*-5, for a total of *R*-24. The Styrofoam may be applied to either the exterior or interior walls. Arguments favoring application to the exterior walls seem to prevail. Rigid Styrofoam applied to the interior walls creates problems with drywall application and finish work. Nail pops are more likely to occur when drywall nails have to pass through the drywall material, followed by 1 in. of foam before wedging into wood. Stud location for drywall nailing is more difficult. Extra trim build-out is needed around door and window openings. Extra-long nails are needed for both drywall and mouldings.

On the other hand, rigid Styrofoam applied to the exterior eliminates most of these problems. In addition, the Styrofoam may be extended over the shoe and floor joists, even extended to cover the foundation to the desired depth, providing a continuous insulating shield. The question as to whether the Styrofoam is applied directly to the studs or over the sheathing must be considered. Structurally, it makes sense to apply sheathing to the studs with the foam between the sheathing and siding. Extra-long nails will be needed for the siding application.

3. *2 × 4–16 in. o.c.* Fiberglass insulation provides *R*-11. Rigid Styrofoam can add *R*-5, for a total of *R*-16.
4. *Double 2 × 4 walls–16 in. o.c. with a separation of 12 in. outside to inside.* 6-in fiberglass applied to each wall provides a total *R*-38 (see Figure 22-1). A variation on this calls for stud spacing of 16 in. o.c. for the exterior wall and 24 in. o.c. for the interior wall. The exterior wall is the support wall. The stud spacing on the two walls may be staggered so that gaps in the insulation are not continuous from inside to outside.

As stated earlier, any method under consideration must be analyzed with respect to cost and resulting energy conservation. Proper insulation must be viewed as an invest-

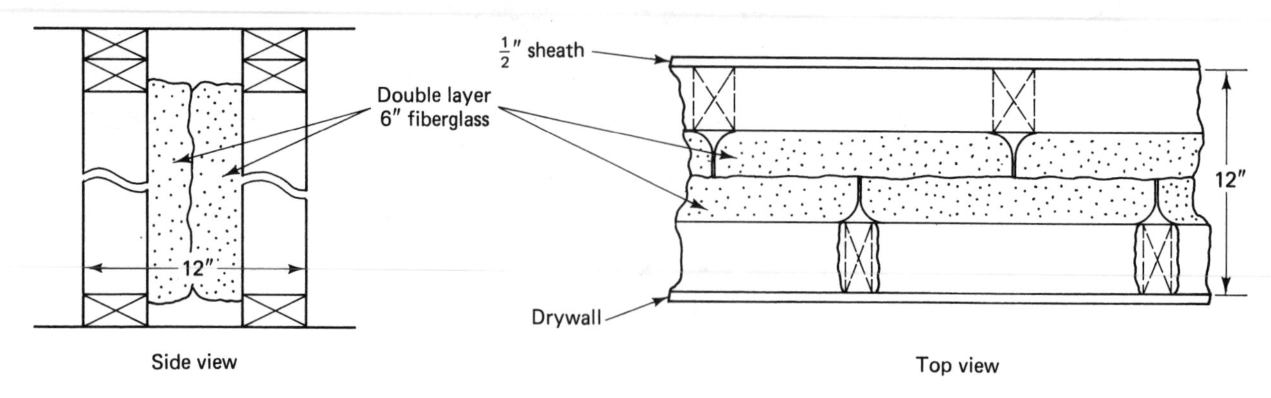

Figure 22-1

ment of dollars. Any investor expects a return. The rate of return is what determines the feasibility of an option. The question simply put is: If a dollar is invested in insulation today, how much money will be saved in future years by reduced heating costs? For example, suppose it has been determined that method B of insulating a house costs \$1000 more than method A. At the same time, based on the type of heating and its cost, heating requirements, and so on, method B will require \$200 per year less in heating costs than method A. This means that method B offers a return of 20% per year (\$200/\$1000 = 0.20 or 20%)—a most attractive yield. Put another way, it will take only 5 years to return the \$1000 to the owner (\$1000/\$200 per year = 5 years). After that, the annual savings is a bonus. Even if the owner has to borrow the extra \$1000, the opportunity is obvious. Real wealth could be acquired if money could be borrowed at, say, 14% and reinvested at 20%.

Certain minimum R-factors seem to have been established for various parts of the country. In the northeast, it is generally agreed that exterior walls should provide a minimum value of R-19 and ceilings R-38. It is safe to assume that there are maximums that should also be considered. Exceeding these results is overkill and wastes money. The amount by which the minimums are to be exceeded should be carefully analyzed with respect to cost-effectiveness.

Turning to one of this chapter's objectives, the amount of insulating material needed is a direct function of the area to be insulated. For ceilings, the ceiling area is used. For walls, the wall area, minus openings, is used. If the foundation is to be insulated, the wall area of the insulated portion is used. If a concrete floor slab is to be insulated, again, area is the basis. One accepted method of insulating a floor slab is to use rigid Styrofoam within 4 ft of its outer edge. When a foundation wall is insulated, it is sometimes recommended that only that portion exposed to the expected frost depth be insulated. Little heat loss will be experienced below the frost level. In any case, area is the required measurement.

Fiberglass insulation is ordered in rolls that cover a specified number of square feet. For example, a roll of foil-faced $6\frac{1}{4}$ in. × $23\frac{1}{2}$ in. insulation will cover approximately 75 ft^2. A roll of foil-faced $6\frac{1}{4}$ in. × $15\frac{1}{2}$ in. fiberglass insulation will cover approximately 50 ft^2. Rigid foam 1 in. thick is available in sheets measuring 2 ft × 8 ft and 4 ft × 8 ft. The estimator should check unit sizes with the supplier when ordering rolls or sheets of insulation.

EXAMPLE 22-1

Specifications for a 28 ft × 44 ft single-story ranch house call for $6\frac{1}{4}$ in. of fiberglass insulation in the exterior walls, 1 in. of rigid insulation wrapped around the exterior walls, extending from the top plate to 5 ft below the foundation top, and $12\frac{1}{2}$ in. of fiberglass over the ceiling. The ceiling insulation will be provided by a double layer of unfaced $6\frac{1}{4}$ in. × $23\frac{1}{2}$ in. fiberglass. The exterior walls are framed with 2 × 6–24 in. o.c. The door and window openings total 210 ft^2. Determine the amount of each type of insulation needed.

***Solution*:** The exterior walls require foil-faced $6\frac{1}{4}$ in. × $23\frac{1}{2}$ in. fiberglass. The wall area equals the perimeter multiplied by the wall height minus the area of the openings. The perimeter is

$$\begin{aligned} P &= 2(28 + 44) \\ &= 2(72) \\ &= 144 \text{ ft} \\ \text{wall area} &= \text{perimeter} \times \text{wall height} \\ &= 144 \times 8 \\ &= 1152 \text{ ft}^2 \\ \text{insulated area} &= \text{wall area} - \text{area of openings} \\ &= 1152 - 210 \\ &= 942 \text{ ft}^2 \end{aligned}$$

Thus 942 ft^2 of $6\frac{1}{4}$ in. $\times$ $23\frac{1}{2}$ in. foil-faced fiberglass is needed for the walls. At 75 ft^2 per roll, the number of rolls required would be

$$\frac{942}{75} = 12.56 \quad \text{or} \quad 13$$

The amount of rigid foam needed for the exterior can be found by adding to the net wall area of 942 ft^2 the area to be insulated below that. This depends on the size of the floor joists and sill plate. Let us assume that they total approximately 1 ft. This 1-ft distance plus the 5 ft of the foundation to be covered totals 6 ft more around the perimeter; that is,

$$6 \times 144 = 864 \text{ ft}^2$$

Added to the net wall area results in a need for

$$864 + 942 = 1806 \text{ ft}^2$$

to be covered. That would require

$$\frac{1806}{16} = 112.875$$

or 113 sheets of 2 ft $\times$ 8 ft rigid foam.

The ceiling area measures 28 ft $\times$ 44 ft. This is an area of

$$\begin{aligned} \text{area} &= 28 \times 44 \\ &= 1232 \text{ ft}^2 \end{aligned}$$

A double layer of $6\frac{1}{4}$-in. insulation is required, which brings the total to

$$2 \times 1232 = 2464 \text{ ft}^2$$

With 75 ft^2 per roll, the order would be placed for

$$\frac{2464}{75} = 32.85 \quad \text{or} \quad 33 \text{ rolls}$$

of $6\frac{1}{4}$ in. $\times$ $23\frac{1}{2}$ in. unfaced fiberglass.

EXAMPLE 22-2

Compare the cost of materials used in framing the exterior walls of the 28 ft $\times$ 44 ft house described in Example 22-1 by each of the following methods: (a) 2 $\times$ 4–16 in. o.c., (b) 2 $\times$ 6–16 in. o.c., and (c) 2 $\times$ 6–24 in. o.c. Assume a lumber cost of \$325 mbf for all sizes used. Ignore studs needed for openings.

***Solution*:** (a) Materials needed for the shoe and double plate total three times the perimeter or

$$3 \times 144 = 432 \text{ linear feet}$$

Placing studs 16 in. o.c. will require

$$\tfrac{3}{4} \times 144 = 108 \text{ studs}$$

Adding 1 stud for each corner and 1 for each starter brings the total to 116 studs. The number of board feet of 2 × 4's used is

For the studs: $\frac{2}{3} \times 108 \times 8 = 576$
For the shoe and plates: $\frac{2}{3} \times 432 = 288$
Total: $576 + 288 = 864$ board feet

The cost is

$$864 \times 0.325 = \$280.80$$

(b) Framing with 2 × 6–16 in. o.c. requires the same number of studs and the same number of linear feet by material as in part (a). In 2 × 6, the number of linear feet equals the number of board feet. The total linear feet of 2 × 6 needed is

$$(108 \times 8) + 432 = 864 + 432 = 1296 \text{ linear feet}$$

With 2 × 6 material 1296 linear feet equals 1296 board feet, resulting in a cost of

$$1296 \times 0.325 = \$421.20$$

(c) Framing with 2 × 6–24 in. o.c. will require the same amount of material for shoe and plates—432 linear feet. The number of studs is

$$\frac{144}{2} = 72$$

plus corners and starters—a total of 80 studs. The linear footage (and board footage) of 80 studs 8 ft in length is

$$8 \times 80 = 640$$

which, added to the shoe and plate needs, totals

$$640 + 432 = 1072 \text{ board feet}$$

The cost is

$$1072 \times 0.325 = \$348.40$$

It can be seen, then, that spending an extra $67 for framing material will allow space for 50% more insulation—a negligible amount of money when considering the overall cost of a house. Any difference in labor costs among the three methods would be negligible.

Next to be considered is the extra cost of insulation material when using $6\frac{1}{4}$-in. as compared to $3\frac{1}{2}$-in. fiberglass. At this writing the price difference between $3\frac{1}{2}$-in. fiberglass insulation and $6\frac{1}{4}$-in. fiberglass insulation is approximately 10 cents per square foot. In Example 22-1 $3\frac{1}{2}$-in. fiberglass insulation would be applied to 942 ft^2. The extra cost for $6\frac{1}{4}$-in. insulation would be

$$\$0.10 \times 942 = \$94.20$$

Therefore, if 2 × 6's–24 in. o.c. framing is used, the extra cost of this type of framing and insulation over 2 × 4's–16 o.c. is

$$\$67.00 + 94.20 = \$161.20$$

It is yet to be established that this is a small amount and represents an excellent investment to be recovered quickly in heating cost savings.

The answers to questions related to the amount to be saved by certain insulating methods lies in results of heat-loss calculations. A heat-loss calculation determines the heating requirements for a structure as measured in British thermal units (Btu). Comparisons of the Btu requirements are made using various insulating methods (or before–after comparisons). The source of heat is priced in terms of the Btu output it provides. These figures can then be projected to approximate annual heating costs.

Several methods for determining heat loss are in use, and unfortunately, each method will likely provide results differing from those determined by other methods. Part of the reason for differing results is that certain assumptions and approximations are used as data in the calculations. Any conclusion drawn can only be as valid as the assumptions leading to those results. Nevertheless, a certain amount of validity lies in each result.

Regardless of which method is used to analyze heat loss from a structure, the basic approach requires a collection of data. Heat loss is caused by a transfer of heat from a heated area to an area of less warmth. One of the assumptions made is the temperature difference between the outside and inside of the structure. A desired inside temperature is assumed, an expected outside average temperature is assumed, and the difference between the two taken as the basis for other calculations. Another assumption is the number of times per unit of time that air in a structure is exchanged with outside air. This occurs because of leakage around openings and the opening and closing of doors and windows. The leakage is referred to as **infiltration losses.** The number of occupants has an effect on the rate of air exchange. The greater the number of occupants, the more frequently the internal house air will be exchanged due to door openings.

The type of materials used in constructing a house, as well as the size and number of door and window openings, have an effect on heat loss. When an older house is analyzed for heat loss, assumptions must sometimes be made as to the type of materials, the stud spacing, and other factors not able to be determined by inspection. For this reason a small amount of error becomes inherent in the conclusions drawn. However, these errors should not seriously affect the conclusions.

All materials used in construction have some insulating capability. The *R*-factors for various materials are available. A list of some is included in Appendix C. When a particular material is under consideration for use in constructing a structure, its *R*-factor may come under scrutiny and be compared with alternatives. Glass is a good example. The window manufacturers have responded to the energy crunch by offering triple-glazed windows. Listed below are the *R*-factors for a single pane, double-paned insulating glass, and triple-paned insulating glass. These are winter figures and assume $\frac{1}{4}$-in. air space between panes.

Single glass	0.89
Two panes	1.55
Three panes	2.13
Low Emissivity (Low E)	3.23

When analyzing these alternatives, the cost of each additional pane must be included. *Remember:* The primary concern lies in the expected return on dollars invested. It is quite possible that the third pane is a questionable investment when the added cost is compared to the slight increase in *R*-value. Depending on the type of window being used, if a third pane is desired, it may be better installed in the form of a separate insert. This would have the advantage over a triple-paned window of providing added resistance to infiltration losses. It should also be mentioned that the amount of space between panes affects the *R*-factor. In the list above, a $\frac{1}{2}$-in. space with three panes increases the *R*-factor to 2.78.

To calculate the heat requirements for a structure, certain facts must be assembled. These include data related to the type of structure and assumptions related to the environment in which the structure is located. The necessary data related to the structure are:

1. Type of interior wall-covering material
2. Type of exterior wall-covering material
3. Type of exterior siding material
4. Wall stud size and spacing
5. Type of foundation material
6. Type of ceiling material
7. Amount and type of insulation in walls, ceilings, and floors
8. Type of vapor-barrier material
9. Building dimensions: length, width, ceiling height
10. Area of openings: doors and windows
11. Adequacy of attic ventilation

The necessary data related to the environment are:

1. Interior design temperature
2. Exterior design temperature
3. Design wind speed
4. Rate of infiltration loss

Once the necessary data have been collected and design assumptions made, calculations of the heat-loss rate are made for each segment of the structure. The heat-loss rate is measured in the number of British thermal units per hour (Btu/hr). The sum of the R factors, R_t, of materials affecting heat loss is found. This sum is then divided into the product of the design temperature difference, dT, and the affected area, A. That is,

$$\text{Btu/hr} = \frac{dT \times A}{R_t}$$

After the calculation of heat loss has been completed for each section, these values are totaled. The result is the total heat loss for the structure as measured in Btu/hr.

EXAMPLE 22-3

Calculate the heat loss for the following structure given the assumptions as stated.

Single-story 28 ft × 44 ft ranch
Foundation: 4-in. slab on grade
Wall height: 8 ft
Exterior framing: 2 × 6–24 in. o.c.
Interior wall covering: $\frac{1}{2}$-in. drywall
Exterior wall covering: $\frac{1}{2}$-in. plywood and $\frac{3}{4}$ in. × 10 in. lapped siding
Ceiling covering: $\frac{1}{2}$-in. drywall
Attic ventilation: adequate
Windows: all double-pane insulated glass with $\frac{1}{4}$-in. air space; total area: 120 ft^2
Doors: two insulated steel doors; total area: 40 ft^2
Insulation: Walls: $6\frac{1}{4}$-in. foil-faced fiberglass
Ceilings: $12\frac{1}{2}$-in. unfaced fiberglass
Slab: 1-in. Styrofoam within 4 ft of perimeter
Vapor barrier: 4-mil polyethylene on ceiling

Design assumptions:

Interior design temperature: 70°F
Exterior design temperature: −10°F
Design wind speed: 15 mph
Slab design temperature: 0°F for outer 4 ft, 35°F center
Infiltration losses: one air exchange per hour

Comments: Temperatures inside a structure will vary from ceiling to floor. The assumption is based on an average between the ceiling and floor temperatures. Infiltration losses vary with the tightness of the structure and the number of people entering and leaving it each hour.

Solution

1. Heat loss due to infiltration

$$\text{Btu/hr} = V \times dt \times (0.02)$$

$$\text{Volume} = 28 \times 44 \times 8 = 9856 \text{ ft}^2$$

$$9856 \times 80 \times 0.02 = 15{,}770 \text{ Btu/hr}$$

Comment: This step represents a deviation from the formula stated above. The factor 0.02 represents the number of Btu per hour per square foot to heat dry air. Calculations for other segments will be done by the Btu/hr formula that uses area.

2. Heat loss through walls

a. Studded portion

Inside air film	0.68
$\frac{1}{2}$-in. drywall	0.45
Stud: $1\frac{1}{2}$ in. by $3\frac{1}{2}$ in.	4.38
$\frac{1}{2}$-in. plywood	0.63
Siding	1.05
Outside air film	0.17
Total (R_t)	7.36

Total area of studs:

$$144/2 \text{ studs} \times 1.5/12 \text{ ft wide} \times 8 \text{ ft high} = 72 \text{ ft}^2$$

$$\frac{dT \times A}{R_t} = \frac{80 \times 72}{7.36} = 783 \text{ Btu/hr}$$

b. Unstudded portion

Inside air film	0.68
$\frac{1}{2}$-in. drywall	0.45
$6\frac{1}{4}$-in. fiberglass	19.00
$\frac{1}{2}$-in. plywood	0.63
Siding	1.05
Outside air film	0.17
R_t	21.98

$$\text{Area} = \text{wall area} - (\text{stud area} + \text{window area} + \text{door area})$$

$$= (144 \times 8) - (72 + 120 + 40) = 920 \text{ ft}^2$$

$$\frac{80 \times 920}{21.98} = 3348 \text{ Btu/hr}$$

3. Heat loss through doors

Inside air film	0.68
Door	7.30
Outside air film	0.17
R_t	8.15

$$\frac{80 \times 40}{8.15} = 393 \text{ Btu/hr}$$

4. Heat loss through windows

Inside air film	0.68
Glass	1.55
Outside air film	0.17
R_t	2.40

$$\frac{80 \times 120}{2.4} = 4000 \text{ Btu/hr}$$

5. Heat loss through ceiling

Inside air film	0.68
$\frac{1}{2}$-in. drywall	0.45
$12\frac{1}{2}$-in. fiberglass	38.00
Outside air film	0.17
R_t	39.30

$$\frac{80 \times 1232}{39.30} = 2508 \text{ Btu/hr}$$

6. Heat loss through slab

$$\text{Total area} = 28 \times 44 = 1232 \text{ ft}^2$$

$$\text{Uninsulated: } 20 \times 36 = \underline{720}$$

$$\text{Insulated: } 512 \text{ (by subtraction)}$$

a. Uninsulated area

Inside air film	0.68
4-in. concrete	0.32
R_t	1.00

$$\frac{40 \times 720}{1} = 28{,}800 \text{ Btu/hr}$$

b. Insulated area

Inside air film	0.68
4-in. concrete	0.32
1-in. Styrofoam	4.17
R_t	5.17

$$\frac{65 \times 512}{5.17} = 6437 \text{ Btu/hr}$$

Total heat loss:

Infiltration	15,770
Walls	783
	3,348
Doors	393
Windows	4,000
Ceiling	2,508
Floor	28,800
	6,437
	62,039 Btu/hr

It is suggested in the Exercises that variations in the heat loss for this structure be further explored.

The question of vapor barriers and permeability has not been discussed here—not because the authors consider the topic unimportant, but because it is not an objective of this chapter. Vapor barriers do play a significant role in proper insulation. References have been included in the Bibliography relevant to this topic.

REVIEW EXERCISES

22-1. Compute the total heat loss in Btu/hr for the house described in Example 22-3, assuming that no insulation has been installed.

22-2. Compute the total heat loss for the house described in Example 22-3, assuming that 1 in. of Styrofoam (*R*-5) has been added to the exterior walls, in addition to the insulation as described.

22-3. Calculate a percentage heat-loss distribution for the house described in Example 22-3 **(a)** with no insulation and **(b)** with insulation as originally described. Determine what percent of the total heat loss is through each of the following: infiltration, walls, doors, windows, ceiling, floor.

22-4. Study the results called for in Problem 22-3: Through what segment of the uninsulated house does the greatest heat loss occur? the least? What priorities should be followed when insulating a house?

The following problems may be assigned as projects to be done outside class.

22-5. Research the heating costs per Btu for various heating sources (oil, electric, wood) and determine the added costs and rates of return for **(a)** insulating as originally described, and **(b)** adding 1 in. of Styrofoam, as compared to an uninsulated house.

22-6. Select a house of your choosing and perform a heat-loss calculation. (You may choose one of the plans in Appendix D, if you wish.)

chapter 23

The Estimating Process

OBJECTIVES

Upon completing this chapter, the student will be able to:

1. Prepare a materials list for construction of a residence.

The purpose of this chapter is to illustrate the process of accurately and efficiently estimating the materials cost for a residence. Concepts and procedures described in preceding chapters are applied.

The estimating process results in figures that do just what the term implies—**estimate** the cost and quantity of materials. A certain amount of error will ordinarily exist in the estimate. It is unfair to both the contractor and the customer to present estimates that grossly vary from the actual amount. It is incumbent on the estimator to minimize error insofar as he has control over the variables.

Despite an accurate forecast by an estimator as to the amount and cost of materials, several factors exist which could alter his figures. Included among these: unforeseen delays in delivery of materials, wasted materials due to poor judgment by workers, materials stolen from the site, defective materials, unforeseen price changes. The effective estimator will consider these factors and prepare the estimate accordingly. To fend off the potential losses incurred by the unforeseen, many estimators use a ''fudge factor,'' a percent multiplier that is added to a total.

Whether a job is being performed under a contract that specifies a price or under a cost-plus agreement, an accurate estimate is to be expected. Actual results should generally fall within plus or minus 5% of an estimate; an outer limit of variance should be 10%.

Accuracy of the estimate will be somewhat assured if the estimator uses good mathematics. A potential problem stemming from omission of materials can be avoided by using a checklist. The checklist should be written so that materials are listed in the order of use. This provides a logical flow for the estimator to follow. In listing materials, the estimator can mentally picture the construction process and allow materials to flow onto paper in the same order in which they will be used on the job site. Figure 23-1 shows a sample list for a single-story ranch house. Certain jobs will require modifications

ESTIMATE SUMMARY

Project
Location ____________

Sheet no. ____ of ____

Item	Size/ number/ length	Bd. ft	Unit cost	Labor	Others	Total
1. Site preparation						
2. Excavation						
3. Footing						
4. Foundation						
5. Drain tile						
6. Foundation coating						
7. Water or well						
8. Septic or sewer						
9. Exterior cellar insulation						
10. Backfill						
11. Gravel						
12. Cellar slab						
13. Girder supports						
14. Girder						
15. Sill seal						
16. Sill plates						
17. Floor joists						
18. F.j. headers						
19. Bridging						
20. Subfloor						
21. Shoe and plates						
22. Studs						
23. Headers and sills						
24. Wall sheathing						
25. Stairs						
26. Cross-ties						
27. Rafters						
28. Roof sheathing						
29. Jack studs						
30. Gable end sheathing						
31. Soffit and fascia						
Framing						
Covering						
32. Drip edge						
33. Asphalt shingles						
34. Ridge vent						
35. Window schedule						

Figure 23-1

Item	Size/ number/ length	Bd. ft	Unit cost	Labor	Others	Total
36. Door schedule (ext.) 37. Exterior trim 38. Siding 39. Interior framing Shoe and plates Studs 40. Wiring 41. Plumbing 42. Insulation 43. Drywall 44. Masonry 45. Kitchen 46. Bath(s) 47. Heating 48. Second-floor covering 49. Door schedule (int.) 50. Molding 51. Paint, caulk 52. Flooring 53. Driveway 54. Landscaping 55. Nails 56. Sales tax 57. Markup						
Totals						

Figure 23-1 *(Continued)*

to this list. Other possible contingencies should be considered as well. An example might be ledge encountered in the excavation process. Extra costs would be incurred for its removal. Clauses in a contract are usually inserted to cover events beyond control of the contractor.

An estimator usually makes his estimate starting with a worksheet on which calculations are made. The careful estimator will keep the worksheets for future reference in checking for errors. The worksheet in Figure 23-2 shows the calculations related to the house shown in Figure 23-3.

WORKSHEET

Project location ________ Sheet no. ____ of ____

1. Site preparation

Figure 23-2

2. Excavation

$$\frac{34 \times 50 \times 6}{27} = 378 \text{ yd}^3$$

3. Footing

$$\frac{8/12 \times 20/12 \times 144}{27} = 5.9 \text{ or } 6 \text{ yd}^3$$

4. Foundation

$$\frac{7 \times 10/12 \times 144}{27} = 31 \text{ yd}^3$$

5. Drain tile and filter paper
   $144 \times 1.4 = 200 \text{ ft}^2$ paper
   144 lf tile
6. Foundation coating
   $5 \times 144 = 720 \text{ ft}^2$ at 100 ft^2 per gallon
   approx. two 5-gal cans
7. Well—as per subcontract
8. Septic system—as per subcontract
9. Cellar insulation
   1 in. Styrofoam to 5 ft below grade, or approx. 6 ft below foundation
   $6 \times 144 = 864 \text{ ft}^2/16 = 54$ sheets 1 in. by 2 ft by 8 ft
10. Backfill
    Suitable excavated site material
11. Gravel
    Crushed stone for drain tile: $\frac{1}{2} \times 1 \text{ ft} \times 1 \text{ ft} \times 144 = 72 \text{ ft}^3$

$$\frac{72}{27} = 2.7 \text{ or } 3 \text{ yd}^3$$

    Cellar slab base: average 6 in. depth

$$\frac{\frac{6}{12} \times 1232}{27} = 22.8 \text{ or } 23 \text{ yd}^3$$

12. Cellar slab

$$\frac{1232 \times \frac{3}{12}}{27} = 11.4 \text{ yd}^3$$

13. Girder supports
    Adjustable steel columns every 8 ft—5 posts
14. Girder
    6×10 built-up; 6/16's, 3/12's
15. Sill seal
    144 lf
16. Sill plates
    2×6: 4/14's, 2/12's, 4/16's
17. Floor joists
    2×10: 70/14's
18. Floor joist headers
    2×10: 4/16's, 2/12's
19. Bridging
    3 ft/space $\times$ 44 spaces $\times$ 2 rows = 264 lf
20. Subflooring
    $\frac{1}{2}$ in. CDX plywood: 1232/32 = 38.5 or 39 sheets
21. Shoes/Plates
    2×6: $3 \times 144 = 432$ lf or 36/12's
22. Studs
    2×6-16 in. o.c. $3/4 \times 144 = 108$ = 4 starters + 4 corners + 18 windows + 11 partitions = 145 studs

Figure 23-2 *(Continued)*

23. Headers/sills
    2 × 6: 7/12's, 4/10's
24. Exterior wall sheathing
    $$\frac{144}{4} = 36 \text{ sheets of } 1/2 \text{ in. CDX}$$
25. Stairs
    2 × 12 risers: 2/14's
    2 × 12 treads: 11 risers × 3 ft = 33 ft or 3/12's
26, 27. Trusses
    2 × 6 rafters: 46/18's
    2 × 4 bottom chords: 46/14's
    2 × 4 tension chords:
    $$T = \frac{S}{2\sqrt{16 + \text{URi}^2}} = \frac{14}{2\sqrt{16 + 5^2}} = 89.6 \text{ in.}$$
    2 × 4 compression chords
    $$C = \frac{T}{2} = \frac{89.6}{2} = 44.8 \text{ in.}$$
    89.6 + 44.8 = 134.4 in.
    Use 1 – 12-ft piece for 1 compression and 1 tension chord: 46/12's
28. Roof sheathing
    $$5/8 \text{ in. CDX: } \frac{13 \times 15.5}{12} = 16.79 \text{ ft rafter length}$$
    $$16.9 \times 2 \times 46 = \frac{1545 \text{ ft}^2}{32} = 49 \text{ sheets}$$
29. Jack studs (gable ends)
    TRi = URi × run = 5 × 14 = 70 in. or 6 ft
    3/4 × 28 = 21 studs – 6 ft or 11/12's
30. Gable end sheathing
    $$\frac{6 \times 28}{32} = 5.25 \text{ or 6 sheets}$$
31. Overhang
    End rafters: 2 × 6: 4/18's
    Overhang framing:
        3 ft × 25 = 75 lf or 6/12's
        Face plate: 6/12's, 2/10's
    3/8-in. A-C exterior plywood
    $$\frac{18 \text{ in.} \times 46 \text{ ft} \times 2}{32} = 4.3$$
    $$\frac{12 \text{ in.} \times 17 \text{ ft} \times 4}{32} = 2.1$$
    Total 6.4 or 7 sheets
32. Drip edge
    Eaves: 46 × 2 = 92 lf
    Rakes: 16.8 × 4 = 67.2 lf
    Total 159.2 lf or 16 lengths
33. Asphalt shingle/felt
    Main: 16.8 × 2 × 46 = 1545.6 ft^2
    Starter: 5/12 × 46 = 38 ft^2
    Total 1584 ft^2 or 16 squares
    15-lb felt: 1584/432 = 3.7 or 4 rolls
34. Ridge vent
    46 ft or 4/10's

Figure 23-2 *(Continued)*

35. Window schedule
    7–2 ft 9 in. × 4 ft 1 in. double hung
    1–2 ft 9 in. × 3 ft 5 in. double hung
    1–9 ft 5 in. × 4 ft 5 in. picture with 2 ft flanking double hung
    1–3 ft 4 in. × 3 ft 4 in. casement
    1–7 ft 10 in. × 5 ft 1 in. bay
36. Exterior door schedule
    1–2 ft 8 in. × 6 ft 8 in. insulated steel entry
    1–8 ft 0 in. × 6 ft 8 in. insulated glass wood-framed slider
37. Exterior trim
    No. 2 pine corner boards: 1 × 5: 4/10′s
    1 × 6: 4/10′s
38. Siding
    Sides: 8.5 × 144 = 1224 ft^2
    Gable ends: 6 × 28 = 168 ft^2
    Subtotal 1392 ft^2
    Less openings
        Windows: 180 ft^2
        Doors: 74 ft^2
        Total 254 ft^2
    Net area 1138 ft^2 or 12 squares
39. Interior framing
    Shoe/plates: 178 ft × 3 = 534 lf or 45/12′s
    Studs: 2 × 4: 534/8′s
40. Electrical—as per subcontract
41. Plumbing—as per subcontract
42. Insulation
    Ceiling: double-layer 6 1/4 in. × 23 1/2 in. unfaced f.g.
        2464 ft^2/75 ft^2 per roll = 33 rolls
    Walls: $6\frac{1}{4}$ in. × $15\frac{1}{2}$ in. foil-faced f.g.

$$(144 \times 7.5) - 254 = \frac{826 \text{ ft}^2}{50} = 17 \text{ rolls}$$

    1 in. Styrofoam, 2 ft × 8 ft sheets:

$$(8 \text{ ft } 8 \text{ in.} \times 144) - 254 = \frac{994 \text{ ft}^2}{16} = 62 \text{ sheets}$$

43. Drywall
    Exterior walls: 144 × 8 = 1152 ft^2
    Ceilings: 1232 ft^2
    Interior walls: 178 × 8 × 2 = 2848 ft^2
    Total area: 5232 ft^2/48 = 109 sheets of 1/2 in. × 4 ft × 12 ft
44. Masonry—as per subcontract
45. Kitchen—as per bid
46. Baths—as per subcontract
47. Heating—as per subcontract
48. Second-floor covering
    5/8-in. particleboard: 1232/32 = 38.5 or 39 sheets
49. Interior door schedule
    Prehung six-panel pine, split jambs: 7–2 ft 6 in. × 6 ft 6 in.
    Slider six-panel pine, no jamb: 2 sets 2 ft 0 in. × 6 ft 6 in.
    Slider six-panel pine, no jamb: 1 set 4 ft 0 in. × 6 ft 6 in.
50. Moldings
    Mopboards: 3 1/2 in. clamshell—638 lf
    Door/window: 2 1/2 in. clam—544 lf
    Window sill: 48 lf
    Opening build-out: select pine—1 × 6—68 lf

Figure 23-2 *(Continued)*

51. Paint/stain—as per subcontract
52. Flooring
    Kitchen/bath: vinyl inlaid—140 ft^2
    Entry: slate—48 ft^2
    Other floors: wall to wall carpet—116 yd^2
53. Driveway—as per subcontract
54. Landscaping—as per subcontract
55. Nails

16d common –	150 lb	1″ roofing	50 lb
8d common	100 lb	4d finish	5 lb
16d galv.	10 lb	6d finish	20 lb
8d galv.	15 lb	8d finish	20 lb
4d shingle	50 lb		

Figure 23-2 *(Continued)*

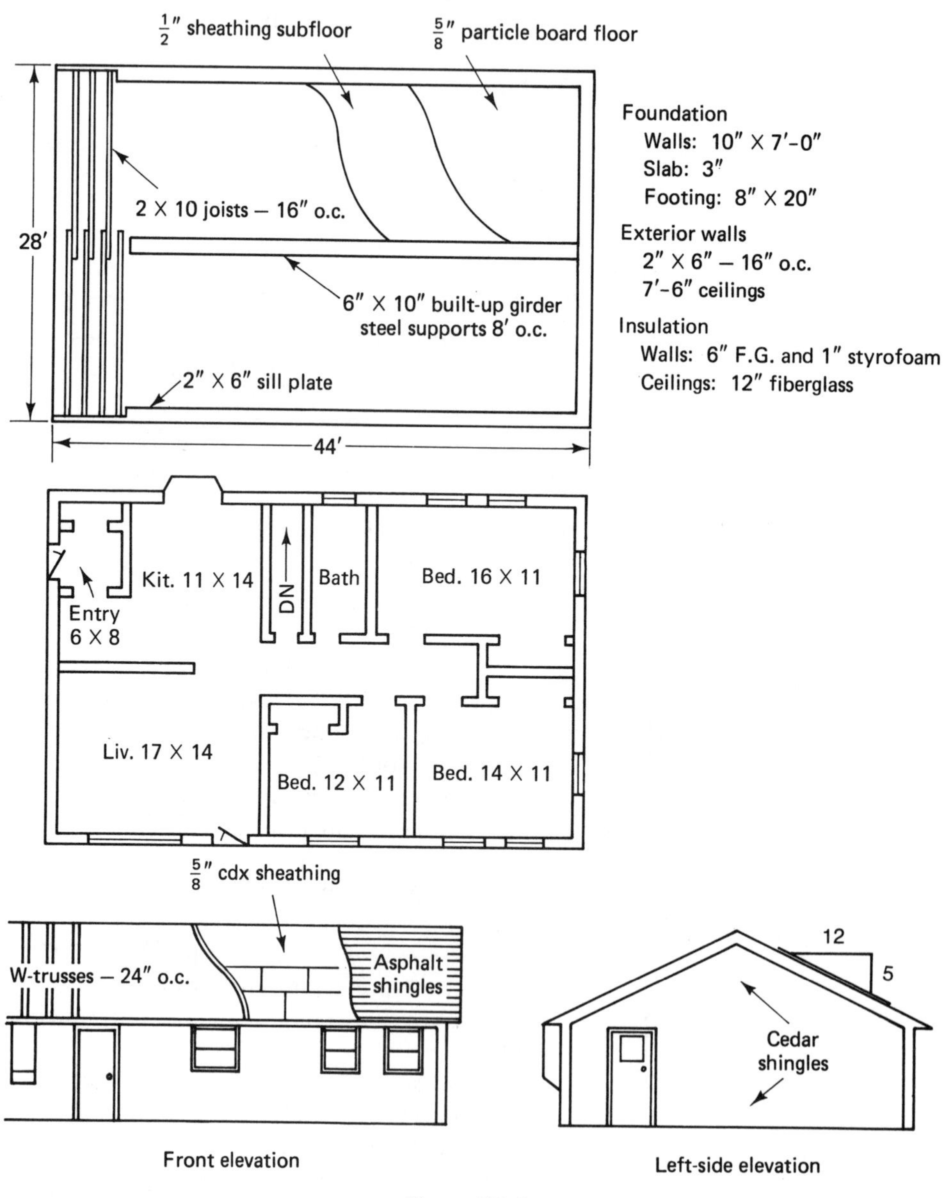

Figure 23-3

Each item has been numbered to correspond to the estimate summary sheet. A partially complete estimate summary sheet is shown in Figure 23-4.

ESTIMATE SUMMARY

Project location ____________ Sheet no. ____ of ____

Item	Size/ number/ length	Bd. ft.	Unit cost	Labor	Others	Total
1. Site preparation					×	
2. Excavation					×	
3. Footing	6 yd^3					
4. Foundation	31 yd^3					
5. Drain tile	144 lf					
Filter paper	200 ft^2					
6. Foundation coating	10 gal					
7. Water or well					×	
8. Septic or sewer					×	
9. Exterior cellar insulation	54 sheets 1 in. × 2 ft × 8 ft					
10. Backfill					×	
11. Gravel	3 yd^3 crushed stone 23 yd^3 gravel					
12. Cellar slab	11.5 yd^3					
13. Girder supports	5 adj. posts					
14. Girder	2 × 10: 3/12's 6/16's					
15. Sill seal	144 lf					
16. Sill plates	2 × 6: 2/12's 4/14's 4/16's					
17. Floor joists	2 × 10: 70/14's					
18. F.j. headers	2 × 10: 2/12's 4/16's					
19. Bridging	1 × 3: 264 lf					
20. Subfloor	1/2 in. CDX: 39 sheets					
21. Shoe and plates	2 × 6: 36/12's					
22. Studs	2 × 6: 145/8's					
23. Headers and sills	2 × 6: 4/10's 7/12's					
24. Wall sheathing	1/2 in. CDX plywood: 36 sheets					
25. Stairs	2 × 12: 3/12's 2/14's					
26. Cross-ties (trusses)						
27. Rafters (trusses)	2 × 6: 46/18's 2 × 4: 46/14's 46/12's					
28. Roof sheathing	5/8 in. CDX: 49 sheets					
29. Jack studs	2 × 4: 11/12's					

Figure 23-4

Item	Size/ number/ length	Bd. ft.	Unit cost	Labor	Others	Total
30. Gable end sheathing	1/2 in. CDX: 6 sheets					
31. Soffit and fascia						
Framing	2 × 6: 4/18′s					
	2 × 4: 12/12′s					
	2/10′s					
Covering	3/8 in. A-C ext:					
	7 sheets					
32. Drip edge	16 pcs.					
33. Asphalt shingles	16 squares					
Felt	4 rolls					
34. Ridge vent	4 pcs.					
35. Window schedule						
7–2 ft 9 in. × 4 ft 1 in. d.h.						
1–2 ft 9 in. × 3 ft 5 in. d.h.						
1–9 ft 5 in. × 4 ft 5 in. pict., 2 ft 0 in. flank. d.h.						
1–3 ft 4 in. × 3 ft 4 in. csmt.						
1–7 ft 10 in. × 5 ft 1 in. bay						
36. Door schedule (ext).						
1–2 ft 8 in. × 6 ft 8 in. insul. steel						
1–8 ft 0 in. × 6 ft 8 in. insul. glass slider						
37. Exterior trim	No. 2 pine:					
	1 × 5: 4/10′s					
	1 × 6: 4/10′s					
38. Siding	Cedar shingles					
	12 squares					
39. Interior framing						
Shoe and plates	2 × 4: 45/12′s					
Studs	2 × 4: 534/8′s					
40. Wiring					×	
41. Plumbing					×	
42. Insulation	6 × 23 unfaced f.g. 33 rolls					
	6 × 15 foil-faced f.g. 17 rolls					
	1 in. Styrofoam 62 sheets					
43. Drywall	1/2 in. × 4 ft × 12 ft: 109 sheets					
44. Masonry					×	
45. Kitchen					×	
46. Bath(s)					×	
47. Heating					×	
48. Second floor						
covering	5/8 in. particleboard: 39 sheets					
49. Door schedule (int.)						
7–2 ft 6 in. × 6 ft 6 in. prehung six-panel pine						
2 sets 2 ft 0 in. × 6 ft 6 in. sliders						
1 set 4 ft 0 in. × 6 ft 6 in. slider						
50. Molding	3 1/2 in. clam: 638 lf					
	2 1/2 in. clam: 544 lf					
	Window sill: 48 lf					
	1 × 6 sel. pine: 68 lf					

Figure 23-4 *(Continued)*

Item	Size/ number/ length	Bd. ft.	Unit cost	Labor	Others	Total
51. Paint, caulk					×	
52. Flooring	Vinyl: 140 ft^2					
	Slate: 48 ft^2					
	W/w carpet: 116 yd^2					
53. Driveway					×	
54. Landscaping					×	
55. Nails						
	16d common: 150 lb					
	8d common: 100 lb					
	16d galv.: 10 lb					
	8d galv.: 15 lb					
	4d shingle: 50 lb					
	1-in. roofing: 50 lb					
	4d finish: 5 lb					
	6d finish: 20 lb					
	8d finish: 20 lb					
56. Sales tax						
57. Markup						
Totals						

Figure 23-4 *(Continued)*

A take-off list summarizing the materials for a project is usually needed for convenience of pricing and for submission to suppliers wishing to bid on a materials package. This list also brings together similar materials that appear in various places throughout the estimate summary sheet. The take-off list is easily generated from the estimate summary sheet. A sample take-off list for the house in Figure 23-3 is shown in Figure 23-5.

TAKE-OFF LIST

Nails and Numbers Construction Company
Termite, Maine

To: Knothole Lumber Co.

Please submit your bid for the following materials at your earliest convenience.

2 × 12: 3/12's, 2/14's
2 × 10: 5/12's, 70/14's, 10/16's
2 × 6: 145/8's, 4/10's, 45/12's, 4/14's, 4/16's, 50/18's
2 × 4: 534/8's, 2/10's, 108/12's, 46/14's

Plywood: 1/2 in. CDX: 81 sheets
5/8-in. CDX: 49 sheets
3/8-in. A-C ext.: 7 sheets
No. 2 pine: 1 × 5: 4/10's
1 × 6: 4/10's
Strapping: 1 × 3: 264 lf

Figure 23-5

Particleboard: 5/8 in.: 39 sheets
Roofing materials:
- 15-lb felt paper: 4 rolls
- 8-in. galv. drip edge: 16 lengths
- Asphalt shingles: 16 squares
- Ridge vent: 4 lengths

Drywall: 1/2 in. × 4 ft × 12 ft: 109 sheets
Cedar shingles: 12 squares
Moldings:
- 3 1/2 in. clam—638 lf
- 2 1/2 in. clam—544 lf
- Sill cap: 48 lf
- Select pine: 1 × 6—68 lf

Insulation:
- 1 in. × 2 ft × 4 ft: 116 sheets
- Sill seal: 144 lf
- 6 1/4 in. × 23 1/2 in. unfaced f.g.—33 rolls
- 6 1/4 in. × 15 1/2 in. foil-faced f.g.—17 rolls

Nails:
- 16d common: 150 lb
- 8d common: 100 lb
- 16d galv.: 10 lb
- 8d galv.: 15 lb
- 4d shingle: 50 lb
- 1 in. roofing: 50 lb
- 4d finish: 5 lb
- 6d finish: 20 lb
- 8d finish: 20 lb

200 ft^2 filter paper
144 lf 4-in. drain tile
10 gal foundation coating
5 adjustable steel columns
Window schedule: All SeeThru brand thermopane
- 7–2 ft 9 in. × 4 ft 1 in. D.H.
- 1–2 ft 9 in. × 3 ft 5 in. D.H.
- 1–9 ft 5 in. × 4 ft 5 in. picture with 2 ft D.H. flanking
- 1–3 ft 4 in. × 3 ft 4 in. casement
- 1–7 ft 10 in. × 5 ft 0 in. bay

Door schedule:
- 1–2 ft 8 in. × 6 ft 8 in. insul. steel entry
- 1–8 ft 0 in. × 6 ft 8 in. insul. glass wood-framed slider
- 7–2 ft 6 in. × 6 ft 6 in. six-panel pine prehung, split jambs
- 2 sets 2 ft 6 in. × 6 ft 6 in. sliding, six-panel pine, no jambs
- 1 set 4 ft 0 in. × 6 ft 6 in. sliding, six-panel pine, no jambs

Figure 23-5 *(Continued)*

Labor hours and related costs are one of the more difficult areas to estimate. Different workers work at different rates. Different types of jobs (framing versus finish work, for example) require different degrees of labor intensity. But it is primarily individual differences that make projections of labor time difficult.

For a contractor working steadily with the same crew, the best source of data on which to base labor requirements is experience. A few minutes at the end of each workday spent recording jobs done and number of labor-hours consumed will provide the most reliable resource for future estimates of labor requirements. For the carpenter working for a contractor, but with the ambition to become a contractor, time spent performing the same record keeping will be time well invested.

Notwithstanding the comments above, references do exist for estimating labor. (The authors would add that many estimating references dismiss the topic as in the two previous paragraphs.) One resource is the Steinberg and Stempel book cited in the Bibliography.

REVIEW EXERCISE

23-1. As a summary project, submit a completed work sheet, estimate summary sheet, and take-off list for one of the plans shown in Appendix D.

chapter 24

Money Matters

OBJECTIVES

Upon completing this chapter, the student will be able to:

1. Determine an amount of simple interest.
2. Determine the size of the monthly payment needed to pay off a loan.
3. Determine the present value of a set of future periodic payments.
4. Determine the amount of a cash discount.
5. Determine the future value of a periodic investment.

Money matters. Indeed it does! Many small businesses have failed—not due to a lack of business activity, but due to an inability on the part of the entrepreneur to manage money. Even in a successful business enterprise, profits can be increased by skillful money management. In this chapter money matters are discussed with an emphasis on how interest is calculated, how discounts work, and what the entrepreneur can do to take maximum advantage of cash flowing through his accounts.

Interest is calculated using either a simple interest or a compound interest approach. In general, when a businessman borrows a sum of money for a relatively short period of time (usually less than 1 year), the money will be used in full for the time period of the loan and then repaid in full. The lender may require periodic payments of interest, but no payments will be made against the *principal* (amount borrowed) during the loan period. Any interest calculation made during the loan period or at the end of the loan period is based on the full amount borrowed. Interest calculations made under these circumstances are called **simple interest** calculations.

On the other hand, when money is borrowed for longer periods with periodic payments made which reduce both the principal and pay the current interest charges, a

compound interest calculation is used. When the amount of principal changes periodically, such as in a loan payoff or when money is left on deposit to earn interest which is periodically added to the principal, the compounding principle is at work.

Simple interest is calculated using the formula

$$I = P \times R \times T$$

where I is the amount of interest, in dollars; P the amount of principal borrowed, in dollars; R the rate of interest, in decimal form; and T the time, in years or fraction thereof.

EXAMPLE 24-1

Determine the amount of simple interest charged if $5000 is borrowed at 15% for 90 days.

***Solution*:** Use the simple interest formula:

$$I = P \times R \times T$$

$$= 5000 \times 0.15 \times \frac{90}{365}$$

$$= \$184.93$$

Thus a business manager who borrowed $5000 for 90 days at 15% interest would be expected to pay $5184.93 at the end of the loan period, with $184.93 of this as interest.

A business works on the principle of turnover. A store merchant hopes to turn over her inventory several times each year. This means that she buys merchandise, retails it, and then reorders for future sales, with this cycle being repeated several times each year. If this businessperson can sell an article for 10% more than her cost and the cycle is able to be repeated four times in a year, the amount of money she invests in that product returns 40% of itself. So even if she must pay 15% interest for working capital, there is profit potential with a 10% markup.

In the same way, a contractor may have to pay 15% interest for borrowed money to use in his business to cover the costs of employee's wages, other business expenses, and purchase inventory. But a 15% markup on his costs means profit potential if he can turn that borrowed money over on several jobs during the course of a year. The difference between interest charged for borrowed money and a percent markup is that the interest charged for a loan is apportioned to the time for which the money is used, while the markup is a flat charge with no reduction for fractions of a year. Therefore, a 15% loan rate would equal a 15% markup on the same amount of money only if the loan was for exactly 1 year and the markup occurred only once during the year.

In the case of a loan repayment when interest is compounded, a different formula is used. (The simple interest formula could be used under some circumstances, but its application becomes very tedious.) To determine the monthly payment required to pay off a loan, the formula used is

$$p = P\left[\frac{(1 + i)^n i}{(1 + i)^n - 1}\right]$$

where p is the monthly payment, P the amount borrowed, i the (annual rate of interest)/12, decimal form, and n the number of monthly payments. This formula may be used to determine the monthly payment necessary to pay off an automobile loan, home mortgage, or any loan that is to be paid off with equal-sized monthly payments. Notice, too, that it could be used to determine the amount that could be borrowed, P, given the ability to make monthly payments of p dollars. Another application would be to determine whether an offered monthly return of p dollars is worth an investment of P dollars.

EXAMPLE 24-2

A carpenter wishes to borrow $3500 to buy equipment for his business. He wishes to pay off the loan with monthly payments made over 3 years. The interest rate is 18%. What is the monthly payment?

***Solution*:** The terms of the formula are

$$P = 3500$$

$$i = \frac{0.18}{12} = 0.015$$

$$n = 36.$$

The formula becomes

$$p = 3500\left[\frac{(1 + 0.015)^{36}(0.015)}{(1 + 0.015)^{36} - 1}\right]$$

A review of algebraic principles will show that exponentiation (raising a number to a power) is done first, followed by multiplication and division, followed by addition and subtraction. Operations within parentheses and terms of a fraction have priority. Following this order, the formula becomes

$$\begin{aligned} p &= 3500\left[\frac{(1.7091395)(0.015)}{(1.7091395) - 1}\right] \\ &= 3500\left(\frac{0.0256371}{0.7091395}\right) \\ &= 3500 \times 0.0361524 \\ &= \$126.53 \end{aligned}$$

If it is desired to know the amount of interest paid on this loan, compare the total of the 36 payments to the amount borrowed. The difference is interest. The total payments amount to

$$36 \times 126.53 = \$4555.08$$

The amount of interest is

$$4555.08 - 3500.00 = \$1055.08$$

EXAMPLE 24-3

A self-employed builder wishes to borrow a sum of money to purchase equipment needed in his business. Analysis of his income and expenses reveals an ability to pay $125 per month over the next 3 years. He can borrow money at 14%. How much can he borrow under these conditions?

***Solution*:** The formula for loan payoff used in Example 24-2 is used with $p = \$125$ and P as the unknown. The periodic rate, i, is 0.14/12 and the number of payments, n, is 36.

$$p = P\left[\frac{(1 + i)^n i}{(1 + i)^n - 1}\right]$$

$$125 = P\left[\frac{(1 + 0.14/12)^{36}(0.14/12)}{(1 + 0.14/12)^{36} - 1}\right]$$

$$125 = P(0.0341776)$$

$$P = \frac{125}{0.0341776}$$

$$= \$3657.36$$

Interest should be thought of as an amount of rent paid for the use of money. People rent money in the same way that they rent other things. The amount of rent paid for money depends on the time for which it is used, the amount used, and supply and demand, the last factor being reflected in the rate of interest charged by lenders. The carpenter in Example 24-2 will pay $1055.08 in rent money to buy equipment. He should determine that this is equipment he will actually put to frequent use during the 3-year period and that this is a better alternative to renting the equipment as it is needed. Included among the decision-making factors are his future needs for this equipment and its life expectancy. Investment tax credits and depreciation are also to be considered. For the business manager unfamiliar with these concepts and their effects, the help of an accountant should be enlisted. As for the interest paid on the loan versus an amount of rental paid for equipment, both are legitimate tax deductions for business purposes. The question of whether to lease or buy is often encountered and should be considered carefully.

Discounts can affect the profit margin of a business in a way greater than is implied by a passing glance at the discount rates usually quoted. Some businesses offer a **cash discount** to encourage prompt payment of monies due them. Cash discounts are quoted as a flat percentage by which the net amount due can be reduced provided that payment is made within a stated time period. For example, a 2% cash discount might be offered for payments made within 10 days of the billing date. Although 2% may seem like a small amount, it should be considered in light of the time factor. The buyer would ordinarily have 30 days in which to pay a bill. By paying within 10 days (ideally on the tenth day) the buyer essentially gives up the use of that money for the remaining 20 days of the billing period. A simple interest calculation will show that 2% gained in 20 days is the equivalent of earning approximately 36% per year. Even if this 2% discount can be gained only once per month, it amounts to 24% per year. In either case, it is no small amount and should be taken if the cash flow of the business permits. Examined another way, if the materials cost of an average-sized house total $35,000, the buyer stands to save $700 on the basis of a 2% discount.

EXAMPLE 24-4

An invoice dated April 5 shows the amount due to be $435.76. A 2% cash discount is offered for payment within 10 days of the billing date. If the bill is paid April 14, how much discount can be taken, and what net amount is due?

***Solution*:** The amount of the discount is

$$0.02 \times 435.76 = \$8.72$$

The amount due is

$$\begin{array}{rr} & \$435.76 \\ \text{less} & \underline{8.72} \\ & \$427.04 \end{array}$$

In summary, timing of payments is important when taking advantage of a discount. There is no advantage to the buyer to pay earlier than the last day on which the discount is allowed. To pay earlier would remove that cash from control and provides no additional discount. If the discount cannot be taken, there is no advantage in paying earlier than the last due date. Again, the idea is to control the flow of cash so as to provide the maximum return to the holder of the cash.

A question related to cash discounts: Is it possible that an advantage can be gained if a business manager borrows funds in order to take advantage of a cash discount? The answer: quite possibly. Consider the following example.

EXAMPLE 24-5

A builder has several accounts that are billed to him on the first of the month. Each offers the terms "2%/10, net 30." This means that a 2% discount may be taken if payment is made within the first 10 days; the net amount (full amount) is due within 30 days. Suppose that these accounts total $3000 due for a particular month and the builder does not have on hand a sufficient amount of cash to make the payments. He does, however, expect to have the necessary amount by the end of the month. Furthermore, he has a line of credit with his bank allowing him to borrow as needed at 15% interest. On the tenth day he borrows enough money to settle the accounts, taking the 2% discount. The amount needed is

$$0.02 \times 3000 = \$60.00 \text{ discount}$$

$$\begin{array}{rr} & \$3000 \\ \text{less} & \underline{60} \\ & \$2940 \end{array}$$

At the end of the month the manager pays off the loan. The interest charges are

$$\begin{aligned} I &= P \times R \times T \\ &= 2940 \times 0.15 \times \frac{20}{365} \\ &= \$24.16 \end{aligned}$$

By following this procedure, the builder was able to save

$$\begin{array}{rrl} & \$60.00 & \text{discount} \\ \text{less} & \underline{24.16} & \text{interest paid} \\ & \$35.84 & \text{saved} \end{array}$$

So here is a case in which it is profitable to pay 15% interest on a loan in order to take advantage of a 2% discount.

There are many other money matters of importance to a businessperson. A self-employed builder must consider more than simply a fair hourly wage when setting prices to be charged to a customer. Included in the costs of doing business are operating expenses for any vehicles used in the business, replacement costs for vehicles and equipment, and insurance costs of several varieties (product liability, health and accident, income protection, etc.). If employees are used, not only are their wages to be considered, but social security taxes must be paid by the employer, as do unemployment security taxes, worker's compensation insurance premiums, and any other benefits the employer wishes to pay on behalf of the workers.

Finally, self-employed persons should plan for their financial future. The government has put into place laws that offer tax advantages to self-employed persons who wish to build their own retirement system. Early planning and consistent investment in a retirement fund will generously reward those who provide for themselves.

As an example of a retirement vehicle, consider the Individual Retirement Account (IRA). This is but one of several ways in which a person can periodically set aside money from earnings to provide for his or her retirement while taking advantage of deferring the payment of income taxes on the earnings.

The laws governing IRAs allow up to $2000 per year (within a certain income range) from earnings to be set aside in a fund, with no income tax paid on either the earnings or the interest accrued until the money is used. Presumably, the taxes charged on these funds will be less in retirement because the individual's income will be less at that time, placing him or her in a lower tax bracket. The law specifies that money set aside in an IRA (together with its accrued earnings) cannot be used before age $59\frac{1}{2}$. Early withdrawal is possible, but a tax penalty may be assessed. A variety of choices exists as to how these funds may be invested. For most, the better choices lie among government and corporate bonds, bank certificates of deposit, stocks, and mutual funds. The IRA may be self-directed or managed for the individual. In the latter case, management fees will be assessed in some form. In either case, the money invested is done through a medium (a bank, stock brokerage firm, insurance company) which guarantees that the funds are used only for their intended purpose.

To illustrate the advantage of an IRA to the individual, compare the amount of money deposited and earned through a lifetime work career of 40 years with and without the use of an IRA. Another finance formula is introduced at this time—that which calculates the amount of an annuity. An **annuity** is any set of equal-sized payments made at regular intervals. In this situation, \$2000 deposited annually for 40 years constitutes an annuity. The formula that calculates the future value of these payments, together with the accumulated interest is

$$S = R\left[\frac{(1 + i)^n - 1}{i}\right]$$

where S is the sum accumulated, R the periodic deposit, i the periodic interest rate, and n the number of deposits made.

Assuming an annual interest rate of 10%, \$2000 deposited each year for 40 years would accumulate to

$$\begin{aligned} S &= 2000\left[\frac{(1 + 0.10)^{40} - 1}{0.10}\right] \\ &= 2000\left[\frac{45.259256 - 1}{0.10}\right] \\ &= 2000 \times 442.59256 \\ &= \$885{,}185.11 \end{aligned}$$

(In practice the accumulated amount would be even greater because the deposits of \$2000 annually would occur at more frequent intervals during the year. Interest would probably be compounded several times during each year, as well. In fact, if \$500 were deposited each quarter and interest at 10% annually were compounded quarterly, the accumulated amount would exceed \$1,000,000!)

Without an IRA an attempt to create a retirement fund would result in far less accumulated wealth because of two drains on the money: the \$2000 would be taxed before being deposited in the account; the earnings on the \$2000 amounts would also be taxed each year. Disregarding the tax on the earnings and assuming an income tax rate of 15%, instead of having \$2000 to deposit each year, only 85%, or \$1700 would be available for investment. The amount accumulated before taxes on the earnings is

$$\begin{aligned} S &= 1700\left[\frac{(1 + 0.10)^{40} - 1}{0.10}\right] \\ &= 1700 \times 442.59256 \\ &= \$752{,}407.34 \end{aligned}$$

Although this is still a considerable amount of money and vividly shows the value of a fixed savings plan, it is, nevertheless, over \$130,000 less than the amount earned in an IRA. Bear in mind, too, that the difference would be increased considerably by the income taxes assessed annually on the interest earned on these funds. An estimate of the income tax paid just on the earnings from the fund during its thirtieth year is nearly \$10,000!

When compared to the retirement income provided by the social security system alone, the income potential of an IRA is indeed impressive. It is extremely important that the self-employed pay themselves well—not only in terms of present earnings, but in terms of funds to provide future earnings as well.

REVIEW EXERCISES

24-1. Find the amount of simple interest and the total amount due on each of the following loans.

- **(a)** \$500 at 12% for 120 days
- **(b)** \$3000 at 14% for 1 year
- **(c)** \$2500 at 15% from April 12 to August 20
- **(d)** \$8500 at 13.75% from July 25 to November 16
- **(e)** \$4500 at 18% from May 6 to October 10

24-2. Find the monthly payment needed to pay off the loan described in each of the following situations.

- **(a)** A \$35,000 mortgage loan at 13.5% for 30 years
- **(b)** A car loan for 48 months in the amount of \$7500 at 11%
- **(c)** A loan to purchase a mobile home for \$15,000 to be paid off in 10 years with 14% interest
- **(d)** A \$50,000 mortgage loan for 30 years at 14%
- **(e)** A loan for land purchase costing \$6000 to be paid off in 36 months with 15% interest

24-3. In each of the situations described in Problem 24-2, determine the total amount of interest paid.

24-4. Suppose you have determined that you can afford to replace the \$325 you are now paying in monthly rent by a mortgage payment of the same amount. The local bank is making mortgage loans for 30 years at 13.5%. How much money could you borrow?

24-5. In Problem 24-4, if the lending bank requires a 20% down payment, how much could be paid for a house?

24-6. Determine (1) the amount of the discount and (2) the net amount due on the following invoice amounts with terms as stated.

- **(a)** Invoice date: May 1
  Terms: 2%/10, net 30
  Invoice amount: \$626.50
  Payment date: May 10
- **(b)** Invoice date: June 12
  Terms: 1%/10, net 30
  Invoice amount: \$4244.76
  Payment date: June 21
- **(c)** Invoice date: July 23
  Terms: $1\frac{1}{2}$%/10, net 30
  Invoice amount: \$823.17
  Payment date: July 31

24-7. A business manager can take advantage of a 2% cash discount only by borrowing the necessary funds at 16%. On the last discount day, she borrows the needed amount, pays the discounted amount, and then pays off the loan 20 days later. The invoice amount was \$3225.84 before the discount was applied. If any, how much did the manager save by this action?

24-8. During the past year a builder had constructed three houses. All materials were purchased from Knothole Lumber Company. Knothole offers a 2% discount on all bills paid by the tenth day of each month. Below is a list of the amounts the builder owed Knothole on the tenth of each month. Assuming that the builder took advantage of the discounts, what was the total amount of his savings during the building season?

Month	*Amount due*
April	$ 5,240
May	8,388
June	12,433
July	28,577
August	34,331
September	19,493
October	5,449

24-9. Assume that the builder referred to in Problem 24-8 had to borrow enough to pay the discounted amounts on the tenth day of each month. He paid 16% interest and paid off each loan after 20 days. What total amount of interest did he pay? Compare this to the amount saved by taking the discounts.

24-10. If $2000 annually is deposited in an IRA that earns 12% per year, what will be the total amount in the account at the end of each of the following time periods?

(a) 10 years
(b) 15 years
(c) 25 years
(d) 35 years

appendix A

The Metric System

The United States is the only major industrialized country in the world still saddled with the cumbersome English system of measurement. (Even the English don't use it.) The metric system is in use almost everywhere else. Not only does using the metric system allow a smooth flow of equipment and products from one country to another, it is also a much simpler and more sensible system to use. To illustrate, here is a comparison of some English and metric equivalents:

1 foot	= 12 inches	1 centimeter	= 10 millimeters
1 yard	= 3 feet	1 decimeter	= 10 centimeters
1 rod	= $5\frac{1}{2}$ yards	1 meter	= 10 decimeters
1 hand	= 4 inches	1 dekameter	= 10 meters
1 fathom	= 6 feet	1 hectometer	= 10 dekameters
1 chain	= 22 yards	1 kilometer	= 10 hectometers
1 mile	= 5280 feet		

Converting from one metric unit to another is accomplished by multiplying or dividing by a power of 10. Recall that multiplying or dividing by 10, 100, 0.01, and so on, can be accomplished simply by moving the decimal point a certain number of places to the left or right.

To convert 1 mile to its equivalent inches, 1 mile must be converted to 5280 ft, and 5280 ft converted to 63,360 in. (5280 ft × 12 in./ft = 63,360 in.). A comparable type of conversion in the metric system would be the conversion from kilometers to centimeters. This simply involves moving the decimal point five places to the right. With a little practice, it can be quickly determined that 8.3 km = 830 000 cm—a conversion that does not require a calculator, or even pencil and paper. Try converting miles to inches in your head!

A-1 METRIC/METRIC CONVERSIONS

In the metric system, each larger unit is 10 times as large as the preceding unit. Hence 10 of the smaller units equal 1 unit of the next larger size. There are three major types of measurements in the metric system: **meters** measure distance or length, **liters** measure volume, and **grams** measure mass. (Mass is interchangeable with weight for most purposes.) Meters will be used to illustrate the method of conversion, but liters and grams are converted in the same manner. A meter, the standard unit of length, is slightly longer than a yard (about 39 in.). Units smaller than and larger than a meter are indicated by a prefix:

Name:	*kilo*meter	*hecto*meter	*deka*meter	meter	*deci*meter	*centi*meter	*milli*meter
Abbrev:	km	hm	dam	m	dm	cm	mm
Size:	1000	100	10	1	0.1	0.01	0.001

The prefixes kilo, hecto, and so on, always mean the same number, regardless of what type of unit is being measured. *Kilo*meter (km) means 1000 meters, *kilo*gram (kg) means 1000 grams, and *kilo*liter (kl) means 1000 liters. A milligram (mg) is 1/1000 or 0.001 gram, a deciliter (dl) = 1/10 or 0.1 liter, and a hectometer (hm) represents 100 meters. To use the metric system, it is necessary to memorize the prefixes given above, what they represent, and in the proper order from left to right. It is then a simple matter to convert from one metric unit to another.

Examples

A. Convert 58.5 cm to meters. Note that meters (m) are *two places* to the *left* of cm. To convert *cm* to *m*, move the decimal point *two places* to the *left*.

$$58.5 \text{ cm} = 0.585 \text{ m}$$

B. Convert 0.35 km to cm. Cm are *5 units* to the *right* of km on the chart. To convert *km* to *cm*, move the decimal point *five places* to the *right*.

$$0.35 \text{ km} = 35\,000 \text{ cm}$$

C. Convert 8 dm to dam. To convert *dm to dam* requires a movement of *two places* to the *left*. A leading zero is shown if there is no whole number to the left of the decimal point.

$$8 \text{ dm} = 0.08 \text{ dam}$$

Exercises A-1A

Study completed examples a, b, and c, and complete the table.

	km	hm	dam	m	dm	cm	mm
a.	8.2 km	⟶	820 dam				
b.		0.053 hm	⟵	5.3 m			
c.			2.5 dam	⟶	⟶	⟶	25 000 mm
1.	? km			47 m			
2.				47 m		? cm	
3.			0.83 dam			? cm	
4.				? m		380 cm	
5.					0.25 dm		? mm
6.					? dm	360 cm	

	km	hm	dam	m	dm	cm	mm
7.	0.048 km					? cm	
8.		? hm		800 m			
9.					46 dm		? mm
10.				? m		820 cm	

Using exactly the same method, it follows that 27 kg = 27 000 g, and 8000 ml = 8 liters. (Note that spaces are used in place of commas in the metric system.) Convert the following values.

11. 18 dg = ________ mg

12. 48 000 cl = ________ kl

13. 0.0058 kg = ________ dg

14. 0.5 liter = ________ kl

15. 5000 mg = ________ g

To convert **square measure,** the number of places the decimal point is moved is *doubled*. To convert **cubic measure,** the number of places the decimal point is moved is *tripled*. Study the following examples to determine the correct method.

Examples

D. Convert 83 000 cm^2 (square centimeters) to square meters.

83 000 cm = ? m

1. First determine the direction and number of places the decimal point would be moved for the linear conversion. In this case the decimal point would be moved *two* places to the *left*.

83 000. cm^2 = 8.3000. m^2

2. Double the number of places moved in square conversion, so the decimal point is moved *four* places to the *left* instead of 2.

83 000 cm^2 = 8.3 m^2

3. These are equivalent square units.

E. Convert 80 000 cc to cubic meters. (Cubic centimeters or cm^3 is often written as *cc*.)

80 000 cm = ? m

1. First determine the direction and number of places the decimal point would be moved for the linear conversion. In this case the decimal point would be moved *two* places to the *left*.

80 000. cc = .080 000. m^3

2. Triple the number of places the decimal point is moved for cubic measure. Therefore, the decimal point is moved *six* places to the *left*.

80 000 cc = 0.08 m^3

3. These are equivalent cubic units.

F. Change 5 000 000 dm^3 to km^3.

5 000 000 dm = ? km

1. Determine the direction and number of places the decimal point would be moved for the linear conversion. In this case the decimal point would be moved *four* places to the *left*.

$5\ 000\ 000\ dm^3$
$= .\underbrace{000\ 005\ 000\ 000}.\ km^3$

2. Triple the number of places the decimal point is moved for cubic measure. Therefore, the decimal point is moved *12* places to the *left*.

$5\ 000\ 000\ dm^3 = 0.000\ 005\ km^3$

3. These are equivalent cubic units.

Exercises A-1B

Using the prefixes for reference, find the equivalences.

km hm dam m dm cm mm

a. $0.0005\ km^2 =$ ___5___ dam^2
b. $20\ 000\ mm^2 =$ ___0.02___ m^2
1. $38\ 000\ cc =$ ________ m^3
2. $0.005\ m^2 =$ ________ hm^2
3. $4200\ hm^3 =$ ________ km^3
4. $0.05\ mm^2 =$ ________ cm^2
5. $82\ 000\ mm^2 =$ ________ m^2
6. $0.05\ m^2 =$ ________ cm^2
7. $48\ 000\ cc =$ ________ dm^3
8. $0.00008\ cc =$ ________ mm^3
9. $41\ m^2 =$ ________ cm^2
10. $18\ 000\ cm^2 =$ ________ m^2

A-2 ENGLISH/METRIC CONVERSIONS

As long as the English system is still in use, converting from metric to English and English to metric will be necessary. The most common conversions are listed below, and it is worthwhile to memorize these. One inch is exactly equal to 2.54 cm. All other conversions are approximate.

1 inch	= 2.54 cm
1 meter	≐ 3.28 ft
1 mile	≐ 1.61 km
1 liter	≐ 1.06 quarts
1 kilogram	≐ 2.2 lb
1 ounce	≐ 28 g

To convert from one system to another, use the unit conversion method discussed in Chapter 4.

Examples

A. What is the weight in kilograms of a 50-lb box of nails?

$$\frac{50\ \cancel{lb}}{1} \times \frac{1\ kg}{2.2\ \cancel{lb}}$$

1. Set up the unit ratio 1 kg = 2.2 lb so that pounds cancel.

$$\frac{50}{1} \times \frac{1\ kg}{2.2}$$

2. Cancel units and perform the division (since 2.2 is in the denominator).

$50\ lb = 22.7\ kg$

B. Gas at a certain station in Canada costs 29.9 cents per liter. What is the price per gallon?

$$\frac{29.9 \text{ cents}}{1 \text{ liter}} \times \frac{?}{?} = \frac{? \text{ cents}}{\text{gal}}$$

1. Set up the given and desired units.

$$\frac{29.9 \text{ cents}}{1 \cancel{\text{liter}}} \times \frac{1 \cancel{\text{liter}}}{1.06 \cancel{\text{qt}}} \times \frac{4 \cancel{\text{qt}}}{1 \text{ gal}}$$

2. Since the conversion factor for liters to gallons is not given, a 2 step conversion is necessary. Convert liters to quarts and quarts to gallons, cancelling the unwanted units of liters and quarts.

$$\frac{29.9 \text{ cents}}{1} \times \frac{1}{1.06} \times \frac{4}{1 \text{ gal}}$$

$$= \frac{112.8 \text{ cents}}{\text{gal}}$$

3. When only the desired units are left, perform the multiplication and division.

29.9 cents/liter = $1.128/gal

4. Most gas stations price to 1/10 cents.

A-3 TEMPERATURE CONVERSIONS

The Celsius scale is used to measure temperature in the metric system. Water freezes at 0°C and boils at 100°C under standard conditions. The Fahrenheit scale, used in the English system, sets the freezing point of water at 32°F and the boiling point at 212°F.

To convert from Fahrenheit to Celsius, use the formula:

$$C = \tfrac{5}{9}(F - 32)$$

To convert from Celsius to Fahrenheit, use the formula:

$$F = \tfrac{9}{5}C + 32$$

Examples

A. A temperature of 48°F is equivalent to what temperature on the Celsius scale?

$$C = \frac{5}{9}(F - 32)$$

1. Formula to convert Fahrenheit to Celsius.

$$= \frac{5}{9}(48 - 32)$$

2. Substitute Fahrenheit temperature into formula.

$$= \frac{5}{9}(16)$$

3. Perform operation in parentheses first.

$$= 8.9°C$$

4. Multiply by 5/9.

48°F = 8.9°C

5. These are approximately equivalent.

B. The thermostat in a house is set at 20°C. What temperature is this on the Fahrenheit scale?

$$F = \frac{9}{5}C + 32$$

1. Formula to convert Celsius to Fahrenheit.

$$= \frac{9}{5} \cdot 20 + 32$$

2. Substitute Celsius temperature.

$= 36 + 32$ — 3. Since there are no parentheses, perform the multiplication first.

$= 68°F$ — 4. Perform the addition.

$20°C = 68°F$ — 5. These are exactly equivalent.

REVIEW EXERCISES

Perform the following conversions.

A-1. 5 lb/meter = ________ lb/foot

A-2. A roof is 14.8 m long. What is its equivalent length in feet and inches, to the nearest $\frac{1}{4}$ in.?

A-3. A wood lathe weighs 462 kg. How much does the lathe weigh in pounds?

A-4. Which is the better buy: nails at $1.25 per pound or $2.85 per kilogram?

A-5. A carpenter has only a metric rule available. If he is framing a wall with studs 16 in. o.c., what length, in millimeters, should he measure on the metric rule? (In Canada, all building layout dimensions are in millimeters.)

A-6. A finish board is 2.6 m long. To the nearest $\frac{1}{4}$ in., how long is the board in feet and inches?

A-7. 2×6 framing lumber weighs 2.1 lb per lineal foot. What is the weight, in kilograms, of an 8-ft-long 2×6?

A-8. Convert 47°C to its equivalent Fahrenheit temperature.

A-9. Convert 85°F to its equivalent Celsius temperature.

A-10. The normal body temperature in a healthy human being is 98.6°F. What is this on the Celsius scale?

appendix B

Right-Triangle Trigonometry

If two right triangles have an equal acute angle, they are similar. In fact, all right triangles with that same acute angle are similar. (Why?) In Figure B-1, $\triangle ABC \sim \triangle ADE \sim \triangle AFG$. ($\sim$ is the symbol for "is similar to.")

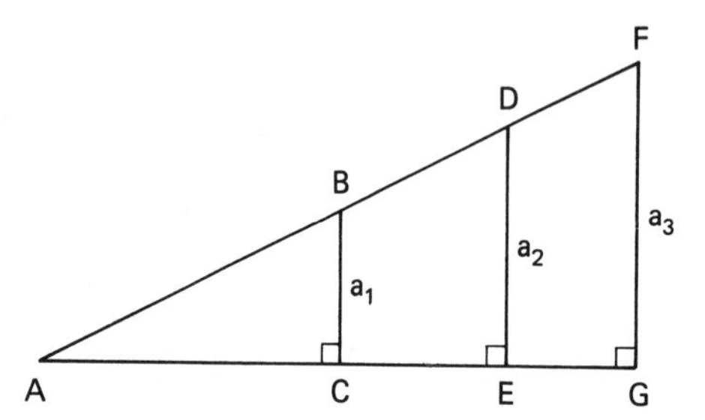

Figure B-1

Recall that similar triangles have a very important characteristic: the ratios of their corresponding sides are equal. Therefore, in Figure B-2,

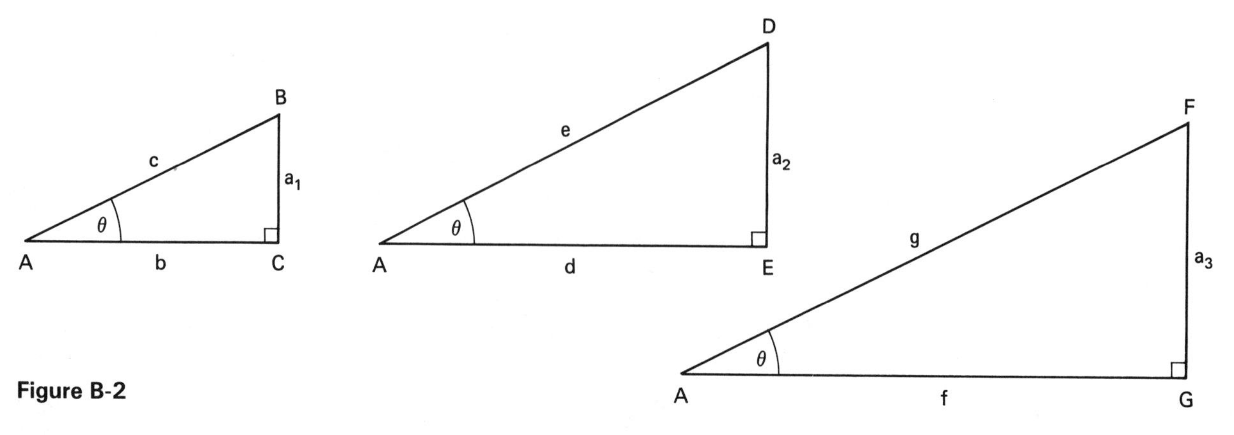

Figure B-2

$$\frac{a_1}{c} = \frac{a_2}{e} = \frac{a_3}{g}$$ These ratios are called the **sine** function when referring to $\angle A$.

$$\frac{b}{c} = \frac{d}{e} = \frac{f}{g}$$ These ratios are called the **cosine** function when referring to $\angle A$.

$$\frac{a_1}{b} = \frac{a_2}{d} = \frac{a_3}{f}$$ These ratios are called the **tangent** function when referring to $\angle A$.

In similar triangles, even though the lengths of the corresponding sides are different, the ratios of corresponding sides are not. These ratios will be constant. Because the ratios are constant, we give them names. In $\triangle ABC$ in Figure B-3, when referring to $\angle A$: side a is the *opposite side*; side b is the *adjacent side*; and side c is the *hypotenuse*. (The

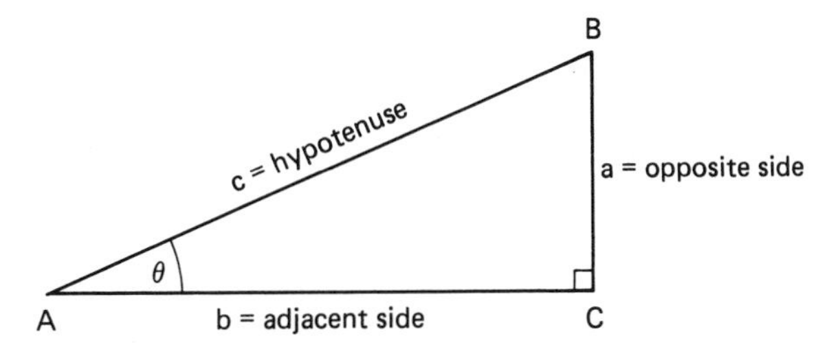

Figure B-3

hypotenuse is always opposite the right angle.) "θ" is used to indicate any acute angle in the triangle. Therefore,

$$\text{sine } \theta = \frac{\text{opposite side}}{\text{hypotenuse}} \qquad \text{abbreviated as } \sin\theta = \frac{\text{opp}}{\text{hyp}}$$

$$\text{cosine } \theta = \frac{\text{adjacent side}}{\text{hypotenuse}} \qquad \text{abbreviated as } \cos\theta = \frac{\text{adj}}{\text{hyp}}$$

$$\text{tangent } \theta = \frac{\text{opposite side}}{\text{adjacent side}} \qquad \text{abbreviated as } \tan\theta = \frac{\text{opp}}{\text{adj}}$$

For $\triangle XYZ$ in Figure B-4, when referring to $\angle X$:

$$\sin X = \frac{x}{z} \quad \left(\frac{\text{opp}}{\text{hyp}}\right)$$

$$\cos X = \frac{y}{z} \quad \left(\frac{\text{adj}}{\text{hyp}}\right)$$

$$\tan X = \frac{x}{y} \quad \left(\frac{\text{opp}}{\text{adj}}\right)$$

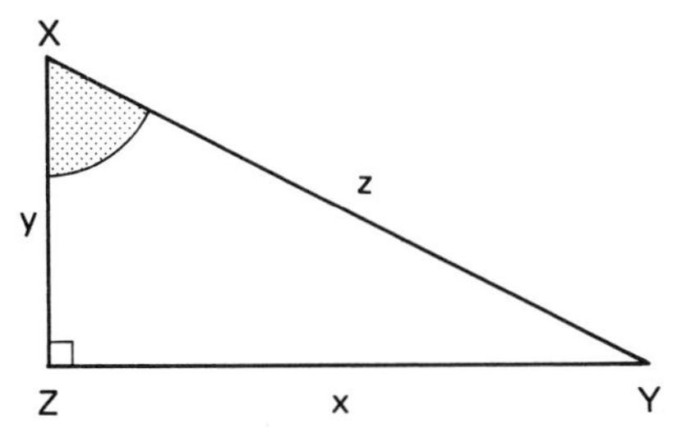

Figure B-4

In reference to $\angle Y$ in Figure B-5, note that x now becomes the adjacent side and y becomes the opposite side. The hypotenuse is always the same side regardless of the angle in reference.

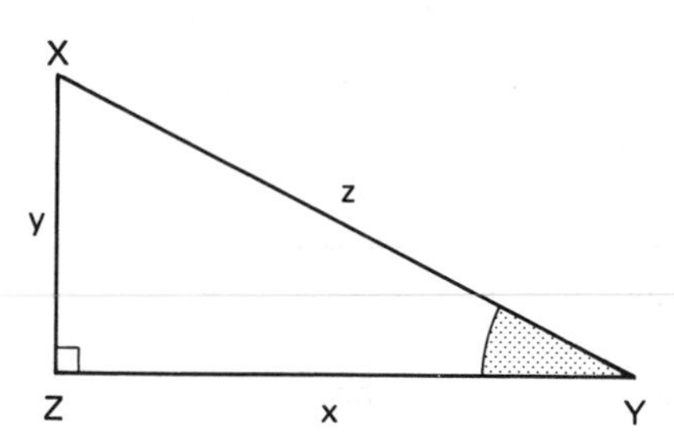

Figure B-5

$$\sin Y = \frac{y}{z} \quad \left(\frac{\text{opp}}{\text{hyp}}\right)$$

$$\cos Y = \frac{x}{z} \quad \left(\frac{\text{adj}}{\text{hyp}}\right)$$

$$\tan Y = \frac{y}{x} \quad \left(\frac{\text{opp}}{\text{adj}}\right)$$

Since every right triangle with an acute angle of 38°, for example, is similar, the ratios of their opposite side to hypotenuse (sin) will be the same: sin 38° = 0.6157. This is also true for the other ratios. No matter what size right triangle, if it has an acute angle of 38°, the ratio of the adjacent side to the hypotenuse will be a constant: cos 38° = 0.7880. Also, the ratio of the opposite side to the adjacent side is a constant: tan 38° = 0.7813. To find these ratios on a calculator with trig functions, use the following procedure:

enter: 38 [sin]

sin 38° = .6157

The number 0.6156615 will then be displayed. This is generally rounded to four decimal places, or .6157. [Be sure that the calculator is in the degree (deg) mode.]

A right triangle with an acute angle of 38° was used as an illustration. A right triangle with an acute angle of any value follows the same principle. For example: If two right triangles of different sizes have an acute angle of 53.1°, they are similar triangles and the ratios of their corresponding sides are equal (Figure B-6).

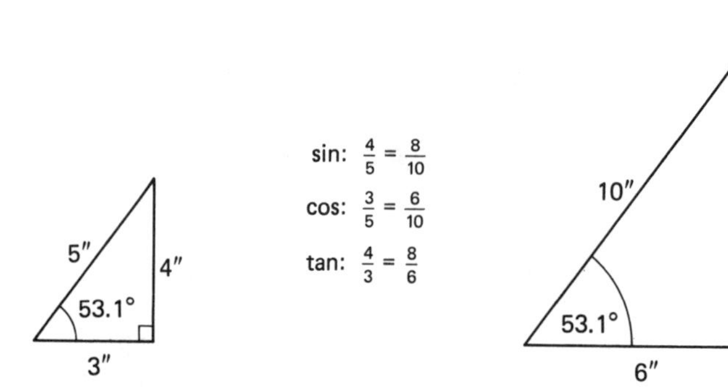

Figure B-6

The following list gives the values of the ratios (called **functions**) for the sine, cosine, and tangent of any right triangle with the specified angles. Using a calculator with trig functions, the student should key in the following angles and functions to ascertain that they are indeed correct. Remember: First enter the angle, then the sin, cos, or tan function.

sin 52° = 0.7880

cos 41° = 0.7547

sin 44° = 0.6947

tan 68° = 2.475

cos 32° = 0.8480

tan 59° = 1.664

The values for the sin and cos of an angle can never exceed 1.000. The tan of an angle can have any value.

Exercise B-1

Complete the table.

	Angle	sin	cos	tan
a.	37.8°	0.6129	0.7902	0.7757
1.	43.8°			
2.	19.26°			
3.	45°			
4.	18.62°			

In the examples given above, the ratio of the sides of a triangle could be found if the angle was known. In reverse form, if the ratio of the sides of a triangle is known, the angle can be found. For $\triangle ABC$ in Figure B-7,

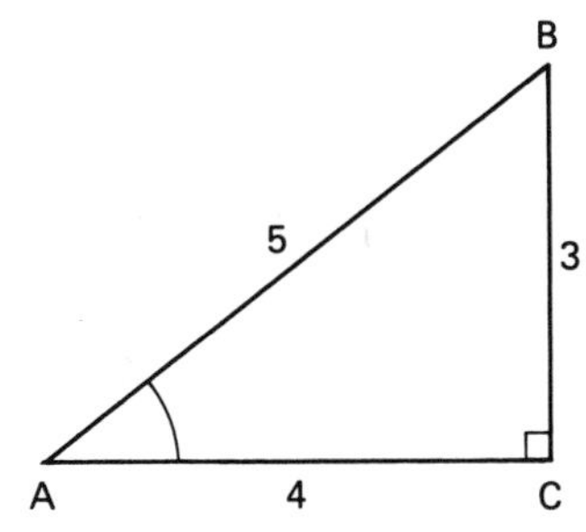

Figure B-7

$$\sin A = \frac{3}{5} = 0.6000$$

$$\cos A = \frac{4}{5} = 0.8000$$

$$\tan A = \frac{3}{4} = 0.75000$$

Knowing that $\sin A = 0.6000$, $\angle A$ can be found by following this sequence on the calculator:

.6 [INV] [SIN] or .6 [2nd] [SIN^{-1}] (depending on calculator)

Knowing that $\cos A = 0.8000$, $\angle A$ can be found by following this sequence:

.8 [INV] [COS] or .8 [2nd] [COS^{-1}]

Knowing that $\tan A = 0.7500$, use this sequence to find $\angle A$:

.75 [INV] [TAN] or .75 [2nd] [TAN^{-1}]

$\angle A = 36.87°$ regardless of which of the three functions was used to find it. Ordinarily, one method is more convenient than the others, and that is the method chosen to find the unknown angle.

Example

A. Find $\angle A$ for the triangle shown.
All three methods of finding $\angle A$ will be shown.

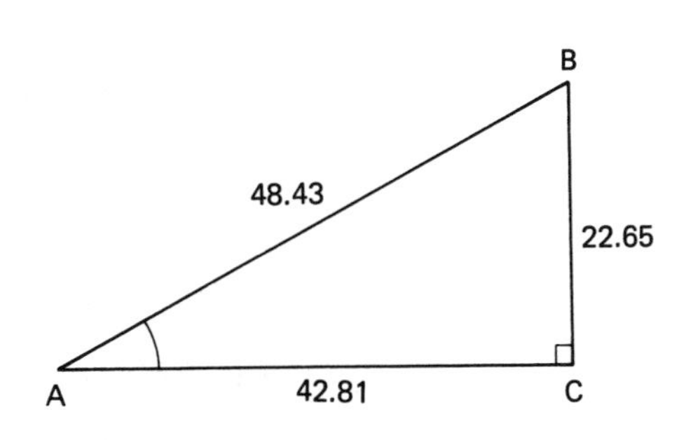

$$\sin A = \frac{22.65}{48.43}$$

$$= 0.4677$$

$$\angle A = 27.88°$$

$$\cos A = \frac{42.81}{48.43}$$

$$= 0.8840$$

$$\angle A = 27.88°$$

$$\tan A = \frac{22.65}{42.81}$$

$$= 0.5291$$

$$\angle A = 27.88°$$

The three sides and three angles of a triangle are referred to as *parts* of the triangle. In a right triangle, if any side plus one other part (another side or an acute angle) are given, the remaining three parts can be found by using trigonometry. This is referred to as *solving* the triangle.

Examples

B. Solve $\triangle ABC$.

$A = 28°$	$a =$
$B =$	$b =$
$C = 90°$	$c = 55$ in.

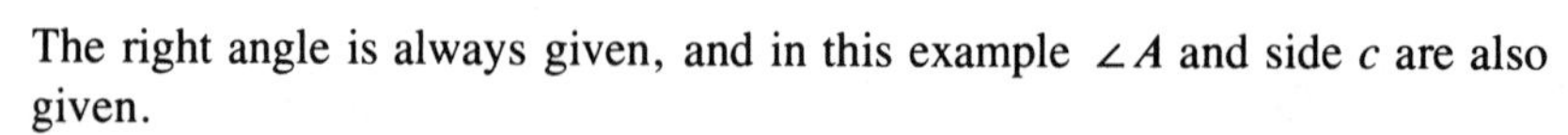

The right angle is always given, and in this example $\angle A$ and side c are also given.

$\angle B = 90° - 28° = 62°$ — 1. The two acute angles of a right triangle always add to 90°.

$$\sin 28° = \frac{a}{55 \text{ in.}}$$

2. Sin $A = a/c$. Set up as a proportion and substitute all known values.

$$\frac{\sin 28°}{1} = \frac{a}{55 \text{ in.}}$$

$$\frac{0.4694716}{1} = \frac{a}{55 \text{ in.}}$$

3. Enter 28 [SIN] to determine value of sin 28°.

$a = 25.8$ in. — 4. Cross-multiply and solve for a.

$$\cos 28° = \frac{b}{55 \text{ in.}}$$

5. Cos $A = b/c$. Set up as a proportion and substitute all known values.

$$\frac{\cos 28}{1} = \frac{b}{55 \text{ in.}}$$

$b = 48.6$ in. — 6. Cross-multiply and solve for b.

$A = 28°$	$a = 25.8$ in.
$B = 62°$	$b = 48.6$ in.
$C = 90°$	$c = 55$ in.

C. Solve $\triangle XYZ$:

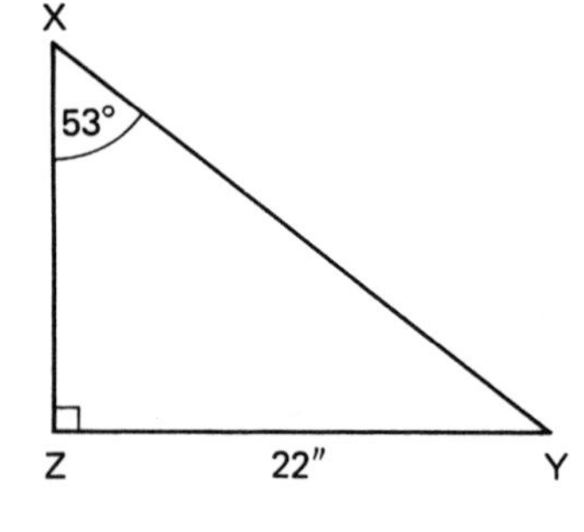

$X = 53°$	$x = 22$ in.
$Y =$	$y =$
$Z = 90°$	$z =$

$\angle Y = 90° - 53° = 37°$ — 1. Find $\angle Y$.

$$\sin 53° = \frac{22}{z}$$

2. Set up sin function as a proportion.

$$\frac{\sin 53°}{1} = \frac{22 \text{ in.}}{z}$$

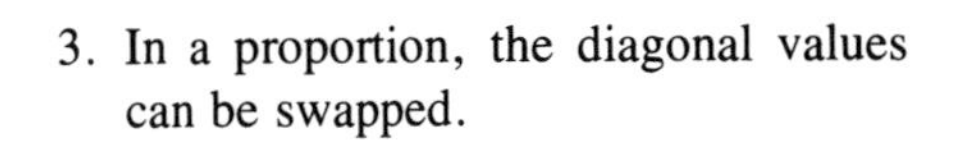

3. In a proportion, the diagonal values can be swapped.

$$\frac{z}{1} = \frac{22 \text{ in.}}{\sin 53°}$$

4. z is easier to solve for in this form.

$$z = 27.5 \text{ in.}$$

$$\frac{\tan 53}{1} = \frac{22 \text{ in.}}{y}$$

5. Set up the tan value to find y.

$$\frac{y}{1} = \frac{22 \text{ in.}}{\tan 53°}$$

6. Swap diagonal values so y can be found more easily.

$$y = 16.6 \text{ in.}$$

$X = 53°$ $x = 22$ in.

$Y = 37°$ $y = 16.6$ in.

$Z = 90°$ $z = 27.5$ in.

On any triangle, the shortest side is always opposite the smallest angle, and the longest side is opposite the largest angle.

Exercise B-2

Solve the triangles.

1. $A = 42°$ $a =$
$B =$ $b =$
$C = 90°$ $c = 38$ ft

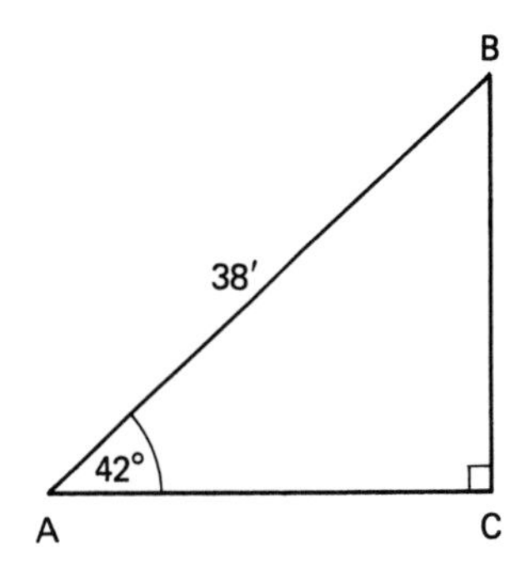

2. $X =$ $x = 15$ ft
$Y = 38°$ $y =$
$Z = 90°$ $z =$

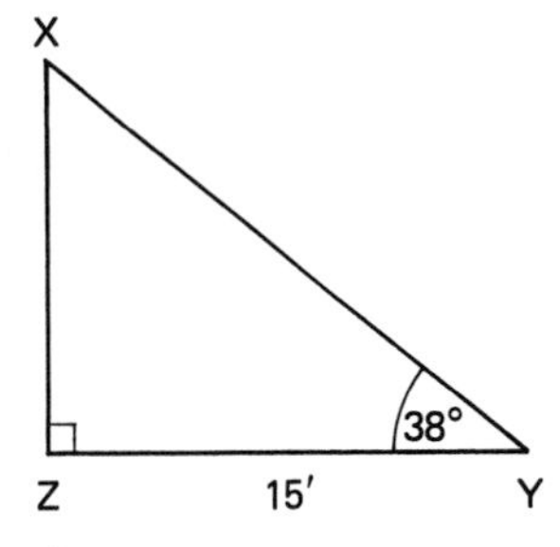

3. $J =$ $j = 46$ in.
$K = 29°$ $k =$
$L = 90°$ $l =$

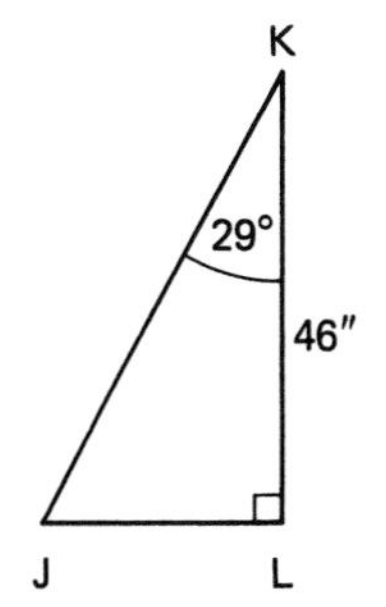

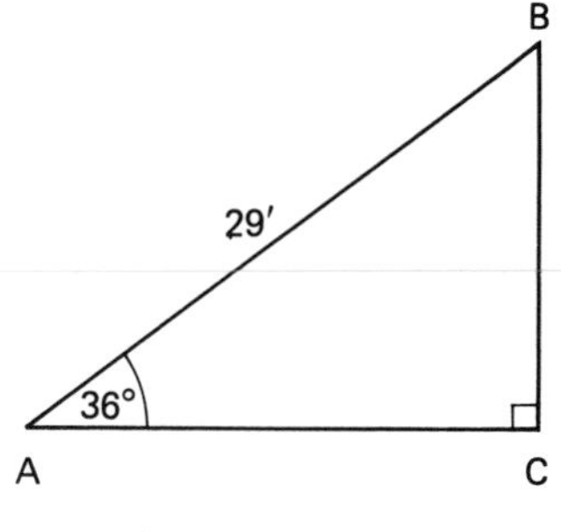

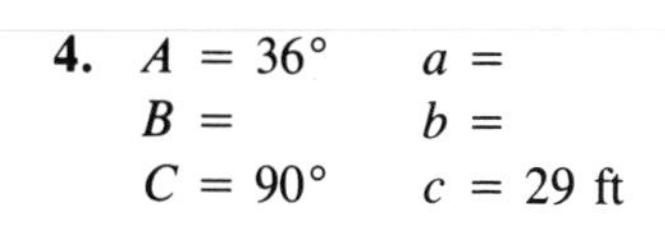

4. $A = 36°$ $\quad a =$
$B =$ $\quad b =$
$C = 90°$ $\quad c = 29$ ft

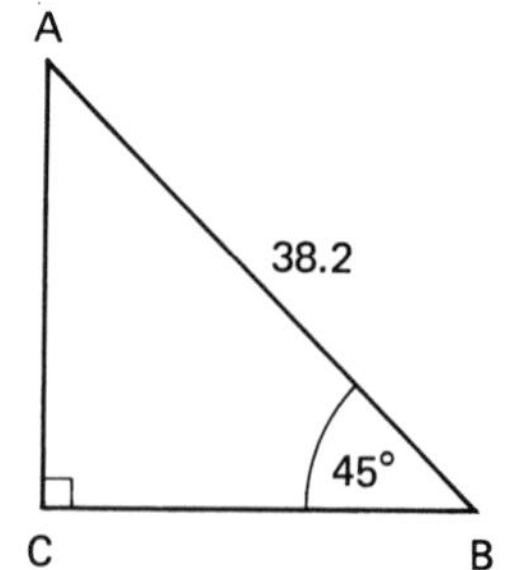

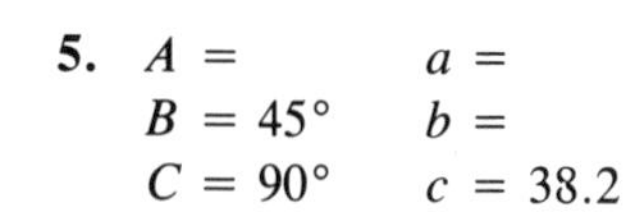

5. $A =$ $\quad a =$
$B = 45°$ $\quad b =$
$C = 90°$ $\quad c = 38.2$

Trigonometry has many useful applications. Study the examples below to learn some of them.

Examples

D. A bridge is to be built across a river and the distance across the river is to be measured at that point. A surveyor sets up stakes 75 ft apart as shown, and with a transit, swings the angle to a stake on the opposite shore. If this angle is 58°, what is the distance across the river?

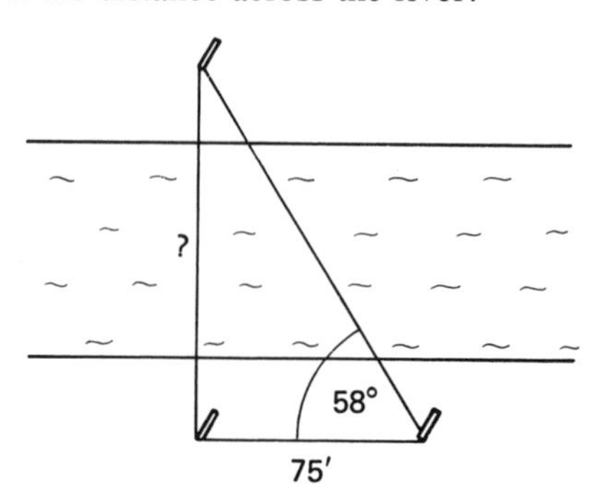

Since the side to be found is opposite the 58° angle, the tangent function is used.

$\tan 58° = \dfrac{x}{75}$ 1. Set up the tangent function.

$\dfrac{\tan 58°}{1} = \dfrac{x}{75 \text{ ft}}$ 2. Cross-multiply to solve for x.

$x = 120$ ft 3. The distance across the river between the stakes is 120 ft.

E. A loading ramp 20 ft long must reach a platform 3 ft 8 in. high. What will be the angle of inclination?

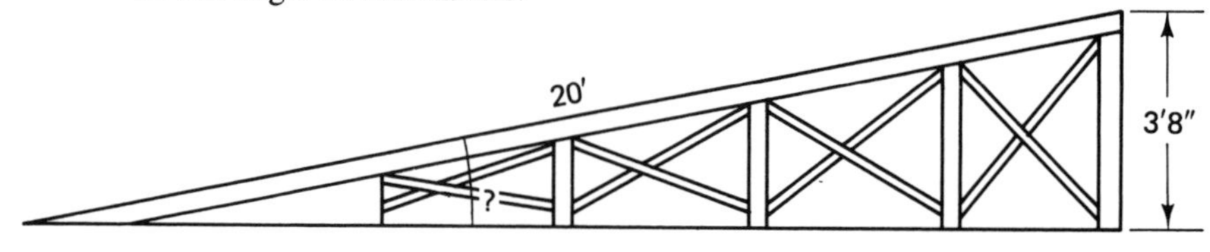

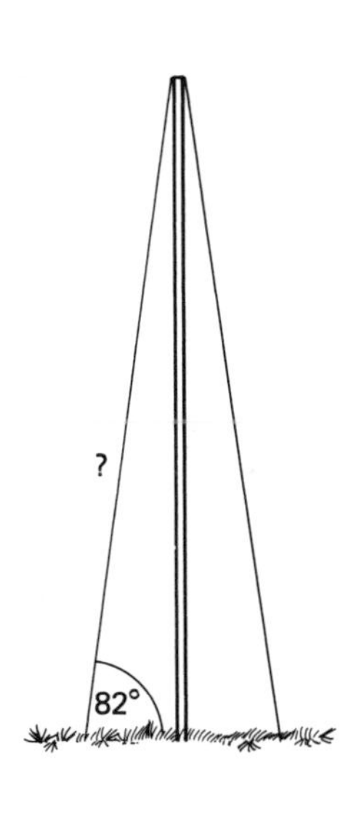

3 ft 8 in. $= 3.\overline{6}$ ft	1. Change the height of the platform to decimal feet.
$\frac{\sin\theta}{1} = \frac{3.\overline{6}\text{ ft}}{20\text{ ft}}$	2. Set up a proportion using the sine function.
$\sin\theta = 0.1833$ $\theta = 10.6°$	3. The angle of inclination is approximately 11°.

Exercise B-3

1. Guy wires supporting a 20-ft pole make an angle of 82° with the ground. How long are the guy wires?

2. A gable roof has a slope of 22°. If the run is 22 ft 6 in., find the rise at its highest point. Give the answer in feet and inches to the nearest inch.

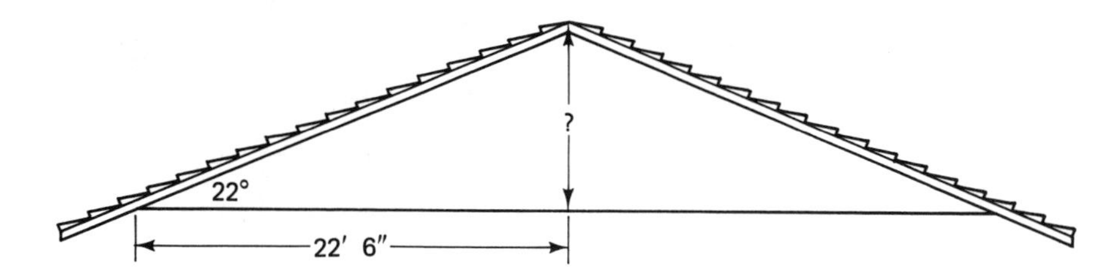

3. A handicapped access ramp is to have an incline of no more than 10°. What should be the length of the ramp if the entrance is 2 ft 9 in. off the ground? Give the answer in feet and inches to the nearest inch.

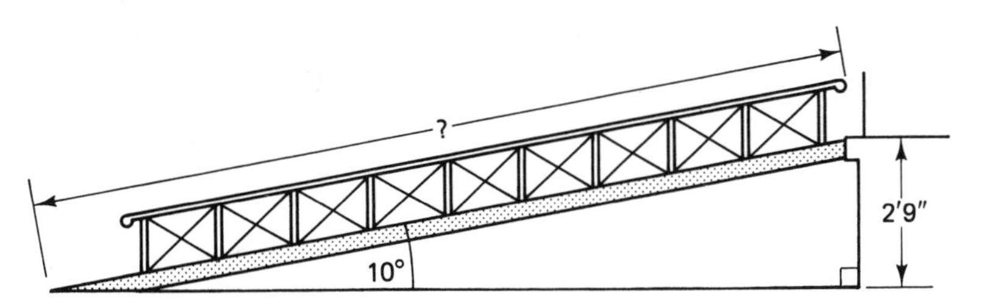

4. A 20-ft ladder is propped against a wall in such a way that the foot of the ladder is 4 ft 3 in. from the wall. What angle does the ladder make with the ground?

5. A surveyor who is 35 ft from the base of a fire tower finds the angle of elevation to the top of the tower to be 71°. How tall is the tower, to the nearest foot, if the man is 6 ft tall?

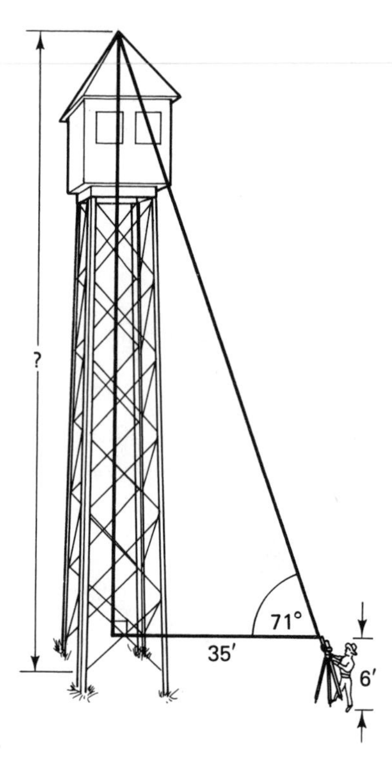

6. A roof has a unit rise of 5 in. (5 in. of rise for every 1 ft of run). What is the slope of the roof?

7. A board that is $5\frac{3}{4}$ in. wide is cut on a miter box at an angle of 37°. What is the length, to the nearest $\frac{1}{32}$ in., of the cut edge?

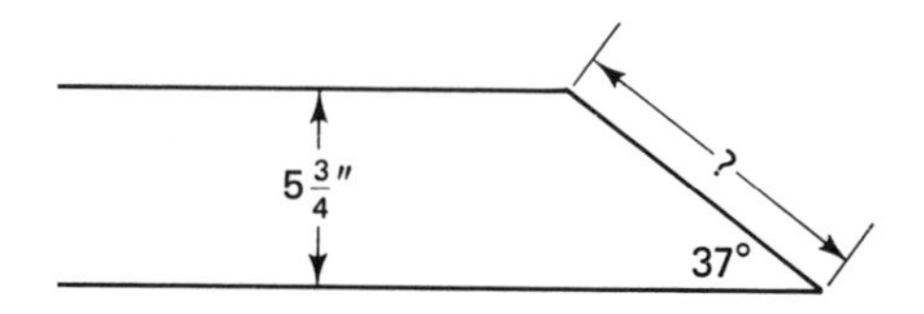

8. On the foundation shown, the diagonals are measured to ensure a perfect rectangle. What should be the angle between the length and the diagonal?

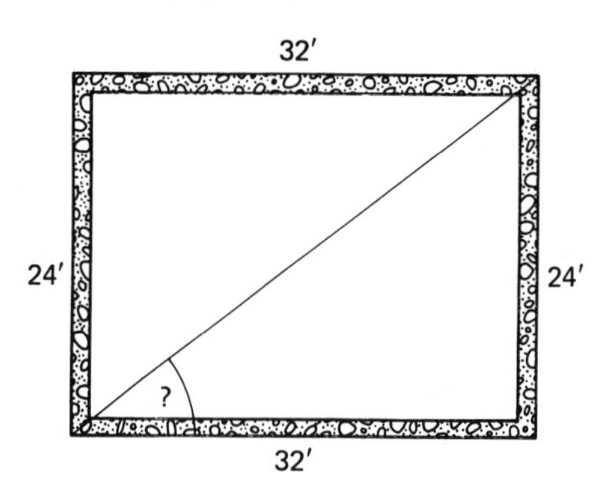

appendix C

Resistances (*R*-Factors) of Various Materials

Note: Values listed are representative of the material, but are not precise. *R*-factors vary with temperature and density. Manufacturers' specifications for specific materials should be consulted.

Material	R-*factor*[a]	*Material*	R-*factor*[a]
Building materials		Air spaces	
Asphalt shingles	0.44[b]	Horizontal	0.87
Brick	0.20	45° slope	0.94
Carpet		Vertical	1.01
Fibrous pad	2.08[b]	Glass[b]	
Rubber pad	1.23[b]	Single	0.89
Clapboards		Insul. (2)	
$\frac{1}{2}$ in. × 8 in., lapped	0.81[b]	$\frac{1}{4}$ in. space	1.55
$\frac{3}{4}$ in. × 10 in., lapped	1.05[b]	$\frac{1}{2}$ in. space	1.72
Concrete	0.08	Insul. (3)	
Door, $1\frac{3}{4}$ in. solid wood	1.96[b]	$\frac{1}{4}$ in. space	2.13
Drywall	0.90	$\frac{1}{2}$ in. space	2.78
Felt paper	0.06[b]	Storm	
Tile, vinyl	0.05[b]	4 in. space	1.79
Wood	1.25		
Plywood	1.25		
Fiber board	2.38		
Insulating materials			
Acoustical tile	2.38		
Expanded polystyrene			
Extruded, 60°F	4.00		
Molded beads, 75°F	3.57		
Expanded polyurethane			
(blown, 50°F)	6.25		
Glass fiber (60°F)	3.00		
Mineral wool (60°)			
Resin binder	3.57		
Loose fill	3.70		
Wood fiber	4.00		

[a]Per inch of thickness.

[b]For material as listed.

appendix D

House Plans

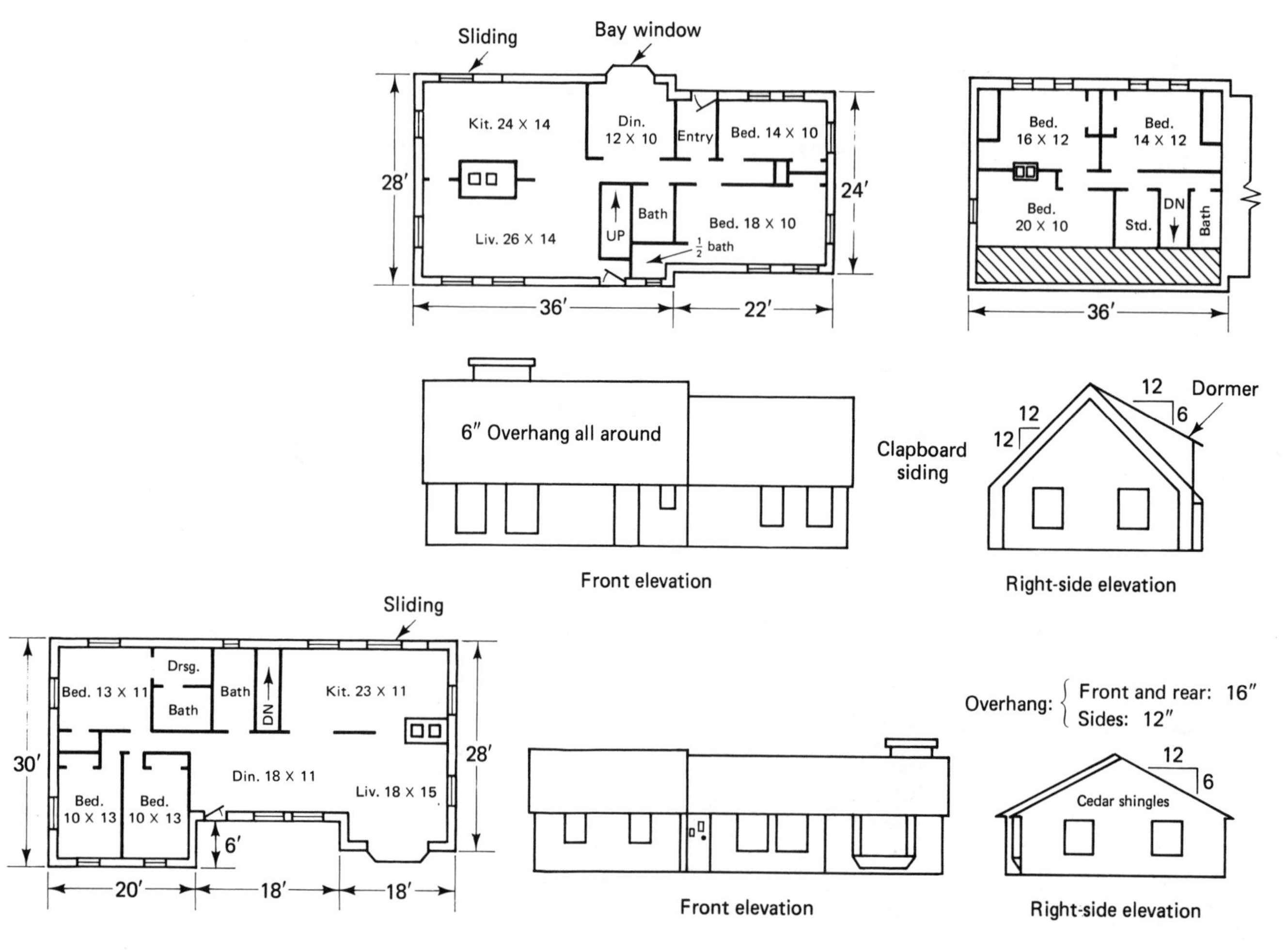

Answers to Odd-Numbered Exercises

CHAPTER 1

Exercise 1-1A

1. ones place
3. hundred-thousands place
5. millions place

Exercise 1-1B

1. 862,450; 862,500
3. 542,000; 540,000
5. 25,478,500; 25,000,000

Exercise 1-2

1. 21
3. 6
5. 22
7. 140 mi

Review Exercises

1-1. $1323
1-3. 1728 in.
1-5. 189 yd^3
1-7. $34,242
1-9. 23,075 plugs
1-11. $366
1-13. 9 in.
1-15. 20 connectors
1-17. 6 pieces
1-19. 8 in.
1-21. 5110 lb
1-23. 154 ft

CHAPTER 2

Exercise 2-1A

1. 7/9
3. 1/2
5. 1/3
7. 4/17
9. 1/8

Exercise 2-1B

1. 15
3. 12
5. 28
7. 36
9. 35

Exericse 2-1C

1. 41/8
3. 15/8
5. 51/8
7. 67/8
9. 27/5

Exercise 2-1D

1. 1 3/5
3. 1 1/16
5. 1 9/16
7. 4 2/15
9. 7

Exercise 2-1E

A. 1/16″
B. 1/4″
C. 9/16″
D. 7/8″
E. 1 1/16″
F. 1 9/16″
G. 1 7/8″
H. 2 3/8″
I. 2 15/16″
J. 3 5/16″
K. 3 3/4″
L. 4 1/8″
M. 4 7/16″
N. 5″
O. 5 5/16″
P. 5 3/4″

Exercise 2-2a

1. 9/24, 20/24
3. 8/12, 9/12
5. 15/24, 16/24
7. 9/64, 6/64
9. 5/18, 12/18
11. 3/4, 2/4
13. 1/9, 6/9
15. 33/36, 20/36
17. 5/6, 2/6
19. 6/16, 5/16

Exercise 2-2B

1. 15/64 in.
3. 11/16 in.
5. 5/8 in.
7. 23/32 in.
9. 51/64 in.
11. 11/64 in., 9/64 in.
13. 63/64 in., 61/64 in.
15. 11/16 in., 9/16 in.
17. 7/8 in., 5/8 in.
19. 7/16 in., 5/16 in.

Exercise 2-3

1. 11/12
3. 1 1/6
5. 7 1/2
7. 14 1/2
9. 9 17/18
11. 382 11/12 ft
13. 4 1/4 in.
15. 7 9/16 in.
17. 4 1/2 in.

Exercise 2-4

1. 1/8
3. 3/16
5. 3 8/9
7. 2 5/8
9. 27/32
11. 39 3/16 in.
13. 46 15/16 in.
15. 36 13/16 in.
17. 1 11/16 in.

Exercise 2-5

1. 2/3
3. 15/28
5. 59 1/2
7. 1 3/4
9. 4 4/5
11. 3 3/4 in.
13. 106 7/16 in.
15. 64 11/16 in.
17. 9 1/6 ft

Exercise 2-6

1. 5/22
3. 18/25
5. 2 4/7
7. 1/48
9. 4/9
11. 38 boards
13. 3 posts
15. 11/16
17. 4 shelves

Review Exercises

2-1. 10 3/8″
2-3. 972 3/16 lbs
2-5. 6-ft lengths
2-7. 4 1/4 in.
2-9. 8 3/4 in.
2-11. 43/64 in., 11/16 in., 25/32 in., 7/8 in.
2-13. 12 1/2 gal
2-15. 1/8 in.

CHAPTER 3

Exercise 3-1

1. 83.41; 83.407
3. 43; 42.6
5. 0.5; 0.50
7. 4.6; 4.60
9. 0; 0.4

Exercise 3-2

1. 31.8
3. 16.47
5. 72.072
7. 74.592
9. 2.033
11. 0.6875 in.
13. 0.375 in.
15. $270.84
17. 0.7567 in.

Exercise 3-3

1. 0.615
3. 0.8067
5. 0.2045
7. 6.365
9. 0.011
11. 0.5502 in.
13. 0.9785 in.
15. 5.372 in.
17. 0.4375 in.

Exercise 3-4

1. 22.68
3. 0.90
5. $29.88
7. 0.305
9. 142.6
11. $16.17
13. 0.625 in.
15. 82.95 ft
17. 274 lbs/min

Exercise 3-5

1. 0.099
3. 2.9
5. 9.12
7. 0.02
9. 1000
11. 240
13. 12
15. 42 ¢/lineal foot
17. 19
19. 83 sheets

Exercise 3-6

1. 830,000
3. 50
5. 48
7. 0.038
9. 62

Exercise 3-7

1. 0.625
3. 0.666. . .
5. 1.875. . .
7. 8.666. . .
9. 19.5625
11. 33/40
13. 41/100
15. 15 17/40
17. 4 47/200
19. 26 6/25
21. 2/3
23. 2 5/11
25. 25 4/9
27. 3 2/3
29. 35 2/9

31.	16″	16″	15 7/8″	15 7/8″	15 29/32″	15 57/64″
33.	7 1/2″	7 1/4″	7 1/4″	7 5/16″	7 5/16″	7 19/64″
35.	7 1/2″	7 3/4″	7 5/8″	7 5/8″	7 5/8″	7 5/8″
37.	48 1/2″	48 1/2″	48 5/8″	48 9/16″	48 19/32″	48 37/64″
39.	13 1/2″	13 1/2″	13 5/8″	13 5/8″	13 5/8″	13 5/8″
40.	58″	58″	58″	58 1/16″	58 1/16″	58 3/64″

41. 73 10/16″ or 73 5/8″
43. 5/32 in.
45. 8 2/8 in. or 8 1/4 in.

CHAPTER 4

Exercise 4-1A

1. 2 ft 9 in.
3. 9 ft 6 in.
5. 7 ft 1 in.
7. 7 ft 1 in.
9. 6 ft 2 in.

Exercise 4-1B

1. 4 ft 2 1/8 in.
3. 7 ft 2 3/8 in.
5. 1 ft 7 3/8 in.
7. 7 ft 1 5/16 in.
9. 4 ft 9 in.

Exercise 4-2A

1. 17 ft 11 in.
3. 2 ft 1 in.
5. 19 ft 7 in.
7. 7 in.
9. 1 ft 8 in.

Exercise 4-2B

1. 43 ft 9 in.
3. 1 ft 2 in.
5. 4 in.
7. 6 in.
9. 10 ft

Exercise 4-3

1. 1.54 miles
3. 70 fathoms
5. 13.75 yds
7. 297 ft

Exercise 4-4

1. 46,656 in.^{3}
3. 2.23 mi^2
5. 1.26 yd^3
7. 2.1 rod^2
9. 17.4 acres
11. 56 yd^3
13. 0.81 acres
15. 80.7 ft^2
17. 15.11 ft^2

Exercise 4-5

1. 132,000 ft/hr
3. 215 lbs/min
5. 0.112 oz per bf
7. 1.53 psi
9. 234 lbs/min
11. 5.6 oz/in.
13. 48 per foot

CHAPTER 5

Exercise 5-1

1. 6 : 1
3. 1 : 1
5. 1 : 3
7. 1 : 6
9. 16 : 1
11. 24 ft × 40 ft
13. 8 : 5
15. $432 and $720

Exercise 5-2

1. 12
3. 15.48
5. 2.148
7. 9/10
9. 0.75
11. 144 mi
13. 53 mi
15. $240.41
17. 200 lbs
19. 13 3/4 ft
21. 33,700
23. 479 rpm
25. 104 hrs
27. 82.08

CHAPTER 6

Exercise 6-1

1. 1.8, 180%
3. 1 41/50, 1.82
5. 1/4, 25%
7. 0.001, 0.1%
9. 1/200, .005
11. 1/40, 2 1/2%
13. $.\overline{6}$, 66 2/3%
15. 1/40, .025
17. 31/50, 62%
19. .0006, .06%

Exercise 6-2A

1. amount
3. base
5. base
7. base
9. base

Exercise 6-2B

1. 12.76
3. 321
5. 625
7. 233
9. 600

Exercise 6-2C

1. 171.6
3. 478
5. 31%
7. 82.3
9. $530.14

Exercise 6-3

1. $4372.50
3. $59.95
5. $125,000
7. $98,884
9. $57.48
11. 22.7%
13. 3′-4″
15. $438.35
17. 5.6%
19. 14.8%

CHAPTER 7

Exercise 7-1A

1. 99° 24′ 59″
3. 20° 3′ 23″
5. 47° 59′ 47″
7. 58° 40′ 35″
9. 79° 19′

Exercise 7-1B

1. 83° 33′ 42″
3. 14.4275°
5. 78° 0′ 14″
7. 121° 24′ 36″
9. 10° 19′ 34″
11. 42° 30′
13. 52.0056°
15. 20° 59′ 59″

Exercise 7-1C

1. 21° 9′ 47″
3. 15° 9′ 15″
5. 124° 18′
7. 12° 49′ 42″
9. 4° 31′

Exercise 7-2

1. $x = 4$, $X = 53°$; $y = 3$, $Y = 37°$; $z = 5$, $Z = 90°$
3. 2.7 ft
5. 146

Exercise 7-3A

1. 67.9
3. 2286
5. 86.5
7. 1764
9. 31.4
11. 5
13. 1
15. 9
17. 4.18
19. 24.5

Exercise 7-3B

1. 8.52
3. 10.4
5. 10.4
7. 47 ft 2 in.
9. 21 ft 8 in.
11. 26 ft

Exercise 7-4

1. $a = 15.4$, $b = 8.92$
3. $j = 26.69$, $l = 53.38$
5. $x = 12.62$, $y = 12.62$
7. $c = 15$ ft
9. $p = 26$ ft
11. 13 ft 10 1/4 in.
13. 18 ft 5 11/16 in.
15. 24 ft

Exercise 7-5

1. 9.23; 21.6
3. 7.99, 19.0
5. 12.00, 30.00
7. 3, 6
9. 7.16, 10.24 sq units
11. 17.2 sq units
13. 29.8 sq units
15. 26.5 sq units
17. 9.45 sq units
19. 4.21 acres
21. 13 ft
23. 6.93 ft^2
25. 17,830 ft^2

CHAPTER 8

Exercise 8-1

1. 26 in.
3. 30 in.
5. 9 ft^2
7. 25 ft^2
9. 66 ft
11. 180 ft^2
13. 4 in.; 40 $in.^2$
15. 47 ft 6 in.; 129.6 ft^2
17. 4 ft^2
19. $16.02
21. $614.90
23. 112 ft
25. 144 ft^2

Exercise 8-2

1. 10 in.; 31.4 in.; 78.5 in^2
3. 13.1 ft; 26.3 ft; 543 ft^2
5. 16 in; 50.3 in; 201 $in.^2$
7. 5 in.; 10 in.; 78.5 $in.^2$
9. 4 in; 8 in; 25.1 in.
11. 201 $in.^2$
13. 254 $in.^2$
15. 11 ft
17. 8 $in.^2$
19. 14 in. pizza

Exercise 8-3

1. 1580 ft^2
3. 1518 ft^2
5. 195 ft
7. 62.8 in.
9. 29.7 ft
11. 54.1 in.; 110 $in.^2$
13. 22 ft 10 1/4 in.
15. 76.3 ft^2
17. 18 $in.^2$
19. 359 ft^2

CHAPTER 9

Exercise 9-1

1. 384 $in.^2$; 432 $in.^2$; 384 $in.^3$
3. 4800 ft^2; 9550 ft^2; 47,500 ft^3
5. 25 yd^3
7. 4620 ft^3

Exercise 9-2

1. 16″, 50.3″, 201 sq in., 1005 in, 251 sq in., 653 sq in.
3. 2.5″, 15″, 19.6 sq in., 98 in., 78.5 sq in., 118 sq in.
5. 11″, 69″, 380 sq in., 5702 in., 1037 sq in., 1797 sq in.
7. 375 cubic inches
9. 295 gal
11. 67.1 sq ft
13. 1508 sq in.
15. 339 sq ft

Exercise 9-3

1. 16″, 50.3″, 2145 cu in., 804 sq in.
3. 8′, 50.3′, 2145 cu ft, 804 sq ft
5. 3.5′, 7′, 179.6 cu ft, 153.9 sq ft
7. 17.2″, 54″, 2664 cu in., 929 sq in.

Exercise 9-4

1. 261.7 ft^2; 306.8 ft^3
3. 452 ft^2; 737 ft^3
5. 15 gal

CHAPTER 10

Review Exercises

10-1. $\frac{2}{3}$; 1; $\frac{4}{3}$; $\frac{5}{3}$; 2
10-3. **(a)** $2853\frac{1}{3}$ bf **(b)** 2240 bf **(c)** 168 bf **(d)** $1013\frac{1}{3}$ bf **(e)** 504 bf.
10-5. 224 lf
10-7. 312 lf
10-9. 35 sheets
10-11. 178 bf

CHAPTER 11

Review Exercises

11-1. \$0.3675/bf
11-3. \$0.259/lf
11-5. **(a)** \$542.19 **(b)** \$306.24 **(c)** \$40.32 **(d)** \$728.00 **(e)** \$235.92 **(f)** \$1852.67
11-7. \$8.96
11-9. \$945 mbf is less by \$0.355/bf

CHAPTER 12

Review Exercises

12-1. 4.9 yd^3
12-3. **(a)** footing: 5.0; walls: 26.3; slab: 10.2; total: 41.5 yd^3
(b) footing: 7.6; walls: 42.6; slab: 14.6; total: 64.8 yd^3
(c) footing: 7.1; walls: 38.9; slab: 14.2; total: 60.2 yd^3

CHAPTER 13

Review Exercises

13-1. Several designs are possible. One is: 8/16′s, 2/8′s, 1/12

13-3.

	(a)	(b)	(c)
Girder	3/18′s	4/8′s, 2/12′s, 7/16′s	2/8′s, 1/10, 7/12′s, 1/14 2/16′s, 1/18
Sill	2/8′s, 6/10′s	3/8′s, 1/10	2/10′s, 8/12′s
Plates	4/12′s, 2/14′s	14/12′s	4/14′s
Floor	2 × 12: 47/20′s	2 × 10: 74/16′s	2 × 10: 56/14′s

13-3. *(continued)*

	(a)	(b)	(c)
Joists	(12 in. o.c.) 2 × 6: 19/10's (12 in. o.c.)	(16 in. o.c.) 2 × 8: 15/10's	36/12's (16 in. o.c.)
F.j. headers	2 × 12: 1/8, 2/10's 3/12's, 2/14's	2 × 10: 8/12's 2/8's	2 × 10: 8/12's 2/10's
Built-up sill plates to be used where necessary to level joists.			
Bridging	330 lf	336 lf	348 lf
Subflooring	35 sheets	52 sheets	48 sheets

CHAPTER 14

Review Exercises

14-1. 23 studs
14-3. **(a)** 480 lf 2 × 4 **(b)** 136 studs **(c)** 1045.33 bf 2 × 4; 1232 bf 2 × 6 **(d)** $313.60 for 2 × 4; $369.60 for 2 × 6
14-5. 498 lf for shoes/plates; 121 studs; total bf: 1466
14-7. $368.40 for 2 × 4-16 in. o.c.; $545.40 for 2 × 6-16 in. o.c.; $439.80 for 2 × 6-24 in. o.c.

CHAPTER 15

Review Exercises

15-1. **(a)** 12.369 **(b)** 13.892 **(c)** 12.816 **(d)** 15.620 **(e)** 15 **(f)** 19.209 **(g)** 16.971
15-3. **(a)** 16 ft 3 in. **(b)** 15 ft 11 1/16 in. **(c)** 17 ft 7 1/2 in. **(d)** 12 ft 7/16 in. **(e)** 7 ft 10 3/4 in. **(f)** 13 ft 3 3/4 in. **(g)** 7 ft 7 11/16 in.
15-5. **(a)** $5\frac{5}{8}$ **(b)** $4\frac{1}{8}$ **(c)** 12 **(d)** $6\frac{7}{8}$ **(e)** $5\frac{3}{4}$
15-7. $2\frac{3}{16}$
15-9. **(a)** front: 90 in.; rear: 129 in.; **(b)** $9\frac{15}{16}$ in. **(c)** front: 15.620; rear: 15.571 **(d)** front: 16 ft $11\frac{1}{16}$ in.; rear: 16 ft $10\frac{7}{16}$ in.
15-11. 2 ft 4 in.

CHAPTER 16

Review Exercises

16-1. **(a)** 6 11/16 in. **(b)** 5 1/4 in. **(c)** 24 in. **(d)** 5 in. **(e)** 6 11/16 in. **(f)** 14 in.
16-3. 17 1/16 in.
16-5. **(a)** 17/12 **(b)** 10 3/16 in. **(c)** 5 11/16 in. **(d)** 4 1/4/12
16-7. 6 3/8 in.
16-9. 1 ft 3 1/4 in.
16-11. 9 7/16/12

CHAPTER 17

Review Exercises

17-1. **(a)** 17 ft $8\frac{5}{16}$ in. **(b)** 22 ft $4\frac{15}{16}$ in. **(c)** 20 ft 3 in. **(d)** 23 ft $11\frac{11}{16}$ in. **(e)** 15 ft $10\frac{7}{16}$ in.
17-3. **(a)** 18 ft $\frac{13}{16}$ in. **(b)** 21 ft $10\frac{15}{16}$ in. **(c)** 17 ft $2\frac{1}{8}$ in. **(d)** 5 ft $8\frac{1}{8}$ in.
17-5. **(a)** $24\frac{1}{16}$ in. **(b)** $24\frac{7}{16}$ in. **(c)** $30\frac{3}{8}$ in. **(d)** $20\frac{7}{16}$ in. **(e)** $20\frac{13}{16}$ in.
17-7. **(a)** no **(b)** 2 11/16/12 **(c)** common: $73\frac{3}{4}$ in.; hip: $103\frac{1}{16}$ in. **(d)** both: $24\frac{9}{16}$ in.
17-9.

	(a)	(b)	(c)
(1) Main	12 ft $3\frac{9}{16}$ in.	16 ft 3 in.	16 ft $9\frac{15}{16}$ in.
Ell	12 ft $3\frac{9}{16}$ in.	13 ft	12 ft $3\frac{1}{2}$ in.
(2) Long	16 ft 6 in.	22 ft $1\frac{3}{8}$ in.	18 ft $7\frac{9}{16}$ in.
Short	16 ft 6 in.	17 ft $8\frac{5}{16}$ in.	18 ft $7\frac{9}{16}$ in.
(3) URi/URu	6/17	5/17	6 15/16/12
(4) Main	$17\frac{7}{8}$ in.	$17\frac{5}{16}$ in.	$33\frac{5}{8}$ in.
Ell	$17\frac{7}{8}$ in.	$17\frac{5}{16}$ in.	$14\frac{1}{16}$ in.

CHAPTER 18

Review Exercise

18-1.

	Rafter	Bottom chord	Tension and compression chords
(a)	2/18's	2/14's	1/12
(b)	2/10's	2/8's	1/14
(c)	2/16's	2/12's	2/12's
(d)	2/14's	1/10, 1/12	2/10's
(e)	2/16's	1/14, 1/12	2/12's
(f)	2/16's	1/14, 1/12	2/12's

CHAPTER 19

Review Exercises

19-1. **(a)** $100\frac{7}{8}$ in. **(b)** 13 **(c)** $7\frac{3}{4}$ in. **(d)** 12 **(e)** 114 in. **(f)** $104\frac{1}{2}$ in. **(g)** 80 in. assumed

19-3. 12; $7\frac{13}{16}$ in.; 12; $8\frac{11}{16}$ in.; 104 in.
19-5. 12; $7\frac{15}{16}$ in.; 11; 9 in. assumed; from top, 6 risers to landing for 71 in., then 7 more risers for 54 in.; from top, 6 risers to landing for 81 in., then 6 more risers for 45 in.

CHAPTER 20

Review Exercises

20-1. **(a)** 9/10's; 4
(b) 10/14's; 6
(c) 19/10's; 7
(d) 25/10's; 10
(e) 29/8's; 10

CHAPTER 21

Review Exercise

21-1.

	(a)	(b)	(c)
(1) Ext. wall Sheathing	47 sheets	43 sheets	52 sheets
(2) Siding			
(i)	1504 ft^2	1376 ft^2	1664 ft^2
(ii)	1274 ft^2	1242 ft^2	1448 ft^2
(3) Drywall			
Walls	91 sheets	110 sheets	116 sheets
Ceilings	39 sheets	44 sheets	59 sheets
(4) Roof sheathing	50 sheets	40 sheets	74 sheets
(5) Shingles	$16\frac{2}{3}$ square	13 squares	$24\frac{2}{3}$ square
(6) Drip edge	17 lengths	15 lengths	19 lengths

CHAPTER 22

Review Exercises

22-1. 183,539 Btu/hr
22-3.

	Uninsul.	%	Insul.	%
Infiltration	15,770	8.6	15,770	25.4
Walls	25,481	13.9	4,131	6.7
Doors	393	0.2	393	0.6
Windows	4,000	2.2	4,000	6.5
Ceilings	75,815	41.3	2,508	4.0
Floor	62,080	33.8	35,237	56.8

CHAPTER 23

Review Exercise

23-1. Students should compare results among classmates.

CHAPTER 24

Review Exercises

24-1. **(a)** \$ 19.73; \$ 519.73
(b) 420.00; 3420.00
(c) 133.56; 2633.56
(d) 365.03; 8865.03
(e) 348.41; 4848.41
24-3. **(a)** \$109,320.40 **(b)** \$1804.32 **(c)** \$12,948.00 **(d)** \$163,278.40
(e) \$1487.64
24-5. \$35,467.58
24-7. \$64.52 discount less 27.72 interest = \$36.80 saved

APPENDIX A

Exercise A-1A

1. 0.047 km
3. 830 cm
5. 25 mm
7. 4800 cm
9. 4600 mm
11. 1800 mg
13. 58 dg
15. 5 g

Exercise A-1B

1. 0.038
3. 4.2
5. 0.082
7. 48
9. 410 000

Review Exercises

A-1. 1.52 lbs/ft
A-3. 950.4 lbs
A-5. 406 mm
A-7. 7.6 kg
A-9. 29°C

APPENDIX B

Exercise B-1

1. sin = .6921, cos = .7218, tan = .9590
3. sin = .7071, cos = .7071, tan = 1.000

Exercise B-2

1. $B = 48°$, $a = 25.4'$, $b = 28.2'$
3. $J = 61°$, $k = 25.5''$, $l = 52.6''$
5. $A = 45°$, 27.0 in., 27.0 in.

Exercise B-3

1. 20.2 ft or 20 ft 2 1/4 in.
3. 15 ft 10 in.
5. 108 ft
7. 9 9/16 in.

Bibliography

DAGOSTINO, FRANK R., *Estimating in Building Construction.* Reston, Va.: Reston, 1973.

JANSEN, TED J., *Solar Engineering Technology.* Englewood Cliffs, N.J.: Prentice-Hall, 1985.

LEWIS, JACK R., *Basic Construction Estimating.* Englewood Cliffs, N.J.: Prentice-Hall, 1983.

OSTWALD, PHILLIP F., *Cost Estimating* (2nd ed.). Englewood Cliffs, N.J.: Prentice-Hall, 1984.

PETRI, ROBERT W., *Construction Estimating.* Reston, Va.: Reston, 1979.

REED, MORTIMER P., *Residential Carpentry.* New York: John Wiley & Sons, 1980.

SHURCLIFF, WILLIAM A., *Superinsulated Houses and Double Envelope Houses.* Andover, Mass.: Brick House, 1981.

SMITH, RONALD C., *Principles and Practices of Light Construction* (3rd ed.). Englewood Cliffs, N.J.: Prentice-Hall, 1980.

"Solar Factsheet," FS111, U.S. Department of Housing and Urban Development, Office of Policy and Development Research, July 1979.

STERLING, R., et al., *Earth Sheltered Housing: Code, Zoning, Financing Issues.* Underground Space Center, University of Minnesota; U.S. Department of Housing and Urban Development, Washington, D.C., 1980.

WASS, ALONZO, *Estimating Residential Construction.* Englewood Cliffs, N.J.: Prentice-Hall, 1980.

Index